Wetlands of the American Midwest

University of Chicago Geography Research Paper no. 241

SERIES EDITORS
Michael P. Conzen
Chauncy D. Harris
Neil Harris
Marvin W. Mikesell
Gerald D. Suttles

Minnesota prairie under snow
Landscape '63. Silk Screen. Lawrence Rosing.

Titles published in the Geography Research Papers series prior to 1992 and still in print are now distributed by the University of Chicago Press. For a list of available titles, see the end of the book. The University of Chicago Press commenced publication of the Geography Research Papers series in 1992 with no. 233.

WETLANDS
of the
AMERICAN
MIDWEST

A Historical Geography of Changing Attitudes

Hugh Prince

THE UNIVERSITY OF CHICAGO PRESS

Chicago and London

HUGH PRINCE is an emeritus reader and honorary research fellow in the Department of Geography at University College London.

The University of Chicago Press, Chicago 60637
The University of Chicago Press, Ltd., London
© 1997 by The University of Chicago
All rights reserved. Published 1997
Printed in the United States of America
06 05 04 03 02 01 00 99 98 97   1 2 3 4 5
ISBN: 0-226-68283-8 (paper)

Library of Congress Cataloging-in-Publication Data

Prince, Hugh C.
    Wetlands of the American Midwest :   a historical geography of changing
    attitudes / Hugh Prince.
        p.   cm. — (University of Chicago geography research paper ; no. 241)
    Includes bibliographical references and index.
    ISBN 0-226-68283-8 (alk. paper)
    1. Wetlands—Middle West.   I. Title.   II. Series.
    GB624.P75   1997
    333.91'8'0977—dc21   97-22226
                                                                                CIP

CONTENTS

FIGURES

TABLES

PREFACE

Research for this study was conducted during and following three visits to the American Midwest in 1954–55, 1966, and 1990. Profoundly different attitudes toward the use and misuse of wetlands prevailed on each of these occasions and influenced the course of the inquiry.

In 1954–55, I spent a year as a graduate student in the geography department at the University of Wisconsin, Madison. My visit was funded by a Fulbright Travel Grant, a University of London Postgraduate Travelling Studentship, and an International Scholarship from the University of Wisconsin, for which I am most grateful. During that period, I studied unsuccessful efforts to drain and reclaim for agriculture peat soils in central Wisconsin. An account of that work appeared in the *Journal of Historical Geography* 21 (1995) 3–22 and is discussed in a wider context in chapter 7 below. Permission from the Academic Press to reproduce illustrations and passages from that article is gratefully acknowledged.

In 1966, I spent the spring semester as visiting professor at the University of Minnesota in Minneapolis. During that period, I visited a prosperous family farm in Kandiyohi County and learned a great deal about the draining of wet prairie soils. The history of draining and farming on heavy soils is discussed in chapter 6.

In 1990, I returned to Minnesota in the fall semester to teach at Macalester College in St. Paul and take part in a graduate seminar in historical geography at the University of Minnesota. On that occasion, I observed some effects of a crisis in corn belt farming and heard from a landowner in Waseca County who was developing wetlands both for farming and conservation. A reappraisal of wetland values under changing social and economic conditions is reviewed in chapter 8.

Since 1991, I have pursued the history of these topics in libraries in London, principally the American history section of

University College London library and collections of printed books, official papers, and maps at the British Library. For taking infinite care to find answers to difficult queries and trace elusive sources of information, I wish to thank Ruth Dar and Anne Oxenham at University College London. Some results of this research were reported in a brief summary of changing attitudes to wetlands at the Eighth International Conference of Historical Geographers at Vancouver in 1992 and in a comment on floods in the upper Mississippi River basin in 1993, published in *Area* 27 (1995) 118–26.

A study such as this is a product of many minds, and my principal debts of gratitude are owed to the authors named in the bibliography. Quite simply, without their inspiration and guidance this study would not have been written. I also wish to thank members of the geography department at University College London for their continuing help over many years. I am particularly grateful to Clive Agnew, Rick Battarbee, Eric Brown, Jacquie Burgess, Nick Clifford, Richard Dennis, Paul Densham, Carolyn Harrison, Ted Hollis, Alun Jones, David Lowenthal, Bill Mead, Richard Munton, John Salt, Julian Thompson, Andrew Warren, and Peter Wood for commenting on different aspects of the project and suggesting fresh lines of inquiry and useful readings. I thank Malcolm Anderson, Alan Baker, Ron Cooke, Terry Coppock, Clifford Darby, John Davis, Joseph Gallagher, Peter Jackson, Jim Johnson, Roger Kain, Edmund Penning-Rowsell, Tony Phillips, Neil Ward, Sarah Whatmore, and Michael Williams for their valuable comments and criticisms on different sections and for recommending further readings. Among many people in the Midwest who helped in collecting material and raising fresh questions I owe special thanks to Michael Bell, John Borchert, Jan Broek, Mark Cassell, Michael Conzen, Charles M. Davis, Siobhan Fennessy, Arlin D. Fentem, Robert W. Finley, Carol and Phil Gersmehl, Douglas Johnson, Hildegard Binder Johnson, Terry Jordan, John Fraser Hart, Miron Heinselman, Leslie Hewes, Jeanne Kay, David Lanegran, Fred Lukermann, Judith Martin, Roger Miller, Robert Moline, James Murray, Joan Nassauer, Clarence Olmstead, Bob Ostergren, William D. Pattison, Jerry Pitzl, Phil Porter, John Rice, Joe Schueler, Richard Skaggs, Roderick Squires, Roger Suffling, Graham Tobin, J. William Trygg, Otto Zeasman, and Mariia Zimmermann.

In the early stages of this project, I was guided by Andrew Hill Clark, who acted as advisor while I was in Wisconsin in 1954–55 and with whom I corresponded during the remaining twenty years of his life. From 1991 onward, Hugh Clout supervised the research. His encouragement and unfailing attention to my early drafts have been an invaluable stimulus to writing and revising. I am deeply endebted to the Cartographic Unit at University College London for drawing the maps and in particular to Bill Mackie, John Bryant, Guy Baker, and Catherine Pyke. I am also grateful to Mike Barnsley, Suse Keay, and Paul Schooling for their help and ingenuity in solving computer problems. Members of the University of Chicago Press have been outstandingly helpful and conscientious. In particular, Carol Saller has not only been a meticulous editor but has also made many constructive suggestions for improving the manuscript.

My greatest thanks are due to Sheila for sharing my faith that one day this project would be completed.

1

Changing Attitudes

The study of change is central to historical geography. It is also a central theme in agricultural and environmental history. Over very long periods of time, physical processes, including glacial deposition, climatic fluctuations, and vegetational changes, have formed and modified wetlands. In Holocene times, beavers dammed streams, muskrats ate cattails, ducks trampled wetland margins, fires repressed tree growth in wet prairies and swamps. During their brief period of occupation, human groups completely transformed the character and appearance of wet prairies and profoundly altered the nature and extent of northern bogs and swamps. In the Midwest of the United States, in the seven states of Ohio, Indiana, Illinois, Iowa, Michigan, Wisconsin, and Minnesota, the most important change was the conversion of wetlands into agricultural land through ditching and tile draining. Artificial draining in wet prairies in Ohio, Indiana, Illinois, Iowa, and southwest Minnesota was most active between 1870 and 1920 and was resumed over large areas in the period from 1950 to 1980. In northern peatlands in Michigan, Wisconsin, and northern Minnesota, draining activity was short-lived and ended disastrously. Where draining failed, wildlife conservation was practiced. I shall describe and discuss these material changes in landscapes in relation to ideas held by contemporary actors and observers.

This study focuses on changes in peoples' minds. Attitudes toward wetlands changed rapidly and fundamentally. Changes in perception were far more radical than changes in action and occurred earlier than changes in landscape. Attitudes were expressed in a great variety of published literature from imaginative writing to official reports on questions of public concern. It is interesting to note how views accepted as common sense at one date were dismissed as nonsense at a later time. Attitudes changed at five levels. At the deepest level, cultural values and moral principles were conceived

1

as fundamentals for the survival and welfare of humanity. New ideologies and political rhetoric sought to change social and economic structures. Changes in knowledge and new directions in research were associated with changes in political outlook. Scientists coined new terminologies and devised new systems of classification; and new institutions and agencies were formed to carry new messages and implement new policies.

Changing Cultural Values and Moral Principles

When exploitation of natural resources was a guiding principle, different cultural appraisals were made of similar environments by hunters and home-seekers. In the eighteenth century, French and British fur traders were attracted to marshes as sources of furbearing animals; in the early nineteenth century, American pioneers were repelled by them as sources of malaria and other dreaded illnesses. When sites for farmsteads were sought, wet prairies were avoided not only from fear of disease but also because they lacked timber with which houses, barns, and fences might be built. Cultural values were born of necessity; they were not adopted by societies and economies as optional extras.

Once farms were established, a new order of priorities emerged. At a time when productivist values were dominant, draining wet soils was regarded as a praiseworthy activity. As early as 1860, Joseph Kennedy, superintendent of the U.S. Census, mentioned draining in his preliminary report: "This important improvement has made great progress in the estimation and practice of our farmers."[1] Enthusiasm was more evident than performance at that date. In 1915, Ben Palmer, a political scientist at the University of Minnesota, advocated draining much of 80 million acres of unproductive swamp and overflowed lands estimated still to remain in the United States. He wrote: "When we consider that these wet lands are so vast in extent, that they are unproductive and an economic waste, and that they are in many states so productive of malarial diseases as to constitute a serious and ever-present menace to the lives and health of the people, the importance of the problem of land

---

1. Joseph C. G. Kennedy, *Preliminary report on the eighth census, 1860.* House of Representatives, 37th Cong., 2d Sess., Ex. Doc. 116 (GPO, Washington, D.C. 1862) 90.

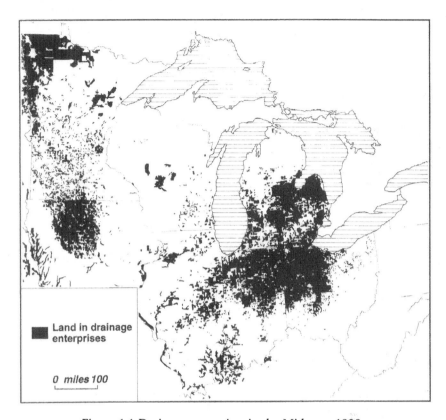

Figure 1.1 Drainage enterprises in the Midwest, 1920

Source: U.S. Bureau of Census, *Fourteenth census of the United States, 1920: Agriculture,* vol. 7, *Irrigation and drainage* (Washington, D.C. 1922).

drainage in the United States is apparent."[2] Extensive draining had been carried out in the Midwest between the 1870s and 1920s (fig. 1.1). In 1930, the *U.S. Census of Agriculture* described drained lands as those that were "actually benefited or made of more value for agricultural purposes by artificial drainage."[3] In 1990, Thomas Dahl of the U.S. Fish and Wildlife Service drew upon similar evidence to compile an inventory of "wetland losses" from the 1780s

2. Ben Palmer, *Swamp land drainage with special reference to Minnesota,* University of Minnesota Studies in Social Sciences 5 (Minneapolis 1915) 1.

3. U.S. Bureau of Census, *Fifteenth census of the United States, 1930: Drainage of agricultural lands* (Washington, D.C. 1932) 2.

to 1980s, portraying agricultural drainage as a threat to wildlife conservation. The report concluded that "wetland acreage has diminished to the point where environmental, and even socioeconomic
benefits (groundwater supply and water quality, shoreline erosion,
floodwater storage and trapping of sediments, and climatic changes)
are now seriously threatened."[4] What was earlier thought desirable
for good husbandry was later considered harmful to a healthy ecosystem. Moral guidelines were reoriented.

Changing Ideologies and Political Rhetoric

Americans advanced westward across prairies and forests in quest of
fortune and freedom. Thomas Jefferson believed that the strength of
the new nation would grow out of enterprise and industry generated
by freehold property owners who would secure the means of subsistence to support their families and exercise democratic powers to
elect federal, state, county, and township governments. In a message
to Congress in 1832, newly elected president Andrew Jackson declared: "Independent farmers are everywhere the basis of society
and the true friends of liberty."[5] By the mid-nineteenth century,
many observers had formed an opinion that sales from the public
domain did not favor independent farmers but fueled land speculation. In 1841, a senator deplored sales of western land "at auction
to bands of speculators and capitalists in large quantities, to lie idle
and unprofitable till they could extort the desired profit from those
whose necessities compel them to have it."[6] Speculators and large
landowners, who were most acquisitive and notorious in wet prairies, were condemned by politicians on all sides as parasites and
public enemies. For most of the nineteenth century, prejudice
against hired hands and tenants was also widespread. In the twentieth century, wetland residents turned their hostility against bankers
and government officials. From the beginning of European settlement in North America, family farms were cherished as stable and

4. T. E. Dahl, *Wetlands losses in the United States 1780s to 1980s* (U.S. Fish and Wildlife
Service, Washington, D.C. 1990) 10.

5. U.S. Congress, House of Representatives, 22d Cong., 2d Sess., Ex. Doc. 2 (1832) 1,
10–11, cited in Roy M. Robbins, *Our landed heritage: The public domain, 1776–1936*
(Princeton University Press, Princeton 1942) 57.

6. U.S. Congress, House of Representatives, 26th Cong., 2d Sess., *Congressional Globe*
(4 January 1841) appendix, 19, cited in Robbins, *Our Landed heritage,* 1942, 81.

cohesive elements in rural society and that sentiment persisted as a myth after most descendants had broken their ties with family lands. Wars of words against speculators and moneylenders, who at times outnumbered genuine settlers, simplified and distorted changing patterns of social division. On the other side, supporters of family farms exaggerated benefits and ignored failings of that revered institution.

The history of wetlands is steeped in ideology and rhetoric. Political slogans were paraded as moral precepts. In 1972, Earl Butz, President Nixon's secretary of agriculture, urged farmers to press ahead with reclamation and plant "from fencerow to fencerow" in furtherance of a productivist policy. Gregg Easterbrook, reflecting on this declaration after world markets had been glutted in the 1980s, suggested that farmers might "attach almost religious significance to what Butz said because it was the one time they were told exactly what they wanted to hear."[7] While Butz was in office, the U.S. Department of Agriculture (USDA) did little to discourage expansion of artificial drainage on wetlands. Following the farm crisis in the 1980s, President Bush set out to reverse this policy. In June 1989, addressing the sixth international waterfowl symposium in Washington, he announced: "It's time to stand the history of wetlands destruction on its head: from this year forward, anyone who tries to drain the swamp is going to be up to his ears in alligators."[8] His message was "no net loss"; every acre drained would have to be matched by an acre restored to wetlands. The intention was clear, but in practice it was unclear how no net loss might be achieved and there was no general agreement about a definition of wetlands. Implementation of a wetlands restoration policy required patient negotiations and lengthy preparations.

Changing Knowledge and Research Interests

At any time, states of knowledge were loosely associated with prevailing value systems, political beliefs, economic and social interests, and current problems. Conversely, topics that were not considered

7. Gregg Easterbrook, Making sense of agriculture: A revisionist look at farm policy, in Gary Comstock (ed.), *Is there a moral obligation to save the family farm?* (Iowa State University Press, Ames 1987) 18.

8. U.S. Department of the Interior, Fish and Wildlife Service, *Wetlands: Meeting the president's challenge* (Washington, D.C. 1990) inside front cover.

relevant to solving administrative, commercial, or technical problems or that might have been regarded as potentially disruptive of established order received little or no attention.

Until the middle of the present century, the study of native American occupation of wetlands was largely ignored. The replacement of the word "Indian" by "native American" reflects an important change in political attitudes and has raised awareness of the historic role of early inhabitants on the continent. The new consciousness has repositioned European Americans as latecomers in a long historical sequence. For much of the nineteenth and twentieth centuries, most historians regarded the native American period as prehistoric. American history was thought to have begun with the entry of Europeans.[9] Frederick Jackson Turner, historian of the frontier, wrote his doctoral dissertation on the character and influence of the Indian trade in Wisconsin. In this account, which dealt mostly with rivalry between the French and English in trade and diplomacy, Indians were shadowy figures. No mention was made of marshes and streams in which beaver were trapped.[10] A much fuller appreciation of the role of Potawatomi in occupying and utilizing the Kankakee Marsh in northern Indiana and Illinois was presented in Alfred Meyer's study in 1935.[11] Most important advances in the study of historical geography of Indian settlements have been made from the late 1970s onward.

Histories that began with the arrival of Europeans tacitly assumed that native Americans had done nothing to modify their environment and left no imprint on the landscape, neither exploiting nor cultivating natural resources. The federal surveys that formed the basis for public land sales recorded species of trees or types of grass or edges of water that marked corners of square-mile sections. These markers were later used to reconstruct vegetation at the time of American entry. The vegetation was not "original" in the sense of

9. Jesse D. Jennings, *Prehistory of North America* (McGraw Hill, New York 1968) deals with the cultural history of Indians as prehistoric and pre-American.

10. Frederick Jackson Turner, The character and influence of the Indian trade in Wisconsin: A study of the trading post as an institution (Johns Hopkins University Studies in Historical and Political Science, Baltimore 1891), reprinted in *The early writings of Frederick Jackson Turner*, introduction by Fulmer Mood (University of Wisconsin Press, Madison 1938) 87–181.

11. Alfred H. Meyer, The Kankakee "marsh" of northern Indiana and Illinois. *Papers Mich. Acad. Sci., Arts and Letters* 21 (1935) 359–96.

being unmodified by human activities, but the distinction between pre-European and pre-human was frequently blurred.[12] It was more difficult to turn a blind eye toward native Americans when discussing first impressions of the country recorded by explorers and pioneers in the early nineteenth century. Many newcomers commented on the presence of native inhabitants and some called for "the removal of the Indian menace."[13] Ignorance of native Americans on the part of later historians left serious gaps in understanding the development of wetlands.

In the mid-nineteenth century, lumbermen expected that northern pineries, situated on or at the margins of wetlands, would continue to yield timber for centuries. By 1915, almost all were exhausted and only cutover stumplands remained. As agricultural settlement advanced into treeless grasslands, demands for timber increased to unprecedented levels. At the same time, pine was put to new uses for constructing railroads, in fencing, in house building, as fuel in homes and factories, and as raw material in manufacturing pulpwood. Earlier forecasts, based on much lower rates of consumption, were rapidly exceeded. Logging and sawmilling speeded up and losses caused by fire escalated. Not until pineries had been cleared were new methods of sustainable forestry investigated and recommendations made for long-term management. Science was called in to find remedies for a disaster.

Research into soil erosion was set going by alarm signals emanating from deeply gullied hillsides in the Southeast and wind-

---

12. P. B. Sears, The natural vegetation of Ohio: I. A map of the virgin forest. *Ohio Journ. Sci.* 25 (1925) 139–49; Sears, The natural vegetation of Ohio: II. The prairies. *Ohio Journ. Sci.* 26 (1926) 128–46; W. B. Dick, A study of the original vegetation of Wayne County, Michigan, *Papers Mich. Acad. Sci., Arts and Letters* 22 (1936) 329–34; D. Finley and J. E. Potzger, Characteristics of the original vegetation in some prairie counties of Indiana, *Butler Univ. Bot. Studies* 10 (1952) 114–18; Francis J. Marschner, *The original vegetation of Minnesota: Compiled from U.S. General Land Office survey notes* ([1930] North Central Forest Experiment Station, St. Paul, Minn. 1974); on the back is an interpretation and critical evaluation by Miron L. Heinselman; R. W. Finley, *Original vegetation cover of Wisconsin, compiled from United States General Land Office notes* (U.S. Department of Agriculture, Forest Service, St. Paul, Minn. 1976); a useful general account is provided by E. A. Bourdo, A review of the General Land Office survey and of its use in quantitative studies of former forests, *Ecology* 37 (1956) 754–68.

13. Douglas R. McManis, *The initial evaluation and utilization of the Illinois prairies, 1815–1840,* Department of Geography Research Paper 94 (University of Chicago, Chicago 1964) 17. In the context of this discussion, attention is drawn to the use of the adjective "initial" in the title of the reference.

eroded plains in the subhumid West, where the dust bowl originated in the spring of 1933. Two years later, a Soil Conservation
Service was established in the USDA. In 1938, the *Yearbook of
Agriculture* reported on the significance of soil studies for agriculture. In discussing effects of draining peat and muck soils, John
Haswell noted that "these deposits shrink and subside after
drainage, and when they are dried out excessively they are a serious
fire hazard. In some cases shrinkage lowers the peat surface nearly
to the outlet level and further drainage by gravity becomes impossible."[14] Drained peatlands in the northern lakes states also suffered
moderate to severe wind erosion. Hard-to-drain clays and silts in
wet prairies, after a tough mat of grass roots had been broken and
underlying soils exposed, were particularly vulnerable to raindrop
splash. Under growing corn, the ground surface was not completely
covered by foliage. Heavy downpours of rain beating on unprotected soil churned it into thin mud which was splashed two feet
high. The splashing of raindrops sifted and washed out of surface
layers soluble salts, clay colloids, and other fine material, leaving a
gritty, silty film which hardened into a lifeless crust or cap and
cracked into polygons as it dried out.[15] Having confronted problems of soil erosion in other parts of North America, scientists discovered that wetland soils in the Midwest were also eroded.

Among solutions proposed for wastage of peat was cessation of
draining and change of land use. In another article in the *Yearbook
of Agriculture* for 1938, F. R. Kenney and W. L. McAtee discussed
the problem of drained areas and wildlife habitats. They reported
that "despoliation has affected almost every kind of wildlife habitat,
but the process has perhaps been most conspicuous and most harmful in relation to aquatic environments. Drainage diversion and pollution have been the principal means of destruction."[16] Soil scientists gave qualified support to early moves toward conserving wetlands for wildlife. The conservation movement advanced much further under leadership from foresters, bird-watchers, game wardens,
and other applied natural scientists.

By 1949, many people were ready to receive an entirely new, at-

14. John R. Haswell, Drainage in the humid regions, in U.S. Department of Agriculture,
Soils and men: Yearbook of agriculture, 1938 (GPO, Washington, D.C. 1938) 726.
15. J. H. Stallings, *Soil conservation* (Prentice-Hall, Englewood Cliffs 1957) 51–61.
16. F. R. Kenney and W. L. McAtee, The problem: Drained areas and wildlife habitats,
in U.S. Department of Agriculture, *Soils and men, 1938,* 79.

titude-changing concept of land and nature. It was propounded by Aldo Leopold, forester, conservationist, and professor at the University of Wisconsin, who lived on a deserted farm in central Wisconsin and died fighting a brush fire along the Wisconsin River. Leopold rejected economists' claims that landowners were free to exploit and dispose of land and natural resources as they wished. His radical concept was that land did not belong exclusively to human beings; it was shared by all living creatures. He wrote: "We abuse land because we regard it as a commodity belonging to us. When we see land as a community to which we belong, we may use it with love and respect."[17] An ideal harmony would prevail when people lived off the land without destroying it. In central Wisconsin, the wetland environment had been severely damaged by draining and ought to have been protected as wilderness.

In theory, Leopold articulated widely shared, deeply felt desires. His "Marshland Elegy" sang the praises of natural wetlands and their secret pleasures, while lamenting the loss of pristine wildness.[18] In practice, solitude was unattainable. People demanded freedom to wander and enjoy wild places. Their exclusion and silence could not be enforced. Areas of special scientific interest were protected; killing of rare species of animals, birds, and fish and collecting of rare plants were prohibited; game hunting was restricted to certain seasons; measures were taken to prevent fires and dumping of litter; but popular consent was difficult to obtain for restrictions on private building development or extensions of public highways. Conflicts between conservationists and recreationists could not be resolved by ecologists or other scientists.

Changing Terminology and Classifications

Concepts and categories of information used by geographers have been changed through time by politicians, journalists, and geographers themselves. Locations and boundaries of regions have been moved, names for different types of wetlands have been revised, characteristics of wetlands in general have been redefined. Of a multitude of terms whose meanings have changed, many are examined in the chapters that follow. This introduction briefly reviews

17. Aldo Leopold, *A Sand County almanac* ([Oxford University Press, New York 1949] Ballantine Books, New York 1990) xviii, 189.

18. Ibid., 95, 101.

changes in usage of five key words: Middle West or Midwest, corn belt, marsh, wet prairie, and wetlands. It is essential to understand changes in definitions of "Midwest" and "wetlands" in reading the title of this study. Changing ideas of "corn belt" as an agricultural region occupying wet prairies and of "marsh" as a generic term for northern wetlands provide means for interpreting narratives of important events.

Changing designation and partitioning of the American West accompanied westward expansion of white settlers from 1780 onward. The name "West" conveyed a mental image of youthfulness, adventure, and opportunity. The East was perceived as densely peopled, powerful, and wealthy but shackled to its colonial past. The West was free. A Northwest Territory, north and west of the Ohio River, was designated in a federal ordinance in 1787 and during the next 61 years, the states of Ohio, Indiana, Illinois, Michigan, and Wisconsin were carved out of this territory. As the West advanced across the Mississippi to drier, short grass plains, the federation was riven by sectional division between slave-owning South and industrial Northeast. The Old Northwest and much of the new West joined forces with the North. After the Civil War, the label "West" referred mostly to the Great Plains region, stretching from the Canadian border to the Gulf of Mexico.

In the 1880s, the term "Middle West" was used in popular writing for the first time. The name was attached to a middle area of the western plains in Kansas and Nebraska, between frontier lands in the Dakotas to the north and the state of Texas to the south. James Shortridge examined literary descriptions of Middle Wests in different locations at different periods. The dominant cultural image of the Middle West situated in Kansas and Nebraska in the period from the 1880s to 1902 was of "a rapidly maturing, mainstream American society."[19] A pastoral ideal of well-cultivated land and established farming families passed an endurance test during years of agricultural depression in the 1890s. The image added stern puritan virtues of thrift and determination to softer pastoral qualities. From 1912 to 1919, in years of high prosperity and intensive draining activity, the name "Middle West" was relocated in the center of the upper Mississippi River basin, taking the places of the obsolete

19. James R. Shortridge, *The Middle West: Its meaning in American culture* (University of Kansas Press, Lawrence 1989) 17.

terms "Old Northwest" and "New Northwest." New Middle Westerners wished to identify with the good image gained by Kansas and Nebraska. Shortridge remarked that literature from that period exuded "an incredible sense of optimism and destiny about the region. A mature agricultural economy was the cultural and economic mainstay, but railroads, steel mills, meatpackers and other industries thrived also, while new universities, libraries, art museums, and political reforms provided a feeling of intellectual accomplishment."[20] Midwesterners viewed themselves as the most American of all Americans. In the 1920s, self-confidence declined. Sinclair Lewis attacked smug complacency in *Main Street* and a worldwide depression shattered a conviction held by commercial farmers that they controlled their own fortunes. From the 1920s to the 1960s, the name and positive image of the Middle West gradually faded. Reports of conflict and disorder from Chicago, Detroit, Cleveland, and other industrial cities cast shadows over the benign image of the Middle West as a peaceful and prosperous region. In the years since 1968, nostalgia for a simple, idyllic rural past has focused on a Middle West relocated in Iowa, Missouri, Kansas, Nebraska, and South Dakota, moved back to where it had been situated nearly a century earlier.[21]

This study takes account of relocations and dislocations of Midwestern consciousness in examining changing attitudes to draining, farming, and conservation.These shifts in regional perception are set within a framework of seven states, Ohio, Indiana, Illinois, Iowa, Michigan, Wisconsin, and Minnesota, whose areas contain the highest proportions of wetlands, drained lands, and conserved lands. A pastoral ideal, materialistic outlook, and nostalgic yearning have played significant parts in land-use decisions taken at different times and different places in this seven-state region called, for convenience, the "Midwest."

The changing form and perception of the corn belt followed a similar course to that of the Midwest. The name "corn belt," in the sense in which the concept is currently understood, was coined in or about 1882.[22] From about 1860, five areas of high corn produc-

---

20. Ibid., 27. Twelve states that adopted the designation "Middle Western" between 1912 and 1919 were Ohio, Indiana, Illinois, Iowa, Michigan, Wisconsin, Minnesota, North Dakota, South Dakota, Nebraska, Kansas, and Missouri, with a focus on Illinois.

21. Ibid., 74.

22. William Warntz, An historical consideration of the terms "corn" and "corn belt" in

tion coalesced to form a single large concentration of corn-growing
and livestock husbandry. The recognition of a distinctive agricultural region in the last two decades of the nineteenth century coincided with a peak period of tile draining activity. The name outlasted important changes in farming practice, economic and social
organization, and rural life. Its boundaries expanded northward in
the early twentieth century to include newly drained prairies in
Michigan, Wisconsin, Minnesota, and Nebraska. A westward expansion of grain-fed cattle production across the plains in Nebraska,
Colorado, Kansas, and Texas began with the introduction of center-
pivot irrigation of crops of corn and soybeans in the 1960s.[23] The
name "corn belt" then applied to a vast area extending from Ohio
to Colorado and from the Red River valley of the North to the high
plains in Texas. It stretched far beyond the core of artificially
drained wet prairies.

Classifications of wetland types were changed when those responsible for making classifications changed. In the first half of the
twentieth century, geologists and soil scientists were firmly in
charge. They identified marshes, swamps, and bogs as different
types of vegetation covering northern peat deposits. Marshes were
open and grass-covered; swamps were tree-covered; bogs were open
and moss-covered. "Marsh" was an inclusive, generic term and the
most common place-name element for wetlands in the northern
lakes states.[24] In the second half of the twentieth century, hydrologists and ecologists attempting to find terms to fit new classifications
considered vernacular words confusing and inadequate. In particular, the word "marsh" was difficult to accommodate in a scientific
classification. In its broader sense, as a place-name element and all-
embracing term for northern wetlands, it was too vague and imprecise to be useful. In its narrower sense, as a term for peatland dominated by grasses and sedges, it was misleading because most aquatic
herbaceous plants grew on permanently waterlogged mineral soils
and did not accumulate peat. A leading textbook by William Mitsch

---

the United States, *Agric. Hist.* 31 (1957) 43, refers to an article in the *Nation* 35 (July 1882)
24.

23. John C. Hudson, *Making the corn belt: A geographical history of middle-western agriculture* (Indiana University Press, Bloomington 1994) 151–88.

24. A. R. Whitson, *Soils of Wisconsin,* in *Wisc. Geol. Nat. Hist. Surv. Bull.* 68 (State of
Wisconsin, Madison 1927) 119–20; Lawrence Martin, *The physical geography of Wisconsin,*
in *Wisc. Geol. Nat. Hist. Surv. Bull.* 36 (State of Wisconsin, Madison 1932) 412, 416–18.

and James Gosselink provided a glossary of common words in which "marsh" was defined as "a frequently or continually inundated wetland characterized by emergent herbaceous vegetation adapted to saturated soil conditions."[25] This expanded and elaborated the vernacular usage of the word.

The nature and extent of wet prairies were difficult to specify. In 1951, Leslie Hewes relied primarily on soil drainage conditions to identify areas of former wetness.[26] At the time Hewes wrote, artificial drainage had almost completely eliminated the distinction between wet and dry prairies: tallgrasses and sedges had been removed, prairie sod was broken, and soils were highly cultivated. In Mitsch and Gosselink's glossary, "wet prairie" was briefly defined and no characteristics were indicated. It was described as "similar to a marsh but with water levels usually intermediate between a marsh and a wet meadow."[27] Marshes were flooded for all or most of the growing season, whereas wet meadow soils were permanently waterlogged below the surface but not covered with standing water for most of the year. In 1971, R. E. Stewart and H. A Kantrud devised a wetland classification for researchers in glaciated prairies in the Midwest. It identified "wetland low prairie" as temporarily flooded.[28] In a systematic classification for the whole of the United States, wet prairies were excluded as "nonwetland."[29] Michael Williams remarked that vast areas of wet prairie "do not appear as wetland by any current classification."[30] Occasionally they were submerged and crops were destroyed or damaged by flooding. In recent descriptions of prairie potholes in the Dakotas, Minnesota, and Iowa, an outer fringe, temporarily flooded, was termed a "low-prairie zone." In all salient features, apart from their

25. William J. Mitsch and James G. Gosselink, *Wetlands,* 2d ed. (Van Nostrand Reinhold, New York 1993) 32.

26. Leslie Hewes, The northern wet prairie of the United States: Nature, sources of information, and extent, *Annals Assoc. Amer. Geog.* 41 (1951) 307–15.

27. Mitsch and Gosselink, *Wetlands,* 1993, 32.

28. R. E. Stewart and H. A. Kantrud, *Classification of natural ponds and lakes in the glaciated prairie region,* U.S. Fish and Wildlife Service Resource Publ. 92 (Washington, D.C. 1971).

29. Lewis M. Cowardin, Virginia Carter, Francis C. Golet, and Edward T. LaRoe, *Classification of wetlands and deepwater habitats of the United States* (U.S. Fish and Wildlife Service, Washington, D.C. 1979) 30.

30. Michael Williams, Agricultural impacts in temperate wetlands, in Michael Williams (ed.), *Wetlands: A threatened landscape* (Blackwell, Oxford 1990) 202.

irregular seasonal hydrological regimes, low prairies were identical with former wet prairies.[31]

The first attempts to formulate scientific classifications of wetlands were made in the 1950s. The generic term "wetlands," spelled as one word, was coined in 1956 by hydrologists S. P. Shaw and C. G. Fredine, working for the federal Fish and Wildlife Service. Circular 39 recognized a need for a national wetlands inventory to determine "the distribution, extent, and quality of the remaining wetlands in relation to their value as wildlife habitat." The inventory described four major categories: inland fresh areas; inland saline areas; coastal freshwater areas; coastal saline areas. Within these categories were grouped 20 wetland types, arranged in order of increasing depth of water or frequency of inundation.[32]

During the 1970s, wetlands became a focus of much political and scientific activity, attracting public attention in adjudication of disputes concerning wetlands to be subjected to or exempted from Section 404 of the Clean Water Acts of 1972 and 1977. In 1975, a federal district court ruled that the U.S. Army Corps of Engineers jointly with the U.S. Environmental Protection Agency had duties under Section 404 of the Federal Water Pollution Control Act Amendments of 1972 to issue or refuse permits for dredging or filling wetland sites. As a result of this decision, the Corps of Engineers quickly issued proposals for regulations defining wetlands by their functions as wildlife habitats, as storage areas for storm and floodwaters, as prime recharge areas for groundwater, and as sedimentation basins. In 1977, the Corps issued a substantially revised definition of wetlands:

Those areas that are inundated or saturated by surface or ground water at a frequency and duration sufficient to support, and that under normal circum-

---

31. H. A. Kantrud, G. L. Krapu, and G. A. Swanson, *Prairie basin wetlands of the Dakotas: A community profile*, U.S. Fish and Wildlife Service Biological Report 85 (Washington, D.C. 1989); Committee on Characterization of Wetlands, William M. Lewis (chair), *Wetlands: Characteristics and boundaries* (National Research Council, Washington, D.C. 1995) 278–83, esp. 280.

32. S. P. Shaw and C. G. Fredine, *Wetlands of the United States: Their extent and their value to waterfowl and other wildlife*, U.S. Fish and Wildlife Service Circular 39 (Washington, D.C. 1956); "wet lands" appear as two words in the title of a paper by historian Richard Lyle Power, Wet lands and the Hoosier stereotype, *Miss. Valley Hist. Rev.* 22 (1935) 33–48; "wet lands" as two words also appear in the passage from Palmer, *Swamp land drainage*, 1915, 1, quoted above.

stances do support, a prevalence of vegetation typically adapted for life in saturated soil conditions. Wetlands generally include swamps, marshes, bogs, and similar areas.[33]

While the Corps of Engineers was defining wetlands in order to regulate Section 404 of the Clean Water Act, the Fish and Wildlife Service was preparing a scientific classification system to replace Circular 39. The new classification arose from "a need to understand and describe the characteristics and values of all types of land, and to wisely and effectively manage wetland ecosystems." In 1979, the Fish and Wildlife Service presented a new definition:

Wetlands are lands transitional between terrestrial and aquatic systems where the water table is usually at or near the surface or the land is covered by shallow water. For the purposes of classification wetlands must have one or more of the following three attributes: (1) at least periodically, the land supports predominantly hydrophytes; (2) the substrate is predominantly undrained hydric soil; and (3) the substrate is nonsoil and is saturated with water or covered by shallow water at some time during the growing season of each year.[34]

In 1990, under "swampbuster" provisions of the Food, Agricultural, Conservation, and Trade Act, the USDA was authorized to refuse federal agricultural loans, payments, and benefits to persons converting wetlands to agriculture. For the purposes of this act, wetland was defined as land that (A) had a predominance of hydric soils; (B) in its undrained state, was saturated, flooded, or ponded long enough during a growing season to develop an anaerobic condition; and (C) supported the growth and regeneration of hydrophytic vegetation.[35]

In 1993, Congress asked the federal Environmental Protection Agency to invite the National Research Council to set up a committee to study the scientific basis for the characterization of wetlands. The committee first addressed the confusion that had arisen through different government agencies emphasizing different parameters, criteria, and indicators for determining wetlands. In 1995, as a means of bolstering confidence, a new uniform reference definition was presented:

33. 42 Fed. Reg. 125–26, 37128–29 (19 July 1977), cited in Committee on Characterization of Wetlands, *Wetlands,* 1995, 51.

34. Cowardin et al., *Classification of wetlands,* 1979, 3.

35. Committee on Characterization of Wetlands, *Wetlands,* 1995, 56.

A wetland is an ecosystem that depends on constant or recurrent, shallow inundation or saturation at or near the surface of the substrate. The minimum essential characteristics of a wetland are recurrent, sustained inundation or saturation at or near the surface and the presence of physical, chemical, and biological features reflective of recurrent, sustained inundation or saturation. Common diagnostic features of wetlands are hydric soils and hydrophytic vegetation. These features will be present except where specific physicochemical, biotic, or anthropogenic factors have removed them or prevented their development.[36]

This study is concerned with the development of all ideas about Midwest wetlands that have influenced management up to the time of the reference definition in 1995. Knowledge of marshes before they were detached from their association with northern peatlands and of wet prairies before they were excluded as nonwetland is as relevant as knowledge of the depth, duration, and frequency of saturation and of plants included on the national hydrophyte list.

Changing Institutions and Agencies

The formation of institutions and transfer of powers from one government agency to another translated attitudes into actions. Inquiries were conducted, opinions debated, measures agreed, and directives issued. Through time, the number of institutions and agencies representing different attitudes and different interests multiplied. This study examines how people living in Midwest wetlands were affected by and helped to influence government action, political programs, and scientific research.

A dream that by going west people would be set free from oppressive government regulation, free from taxation, free from service in foreign wars, free from restrictions on exploitation of natural resources, did not come true. From the beginning of American settlement in the west, federal agents pursued westerners to the farthest frontier. The Treasury Department sold lands from the public domain, the army manned forts, and the Corps of Engineers dredged and embanked channels of navigable waterways. In 1849, a newly established Department of the Interior took over the General Land Office, was made responsible for the Office of Census and was given charge of a recently formed Indian Bureau. In 1862, a Department

36. Ibid., 3.

of Agriculture was established to promote agricultural development by diffusing useful information on subjects connected with agriculture. In 1881, it acquired a new forestry division; in 1883, a veterinary division; and in 1898, it set up a soil survey.[37] In 1879, responsibility for topographical surveying was transferred to the U.S. Geological Survey.

Nineteenth-century attitudes to wetlands were dominated by a Jeffersonian ideal that the rightful occupiers of the soil were independent American farmers. As property owners their freedom to do what they liked with the land was almost unlimited. Few restraints were imposed on destructive exploitation of natural resources and there was little dispute about an owner's right to alter the appearance of landscape. Only where untouched wilderness lay in the public domain were questions of care and guardianship contested by nonresident town dwellers. After 1872, national parks and national forests in the far west of the United States were kept in public ownership for the benefit of people seeking outdoor recreation.

In the twentieth century, nonfarm interests increasingly challenged an absolute right claimed by private owners to destroy natural environments. Progressively, powers to regulate and control land use were taken away from individual occupiers and transferred to more and more distant government agencies. County boards of commissioners had powers removed by state legislatures, states lost powers to federal government, and U.S. sovereignty was gradually curtailed by international agreements on tariffs and trade, affecting farm price support, and on wildlife conservation, through worldwide conventions protecting migratory birds and endangered species.

Since 1920, economists, historians, geographers, hydrologists, ecologists, politicians, and government officials have changed their views about wetlands and have revised public policies toward wetland management. Up to the 1920s, agricultural expansion and draining were foremost concerns of scientists and were actively encouraged by politicians. In the 1930s, soil conservation and financial security were at the top of the political agenda and interest in draining declined. In the period from 1945 to 1970, farm enlarge-

37. Gladys L. Baker, Wayne D. Rasmussen, Vivian Wiser, and Jane M. Porter, *Century of service: The first 100 years of the United States Department of Agriculture* (USDA, Washington, D.C. 1963) 13, 22, 51, 55, 83.

ment and investment in buildings and machines again took priority
and interest in draining revived. From the 1960s, outdoor recreation and wildlife conservation were pressing issues and, in the
1980s, government, with support from scientists, took steps to reduce surplus agricultural production and stop further loss of wetlands. The views of nonfarm interests were now very strong, clam-
orous, and insistent.

A decisive shift in government policy was marked by ending the
publication of detailed statistics on a county basis relating to
drainage of agricultural land. Areas of land in drainage enterprises
and of drained farmland had been collected for national censuses
every ten years from 1920 to 1960, the figures for drained land in
1950 and 1960 being derived by subtracting unreliable estimates of
areas of unimproved land (which included undrained and other uncultivated land) from total areas of land in drainage enterprises.[38]
Later censuses in 1967 and 1978 provided only rough estimates of
drained land. In 1940, the U.S. Fish and Wildlife Service was established within the Department of the Interior. It began to collect data
on the extent of wetlands in 1956 but did not recognize wet prairie
as a wetland category and failed to acknowledge that tile draining
extended beyond the limits of soils characterized as formerly waterlogged.[39] The Fish and Wildlife Service did not monitor changing
acreages of wetlands effectively until a National Wetlands Inventory
was inaugurated in 1983.[40] Maps for the inventory were drawn
from aerial photographs at scales ranging from 1:60,000 to
1:130,000. Photo interpretation and field reconnaissance defined
wetland boundaries according to the 1979 classification devised by
Lewis M. Cowardin and others. By 1991, inventory maps covered
about 70% of the coterminous states, including the whole of
Indiana and Illinois, and most of Minnesota, Michigan, Ohio, and
Iowa. Mapping had made least progress in Wisconsin.[41] The
National Wetlands Inventory project also published reports on the

38. U.S. Bureau of Census, *Census of agriculture, 1959: Drainage* (GPO, Washington,
D.C. 1961), vol. 4, 9–23, esp. 18.

39. Shaw and Fredine, *Wetlands of the United States,* 1956.

40. R. W. Tiner and B. O. Wilen, *The U.S. Fish and Wildlife Service National Wetlands
Inventory project* (U.S. Fish and Wildlife Service, Washington, D.C. 1983).

41. B. O. Wilen, Fact sheets and information (National Wetlands Inventory, St. Petersburg, Fla. 1991).

status of and trends in wetlands and deepwater habitats.[42] In western parts of the Midwest, substantial losses of wetlands in the 1970s and 1980s were attributed to agriculture.

Farmers began to feel isolated from many different government agencies pursuing divergent policy objectives. In the 1960s, farmers in south central Minnesota who wanted help with draining wetlands would call a county agent of the Agricultural Advisory Service and draw up a plan. By the end of that decade, some farmers were beginning to regard draining wet prairies as "improper management of the resource." In 1969, Robert Moline reported that "those opposed to drainage are capturing a significant audience and may justifiably claim that wetland preservation rather than wetland drainage constitutes the public benefit."[43] In the 1990s, farmers planning to lay new drains had to seek permission from the federal Natural Resources Conservation Service, the Agricultural Stabilization and Conservation Service, and the Environmental Protection Agency. If a wet patch had value as a wildlife habitat they needed approval from the U.S. Fish and Wildlife Service and had to consult the Forest Service. If their drains flowed into a stream, they had to satisfy the Army Corps of Engineers about the quantity of water to be discharged. In addition, a number of state agencies had to be consulted. In the state of Minnesota, the Department of Conservation, Game and Fish Division, Division of Forestry, Lands and Minerals Division, and Minnesota Outdoor Recreation Resources Commission had powers to make inquiries, recommend modifications, and request the withdrawal of proposed drainage schemes. Farmers who wanted to drain wet prairies felt frustrated by increasing numbers of government regulations.

Producers of crops and livestock also felt that they had lost their financial independence. From the time of early agricultural settlement in the Midwest, landowners and occupiers borrowed money from sources outside their families. In the nineteenth century, railroads, cattlemen, lumber companies, land companies, and banks were financed partly by New England and New York merchants. In

42. Dahl, *Wetlands losses,* 1990; W. E. Frayer, *Status and trends of wetlands and deepwater habitats in the coterminous United States, 1970s to 1980s* (Michigan Technological University, Houghton 1991).

43. Robert T. Moline. *The modification of the wet prairie in southern Minnesota,* Ph.D. diss., Geography, University of Minnesota 1969, 188.

the twentieth century, multinational corporations manufacturing agricultural machinery and chemicals as well as giant combines of food packers and processors lent money to producers and acquired vested interests in agriculture. Growers became increasingly dependent on outside capital. As their debts mounted, they became less and less free.

Changing economic, social, and cultural aspects of rural life were reflected in a proliferation of new organizations and new publications. During the twentieth century, new academic disciplines, new scientific societies, and periodicals were launched and new approaches were opened to the study of wetlands and other environmental topics. Special interest groups took initiatives in raising questions about conservation of nature. Concerns for mountains, forests, and wildlife were championed by the American Forestry Association founded in 1875, the Sierra Club established in 1892, the National Audubon Society founded in 1905 and the Izaak Walton League organized in 1922.[44] Conservation was not a populist movement; it was led by scientists and technologists, including biologists, geologists, hydrologists, foresters, agronomists, and anthropologists; "its role in history," wrote Samuel Hays, arose "from the implications of science and technology in modern society."[45] It was a movement toward centralized planning and efficient management of natural resources, away from wasteful, competitive exploitation.

Important information on past and present economic and social questions was collected and examined in new journals, notably *Agricultural History*, which first appeared in 1927, and *Rural Sociology*, first published in 1935. Studies described histories of draining, farming, land tenure, and finance, together with attendant crises and failures. Historians and geographers both had long-standing commitments to research on the changing character and extent of wetlands. They traced the progress of draining and the development of conservation. In the 1970s, studies of changing attitudes toward nature, changing representations of landscapes, and changing environmental images drew both environmental historians and

44. Roderick Nash (ed.), *The American environment: Readings in the history of conservation* (Addison-Wesley, Reading, Mass. 1976) ix–xvii.

45. Samuel P. Hays, *Conservation and the gospel of efficiency: The Progressive conservation movement, 1890–1920* (Harvard University Press, Cambridge 1959) 265.

historical geographers together, in debates more than in collaboration. Some of these discussions appeared in two new journals, the *Journal of Historical Geography*, founded in 1975, and the *Environmental Review*, inaugurated in 1976. In a survey of common ground between the two subdisciplines, Michael Williams welcomed the emergence of environmental history as "one of the most exciting things to happen in American history this century."[46] Both environmental historians and historical geographers contributed to studies on four themes: human modification of the earth; global expansion and capitalist economy; the place of humans in nature; and interrelations between habitat, economy, and society. Recent research has begun to explore changing attitudes toward wetlands and has reached out to examine ecological changes in waterlogged environments.[47]

Urgent problems of planning and managing wetlands withdrawn from agricultural use have engaged engineers, hydrologists, and biologists. In March 1980, a group of 35 wetland scientists and managers chartered a new professional society of wetland scientists. In 1995, the society's membership exceeded 4,000 and its quarterly journal, *Wetlands*, reached its fifteenth year of publication.[48] In the 1980s, universities, government agencies, and consultancies began to appoint professional wetland scientists. Employment opportunities for trained natural scientists were greater than for those skilled in the practice of wetland crafts. In the 1990s, occupiers of wetlands were at the center of expanding networks of expert advisors, consultants, researchers, social scientists, federal and state officials, lawyers, accountants, salesmen, lobbyists, politicians, journalists, birdwatchers, artists, vacationers, all of whom expressed different views about food production, conservation, rural life, government expenditure, and other issues related to wetlands.

46. Michael Williams, The relations of environmental history and historical geography, *Journ. Hist. Geog.* 20 (1994) 3.

47. John A. Jakle, *Images of the Ohio valley: A historical geography of travel, 1740 to 1860* (Oxford University Press, New York 1977) 3–20; Roger A. Winsor, Environmental imagery of the wet prairie of east central Illinois, 1820–1920, *Journ. Hist. Geog.* 13 (1987) 375–97; Janel M. Curry-Roper and Carol Veldman Rudie, Hollandale: The evolution of a Dutch farming community, *Focus* 40 (1990) 13–18; Gordon G. Whitney, *From coastal wilderness to fruited plain: A history of environmental change in temperate North America, from 1500 to the present* (Cambridge University Press, Cambridge 1994) esp. 271–82.

48. Society of Wetland Scientists, *1995 membership directory and handbook*, Supplement to *Wetlands* 15 (1995) iii.

Critical Interpretation of Literary Evidence

This study of changing attitudes is based on reading and interpreting accounts of past observers and actors. Material has been drawn from reports and reflections mostly composed by well-informed, articulate observers but also from popular commentators and writers of fiction. Where possible, descriptions have been checked against archaeological, cartographic, statistical, and scientific evidence. Much reliance has been placed on previous critical studies by geographers, historians, sociologists, and ecologists. The originality of this research lies in bringing together fragments of information from different sources that shed light on preferences, prejudices, and social, economic, and cultural outlooks. It has been difficult to avoid surmise and conjecture but inferences have been based on the clearest and most explicit expressions of opinion. Later events showed that many opinions were well-informed and progressive, but some proved to be misguided or even harmful. In the United States, in the nineteenth and twentieth centuries, the recording of historical facts was voluminous and generally accurate, conflicts of interest were boldly presented and openly contested, differing opinions were freely discussed and eloquently argued. Free thoughts, sharply debated and widely published, offered a wealth of material for documenting changing attitudes.

Some voices have remained silent: native Americans removed from their villages, farmers who quit, drainage districts that went bankrupt left few records. On balance, this study does not adequately represent views of individual farmers, draining contractors, conservation officers, and other local residents closest to wetland sites. I did not conduct questionnaire surveys, and I have used notes of my own visits to wetlands only as background impressions in interpreting statements by other writers. Some of the most valuable evidence for opinions of developers and prospective settlers in the first quarter of the twentieth century I found in files of correspondence and press statements from E. R. Jones, state drainage engineer of Wisconsin. In 1954, those files were kept in the College of Agriculture at the University of Wisconsin in Madison. Although I visited state historical societies and museums in Madison, Minneapolis, and St. Paul, I did not inspect local archives. On the other hand, many published local studies contain first-hand observations, and I have gathered numerous expressions of wetlanders' attitudes from biographies of early settlers, edited records, and correspondence of cattlemen, farmers, and travelers published by state and local history societies. I have also drawn upon recent interviews recorded by sociologists.

Chronological and Regional Approaches to Representations of Wetlands

This study traces changing attitudes to lands regarded as wet, water-logged, or inundated at different periods. Chapter 2 does not add anything new to knowledge of the physical geography of wetlands but attempts to connect earlier, geological-pedological-vegetation-association views of wet prairies and bogs with later hydrological-biogeochemical-ecosystem views of a hierarchy of wetlands. It puts different pieces of knowledge together to form new patterns that relate closely to changes in social and economic appraisals of wetland values. At different periods, natural scientists responded to questions posed by society and the economic system, and they themselves raised new problems for society and the economy to solve.

Succeeding chapters follow roughly a chronological sequence, comparing developments in southern wet prairies with northern bogs and swamps. Chapter 3 discusses the extent to which native Americans modified wetlands and possibly assisted their expansion. It attempts to assess how changes in indigenous populations and economies affected wetlands, examines the impact of the fur trade on native societies, and reviews evidence for removal of remains of native American occupation in the nineteenth century.

Chapter 4 discusses early American attitudes toward wetlands. Initial shock and fear of fevers, floods, and fires repelled all but the most intrepid pioneers. Hesitantly, a few ventured to colonize small prairies close to groves of timber. The influence of federal land survey plats and notes on the selection of land by individual purchasers and by states applying for grants of swamplands is examined. Reasons for hostility toward speculators and ineffectiveness of measures intended to encourage settlement by family farmers are reviewed.

Chapter 5 asks why successive attempts to overcome early prejudices against wet prairies did not result in establishing flourishing enterprises. Perceptions and aims of large landowners, cattlemen, railroad promoters, and manufacturers of steel plows and barbed wire are examined. Many entrepreneurs who occupied wet prairies in the middle decades of the nineteenth century were incorrigible optimists. Land prices did not rise as fast as on other types of soil; losses on investment in construction were sustained repeatedly yet a spirit of daring was not quenched. Members of families who suffered setbacks returned after financial panics to buy more land and embark on new ventures. Changes in the organization of large estates and methods of funding improvements are discussed in the light of continuing antagonism toward speculators.

Chapter 6 traces the growing confidence in prospects for wet prairies through drainage improvements. Progressively more effective methods of draining were introduced and these boosted land values. Ultimately, investments in drainage works had to be paid for out of increased production, and corn belt farming was made more efficient. A spiral of success raised expectations and led some to consider whether northern peatlands might be converted into productive farmlands.

Chapter 7 recounts a history of recurring failures and disappointments in attempting to reclaim northern bogs and swamps. Railroads, backed by federal and state land grants, attempted to attract settlers; lumbermen rapidly cleared northern pineries and tried to dispose of cutover stumplands; and other agencies endeavored to sell peatlands to immigrants. Many hoped that draining might transform the condition of peat soils, but draining caused peat to shrink and burn and did little to lengthen a perilously short growing season. Worse still, drainage enterprises were too ambitious, badly organized, and undersubscribed. In the financial crisis of the 1920s most went bankrupt and farming was abandoned. Searches for alternative uses met with little success. Even as forest reserves and wildlife refuges they were too costly and did not offer sufficient employment to sustain permanent communities.

In the 1920s a mood of despondency settled on farmers in a hitherto prosperous section of the corn belt. Chapter 8 discusses changing attitudes to a slump in farm prices and a fall in investment in draining. Farmers clung to their land during the depression, and from 1945 to 1970 they were rewarded by rocketing land values and enormous increases in yields of corn, soybeans, and livestock. Despite economic recovery, they felt increasingly insecure. Endebtedness rose, family succession broke down, local social provision declined. Government officials blamed farmers for eroding the soil, discharging toxic agricultural chemicals into sources of water supply, and destroying wildlife habitats by draining. In the 1980s, wet prairies were hit by a severe farm crisis; in the 1990s, disastrous floods followed a prolonged drought. Grit and determination were not sufficient to enable farmers to pull through. Those who remained hopeful diversified their enterprises, conserved lakes and wildlife, and catered for recreational users of wetlands. The restoration of wetlands raised new questions about the functioning of ecosystems, the authenticity of artificially created natural environments, the legitimacy of ownership, and the appropriateness of management of land and water.

Changes in perceptions and representations of wetlands led to changes in action. Chapter 9 reviews ideas that prevailed in the past and the legacy of those ideas in recent thinking. A major change in wetland values followed an assertion that humans were members of biotic communities: they did not stand outside nature and they could not run away. The survival of wetlands depended on continuing human thought and intervention.

2

Physical Characteristics of
Wet Prairies and Bogs

Looking north from the edge of the Bloomington moraine in 1812, General William Henry Harrison declared: "The country beyond that is almost continued Swamp to the Lakes"[1] At the present time there is scarcely a vestige of wet land between the Bloomington moraine and the shore of Lake Michigan, almost all the country being well-drained and highly cultivated. Indeed, what Harrison described as swamp was not and never had been what would now be called swamp, but he and others among his contemporaries made no distinction between wide expanses of treeless wet prairie in the southern Midwest and swamp forest, open bog, and other types of peatland in the northern lakes country. When he wrote, the distinction was of no importance; today it divides a region of almost uninterrupted farmland from another where forests and farms are interspersed with uncultivated wetlands. For convenience, the southern type of wetland is termed here "wet prairie" and the northern type "bog."

It is not denied that the terms "wet prairie" and "bog" are imprecise and fail to represent a rich diversity of physical characteristics of wetlands in the southern and northern Midwest, but they serve their purpose as labels pointing to important differences. This account compares and contrasts southern wet prairies with northern bogs in terms of landforms, soils, and vegetation. Descriptions of hydrology, biogeochemistry, and ecosystems deal more fully with northern peatlands than with now nearly extinct wet prairies. Studies by Leslie Hewes inscribed the name "wet prairie" on large tracts of tall grasslands rooted in stiff, impervious mineral soils, situated on flat upland till plains on older drift in Ohio, Indiana, Illi-

---

1. Logan Esarey (ed.), *Messages and letters of William Henry Harrison,* Indiana Historical Collections 9 (Indianapolis 1922) 214.

nois, southeastern parts of Wisconsin, and southwest Minnesota.[2]

In a narrow sense, "wet prairie" was a name given to a type of vegetation, so an appropriate designation for its northern counterpart ought to be a name for another vegetation type. "Marsh," by far the most frequent place-name element for wetlands in the northern lakes states of Michigan, Wisconsin, and Minnesota, must be ruled out because marsh soils are predominantly mineral. Confining attention to organic soils, no fewer than three different types of vegetation may be identified in northern peatlands. "Swamp" is dominated by trees, characteristically tamarack (*Larix*); and shrubs, frequently bog birch (*Betula pumila*), growing on firm peat. "Fen" is peat-accumulating wetland receiving some drainage from surrounding mineral soils. Typically it is alkaline or only mildly acid and supports sedge (*Carex*), and willow (*Salix*). "Bog" is peat-accumulating wetland sited in a closed basin that has no significant inflows or outflows. It is acid and is dominated by acid-loving mosses, particularly *Sphagnum*. Fens are much less common than bogs and bogs are much more characteristic of true peat soils than swamp forests. A special legal meaning attaches to the term "swamp," defined in the Swamp Land Acts of the nineteenth century, whereas "bog" has no such restricted usage. Engineers and hydrologists use "bog" as a collective name for northern wetlands and it will serve as a term of contrary meaning to "wet prairie."

Changing Objectives in Scientific Studies of Wetlands

The physical geography of waterlogged areas in the Midwest has been completely rewritten in the past forty years. From the late nineteenth century until the mid-twentieth century "wet lands," phrased as two words, were regarded as lands that were poorly drained. The emphasis was on land. Poor soil drainage was an impediment that prevented land from being utilized productively. Wet lands were described and studied by earth scientists: geologists, geomorphologists, and soil scientists, and by engineers concerned with land

---

2. Leslie Hewes, Some features of early woodland and prairie settlement in a central Iowa county, *Annals Assoc. Amer. Geog.* 40 (1950) 40–57; Hewes, The northern wet prairie of the United States: Nature, sources of information, and extent, *Annals Assoc. Amer. Geog.* 41 (1951) 307–23; Hewes, with Phillip E. Frandson, Occupying the wet prairie: The role of artificial drainage in Story County, Iowa, *Annals Assoc. Amer. Geog.* 42 (1952) 24–50; Hewes, Drained land in the United States in the light of the Drainage Census, *Professional Geographer* 5 (1953) 6–12.

drainage in schools of agriculture and forestry. Since the 1950s, the term "wetlands," spelled as one word, has gained currency. Wetlands are studied by wetland scientists: hydrologists, biogeochemists, biologists, and ecologists concerned with the conservation of plants and animals. The emphasis has shifted away from land to wetness. Indeed, the U.S. Fish and Wildlife Service classification of wetlands includes not only lands at least periodically saturated or covered by water but also deepwater habitats where water is more than 2 meters deep.[3]

Wetlands are described as distinctive habitats transitional between terrestrial and aquatic environments, places where hydrophytic plants grow and also where hydric soils form. Undrained areas having both hydrophytic vegetation and hydric soils are normally accepted as wetlands, but where either vegetation or soils are not characteristically hydric, scientists "have difficulty defining the *minimum* threshold of inundation and/or saturation necessary to create and maintain wetlands."[4] By observing hydrological conditions, Ralph Tiner would identify as wetlands areas that are either flooded by flowing water for more than one week during the year in most years or saturated near the surface by surface water or groundwater for more than four to six weeks during the year in most years.

Disappearing Wet Prairies

A profound change in approach to the study of waterlogged areas is reflected in the replacement of earth science terminology by wetland science terminology and by new maps delineating wetland areas much smaller in extent than were recorded on earlier maps. Early steps in the formulation of the U.S. Fish and Wildlife Service classification can be traced from a typology of vegetation based on common names (wet meadow, swamp, fen, marsh) to a hierarchical structure (system, subsystem, class) that avoids common names.[5] In

3. Lewis M. Cowardin, Virginia Carter, Francis W. C. Golet, and Edward T. LaRoe, *Classification of wetlands and deepwater habitats of the United States* (U.S. Fish and Wildlife Service, Washington, D.C. 1979) 3.

4. Ralph Tiner, How wet is a wetland? *Great Lakes Wetlands* 2 (1991) 1.

5. Cowardin et al., *Classification of wetlands,* 1979, 27–31; A. C. Martin, N. Hotchkiss, F. M. Uhler, and W. S. Bourn, *Classification of wetlands of the United States,* U.S. Fish and Wildlife Service, *Spec. Sci. Rep. Wildlife* 20 (Washington, D.C. 1953); S. P. Shaw and C. G. Fredine, *Wetlands of the United States,* U.S. Fish and Wildlife Service Circular 39 (1956); R. E. Stewart and H. A. Kantrud, *Classification of natural ponds and lakes in the glaciated*

the course of revision, Stewart and Kantrud's zone of "wetland low prairie" has been deleted from the current classification because it is now defined as "nonwetland."[6] Three types of wetland distinguished by separate names in popular speech—marshes, wet meadows, and fens—have been reduced to a single class designated as "emergent wetlands."[7] The inclusion of shallow lakes as wetlands in the classification has been more than offset by what is described as a "loss" of wetlands, mostly resulting from agricultural drainage.[8]

The deletion of the category "wet prairie" follows extensive reclamation for agriculture by the digging of ditches and laying of tile drains. The area provided with artificial drainage exceeds the area that, before draining, was intermittently, seasonally, or semi-permanently flooded.[9] Only in exceptionally wet years such as 1973 and 1993 are these lands flooded during the summer.[10] On the western margin of tallgrass prairies, in western Minnesota, South and North Dakota, and in the prairie provinces of Canada, small shallow pools known as "prairie potholes" are surrounded by temporary or seasonal wetlands, but these have also been encroached upon by artificial drainage[11] (fig. 2.1). Wet prairies are now farm-

prairie region, U.S. Fish and Wildlife Service, Resource Publ. 92 (Washington, D.C. 1971); F. C. Golet and J. S. Larson, Classification of freshwater wetlands in the glaciated Northeast, U.S. Fish and Wildlife Service, Resource Publ. 116 (Washington, D.C. 1974).

6. Cowardin et al., Classification of wetlands, 1979, 30.

7. Ibid., 28.

8. Thomas E. Dahl, Wetlands losses in the United States 1780s to 1980s, U.S. Fish and Wildlife Service (Washington, D.C. 1990); R. W. Tiner, Wetlands of the United States: Current status and recent trends, U.S. Fish and Wildlife Service, National Wetlands Inventory (Washington, D.C. 1984); G. A. Pavelis (ed.), Farm drainage in the United States: History, status, and prospects, U.S. Department of Agriculture, Economic Research Service, Misc. Publ. 1455 (Washington, D.C. 1987).

9. Robert T. Moline, The modification of the wet prairie in southern Minnesota, Ph.D. diss., University of Minnesota 1969, 83, remarks on the disappearance of the name "wet prairie" and attributes this to the elimination of poor drainage from mineral soils in the prairie region.

10. C. Parrett, N. B. Melcher, and R. W. James, Flood discharges in the upper Mississippi River basin 1993, U.S. Geological Survey Circular 1120-A (Washington, D.C. 1993); Hugh Prince, Floods in the upper Mississippi River basin, 1993: Newspapers, official views, and forgotten farmlands, Area 27 (1995) 118–26.

11. William J. Mitsch and James G. Gosselink, Wetlands, 2d ed. (Van Nostrand Reinhold, New York 1993) 57–59; M. W. Weller, Freshwater marshes: Ecology and wildlife management, 2d ed. (University of Minnesota Press, Minneapolis 1987); J. A. Leitch and L. E. Danielson, Social, economic, and institutional incentives to drain or preserve prairie wetlands (Department of Agricultural and Applied Economics, University of Minnesota, St. Paul 1979); hydrology, geochemistry, and soils are discussed in J. L. Richardson and J. L.

lands subject from time to time, here and there, to residual problems of inadequate soil drainage. They no longer support hydrophytes and are not classed as wetlands.

Extent of Northern Peatlands

Sketch maps drawn in the 1970s represent the southern edge of the northern peatlands just reaching into northern Minnesota, but many classic sites described in ecological reports lie far to the south of that line.[12] Important studies describe peatlands in Minnesota, Wisconsin, Michigan, and some sites further south in Ohio, Indiana, and Illinois.[13] The demarcation of the regional boundary is based on a redefinition of peatlands. Neither the Fish and Wildlife Service classification nor the Soil Conservation Service taxonomy identifies peatland or peat as a distinctive category.[14] Scientific classification substitutes "histosols" or "hydric organic" soils for common names "peat" and "muck," while "palustrine moss-lichen, scrub-shrub, forested and emergent wetlands" are substituted for what are commonly called "bogs," "swamps" and "marshes."[15]

Not only has scientific terminology changed, but peat itself has disappeared from large areas as a result of digging, draining, decomposition, and fire. In Wisconsin, in 1976, alluvial soils of stream bottoms and peat and muck soils of major wetlands covered 2,850,000 acres or about 8.2% of the state's land area. It was estimated that the actual area on the ground might be double that, since "many small areas of these soils cannot be shown on the soil map

Arndt, What use prairie potholes? *Journ. Soil and Water Conservation* 44 (1989) 196–98.

12. P. H. Glaser, *The ecology of patterned boreal peatlands of northern Minnesota: A community profile,* U.S. Fish and Wildlife Service, Report 85 (7.14) (Washington, D.C. 1987); J. Terasmae, Postglacial history of Canadian muskeg, in N. W. Radforth and C. O. Brawner (eds.), *Muskeg and the northern environment in Canada* (University of Toronto Press, Toronto 1977) 9–30.

13. R. L. Lindeman, The developmental history of Cedar Creek Lake, Minnesota, *Amer. Midl. Naturalist* 25 (1941) 101–12; M. L. Heinselman, Forest sites, bog processes, and peatland types in the glacial Lake Agassiz region, Minnesota, *Ecol. Monographs* 33 (1963) 327–74; J. A. Larsen, *Ecology of the northern lowland bogs and conifer forests* (Academic Press, New York 1982).

14. U.S. Soil Conservation Service, *Soil taxonomy: A basic system of soil classification for making and interpreting soil surveys,* USSCS Agricultural Handbook 436 (Washington, D.C. 1975); U.S. Soil Conservation Service in cooperation with National Technical Committee for Hydric Soils, *Hydric soils of the United States* (Washington, D.C. 1987).

15. Cowardin et al., *Classification of wetlands,* 1992, 5.

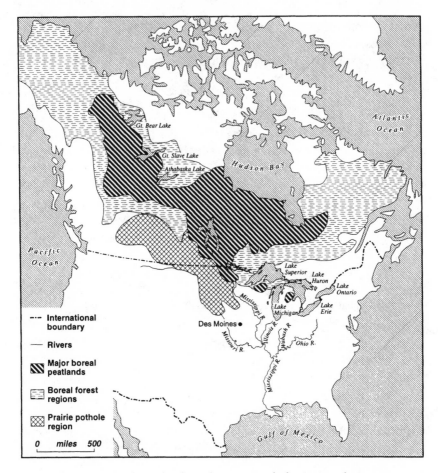

Figure 2.1 Boreal peatlands and prairie potholes in North America.

Sources: William J. Mitsch and James Gosselink, *Wetlands*, 2d ed. (Van Nostrand Reinhold, New York 1993) 370; Jon H. Goldstein, *Competition for wetlands in the Midwest: An economic analysis* (Resources for the Future, Washington, D.C. 1971) frontispiece; Committee on Characterization of Wetlands, William M. Lewis (chair), *Wetlands: Characteristics and boundaries* (National Research Council, Washington, D.C. 1995) 278, outlines area of prairie potholes.

because of its scale."[16] The diminishing area of organic soils, pre-viously known as peat and muck, is not accounted for solely by a

16. Gerhard B. Lee, Soil Region J: Soils of stream bottoms and major wetlands, in Francis D. Hole (ed.), *Soils of Wisconsin,* in *Wisc. Geol. Nat. Hist. Survey Bull.* (University of Wisconsin Press, Madison 1976) 115.

restricted definition or by omission from the map of areas too small to plot, but is represented in part by actual wastage and removal of peat. Gerhard Lee mentions the exploitation of several thousand acres of sphagnum peat in Jackson and Juneau counties for horticultural purposes. In addition, agriculture has claimed 53,500 acres or about 2% of the total area of peat and muck in the state. The full extent of shallow peat lost by slow oxidation or burning after drainage is not known. Lee remarks that "some areas mapped as peat in early surveys in Price county are wet mineral soils today." Uncontrolled fires in bogs in north central Wisconsin continued to deplete peat resources.[17]

Fifty years earlier, in 1926, another soil survey of Wisconsin estimated the area of peat alone to be 2,540,800 acres or 7.2% of the state's land area. Other wet soils, including muck, Clyde, Dunning, and various mineral soils, accounted for an additional 2,339,200 acres or 6.6% of the state. The total area of major wetlands amounted to 4,880,000 acres, about 13.8% of the state.[18] Thousands of small kettle holes and numerous other pockets of waterlogged soils were too small to record on the map, but the difference between areas surveyed in 1976 and 1926 also reflects a large amount of peat lost through shrinkage, cultivation, exploitation, and combustion. A. R. Whitson reported that peat had been burned over "quite extensive areas" in central Wisconsin.[19] In 1915, F. W. Huels observed that forest and "marsh fires, frequently smoldering for months, burn great holes deep into the deposit" and "many peat areas in northern Wisconsin have been partly consumed by such fires."[20] In 1903, Samuel Weidman noted that large volumes of organic soils had been destroyed by repeated burning in cutover forests in north central districts.[21] In 1907, A R. Whitson and E. R. Jones included stiff, hard-to-drain clays in their estimated total of 7,360,000 acres of wetland, nearly 21% of the state's surface.[22]

17. Lee, Soil Region J, 1976, 120.

18. A. R. Whitson, *Soils of Wisconsin*, in *Wisc. Geol. Nat. Hist. Surv. Bull.* 68 (State of Wisconsin, Madison 1927) 33, 83, 133.

19. Whitson, *Soils*, 1927, 133.

20. F. W. Huels, *The peat resources of Wisconsin*, in *Wisc. Geol. Nat. Hist. Surv. Bull.* 45 (Madison 1915) 241.

21. Samuel Weidman, *Preliminary report on the soils and agricultural conditions of north central Wisconsin*, in *Wisc. Geol. Nat. Hist. Surv.* (Madison 1903) 46.

22. A. R. Whitson and E. R. Jones, *Drainage conditions of Wisconsin*, in *Univ. Wisc. Agric. Expt. Sta. Bull.* 146 (Madison 1907) 46 (table ix).

Assuming that mineral soils were more extensively converted to
nonwetland through artificial drainage, the area of peat in Wisconsin must have been considerably greater in 1907 than in 1926 or
1976.

In Michigan, in 1900, peat and muck soils were estimated as
covering 5,200,000 acres or about 11.6% of the surface of the state,
but that probably accounted for less than half the area of wetlands at
the time of entry by pioneer settlers.[23] In the 1980s, the wetland
area was estimated at 5,583,400 acres or 15% of Michigan's surface
area. An apparent increase in the extent of wetlands in Michigan between 1900 and the 1980s may be accounted for partly by excluding shallow peat deposits from the earlier survey and partly by including lakes and inland waters in the total area of the state.[24] In
Minnesota, a report in 1913 made by the state drainage commission
estimated an original extent of swamp, wet, and overflowed lands as
10,112,720 acres or 18.8% of the surface area of the state.[25] In the
1980s, the wetland area was estimated at 8,700,000 acres or 16.2%
of the state's surface.[26] The total losses of peatlands in Wisconsin,
Michigan, and Minnesota are not as extensive as in wet prairie regions. The plow and drain tile did not make deep or lasting inroads
into bogs and swamps, but the ax of the lumberman and subsequent
burning destroyed large areas of peat.

Wastage of Peat by Fire

Great conflagrations in the northern woods in the late nineteenth
century did not spare exposed peat deposits. Unquenchable fire-
storms that raged across the Midwest in October 1871 from Indiana
to North Dakota destroyed small towns in northern Michigan, and

23. J. D. Towar, *Peat deposits of Michigan,* in *Mich. State. Agric. Coll. Expt. Sta. Bull.* 181
(East Lansing 1900) 157, cited in Charles A. Davis, *Peat: Essays on its origin, uses, and
distribution in Michigan,* Report State Board of Geol. Survey Michigan for 1906 (East
Lansing 1907) 97–395; Michigan Department of Natural Resources, Recreation Division,
Michigan's 1987–88 Recreation Action Program (East Lansing 1988).

24. The total area of Michigan in 1900 was recorded as 44,827,580 acres; in 1980, it was
37,583,400 acres; U.S. Department of Interior, Fish and Wildlife Service, National wetlands
inventory, unpublished data, St. Petersburg, Fla. 1983.

25. Appendix 4, table A, in Ben Palmer, *Swamp land drainage with special reference to
Minnesota,* University of Minnesota Studies in Social Sciences 5 (Minneapolis 1915) 122;
E. K. Soper, *The peat deposits of Minnesota,* in *Minn. Geol. Surv. Bull.* 16 (1919).

26. University of Minnesota Center for Urban and Regional Affairs, *Thematic map:
Available wetlands for bioenergy purposes* (University of Minnesota, St. Paul 1981).

Peshtigo in Wisconsin, and reduced to ashes swampy districts in Chicago. In Wisconsin, it swept over Lake Horicon or Winnebago Marsh which, according to an eyewitness report, "was itself on fire."[27] Fires in Michigan in 1881 were even more destructive, and in 1894 conflagrations again swept across Michigan and Wisconsin and devastated Hinckley in Minnesota. In 1908 fires returned yet again to Metz in Michigan, to Minnesota, and to Wisconsin, where Frederick Huels observed that Peshtigo Marsh "has many burned stumps. At the time of visit the marsh was very dry and burning. Forest fires were raging round about the marsh and prospecting was done under difficulties."[28] Swamps in Chippewa, Vilas, and Barron counties were extensively burned over, and on the Door peninsula, Sturgeon Bay Bog was burning in the summer of 1908. At the time of prospecting, "the dry vegetation had been pretty well burned off. The peat itself was on fire."[29] Yet more fires broke out in 1910, 1911, and 1918, ravaging large areas in Michigan and Minnesota. Stephen Pyne concludes that late-nineteenth-century conflagrations inflicted enormous damage on northern peatlands: "In former bogs as much as a dozen feet of organic mat had been burned away, leaving bare soil and rock as a residue. In many former pine sites only sand remained."[30] However destructive individual outbreaks might have been, their cumulative effect was not as serious as the effect of draining on wet prairies. Total losses of peat attributable to nineteenth- and twentieth-century fires must be measured in tens of thousands, not millions, of acres. Vast areas of peat deposits still remain far to the south of the Canadian muskeg.

A new definition of the southern edge of northern peatlands rests more heavily on ecology, on an association of boreal plants, than on the presence or absence of organic soils. The limit is thought to be determined by the intensity of solar radiation in summer months when precipitation and humidity are otherwise adequate to support bogs further south. In sum, the distinctively boreal character of northern peatlands results from the incidence of long,

27. Stephen Pyne, *Fire in America: A cultural history of wildland and rural fire* (Princeton University Press, Princeton 1982) 38, 200, 203; C. D. Robinson summarized the effects of the October 1871 conflagration in Wisconsin. His report is reprinted in Franklin Hough, *Report on Forestry* 3 (GPO, Washington, D.C. 1882) 231–36.

28. Huels, *Peat,* 1915, 140.

29. Ibid., 121, 127, 134, 139.

30. Pyne, *Fire,* 1982, 217.

hard winter frosts and very short growing seasons in summer.[31] The move of the boundary northward is not associated with global warming or other physical changes in the environment. Vegetation has been stripped and peat has been thinned by fire, by lumbering, and to a small extent by clearing for cultivation, but these losses have been much smaller than losses suffered by wet prairies in the south.

Reconstructing Past Physical Geographies

Almost all vestiges of wet prairie have disappeared as a result of agricultural drainage. They were drained before wetland scientists had an opportunity to observe them and study their hydrology and ecology. In the northern lakes states of Michigan, Wisconsin, and Minnesota, extensive areas of peat remain but their character has been modified by widespread clearing of trees and repeated burning. At present, wetland scientists are more interested in less disturbed ecosystems north of the Canadian border.

A history of artificial drainage starts from an earlier physical basis, profoundly different from that on which a history of wetland conservation is based. A history of native American utilization of wetland resources relates to a physical geography formed at an even earlier period. The reconstruction of these early physical geographies draws upon surveys and notes made by explorers and scientists who viewed the physical environment in ways that differ from contemporary observations. Many nineteenth- and early twentieth-century land surveyors, soil scientists, and agricultural writers discussed how wetlands might be improved by artificial drainage and reclaimed for agriculture and human settlement. Many modern wetland scientists discuss how wetlands may be preserved as wildlife habitats, contribute to flood protection, or serve to reduce the discharge of agricultural pollutants, industrial effluents, and domestic sewage into streams.

Currently, geomorphologists, hydrologists, soil scientists, biologists, and systems ecologists are contributing to the description and classification of wetlands. For the purpose of implementing Section 404 of the Clean Water Act of 1977, the U.S. Army Corps of Engineers was required to draw up a legal definition of wetlands that

---

31. Larsen, *Ecology,* 1982, cited in Mitsch and Gosselink, *Wetlands,* 1993, 371.

were to be protected by the refusal of permits for dredge-and-fill operations.[32] The primary objective of the present regulation is to safeguard aquatic vegetation. A definition drawn up in 1975 laid somewhat greater emphasis on waterlogged soils: "those areas that normally are characterized by the prevalence of vegetation that requires saturated soil conditions for growth and reproduction."[33] That definition did not cover mats of aquatic plants floating on shallow water and it was also difficult to apply to areas from which aquatic vegetation had been removed or destroyed. For managing and regulating wetlands, the new revised U.S. Army Corps of Engineers' definition specifies vegetative cover as a primary indicator. The 1979 U.S. Fish and Wildlife Service definition, taking account of wetland hydrology, wetland soils, and hydrophytic vegetation, provides a broader framework for ecological and socioeconomic studies and will be adopted in the following description.

Terrain

The salient characteristics of five roughly differentiated types of wetlands in the American Midwest are summarized in table 2.1. For comparative purposes, characteristics of individual features are grossly simplified, but looked at together they provide an indication of how different ecosystems function. Landforms are the first features to be identified. In 1907, Charles A. Davis recognized that bogs in Michigan occupied low-lying (lowl, in table 2.1) situations. Many were located in shallow lake basins; others formed in river floodplains (FP, in table 2.1). Davis also noted that some bogs developed a buildup of organic matter and sediment from the bottoms of lakes; others developed by growth of aquatic vegetation inward from the water's edge.[34] In most lakes both processes were at work.

Almost all wetlands in the Midwest lie within the glaciated area. In Ohio and Indiana the southern limit of wet prairies corresponds exactly with the limit of glaciation and in other states the correspondence is very close[35] (fig. 2.2). South of the glacial margin, most extensive tracts of wetlands are bottomlands or floodplains in

32. 33 CFR328.3(b) 1984, cited in Mitsch and Gosselink, *Wetlands,* 1993, 27.
33. 42 *Fed. Reg.* 37128, 19 July 1977, cited in Mitsch and Gosselink, *Wetlands,* 1993, 27.
34. Davis, *Peat in Michigan,* 1907, 97–395.
35. Guy Harold Smith, The relative relief of Ohio, *Geog. Rev.* 25 (1935) 282.

TABLE 2.1 Comparison of wet prairie, marsh, fen, swamp, and bog

		Prairie	Marsh	Fen	Swamp	Bog
Terrain	Situation	upld	lowl	lowl	lowl	lowl
	Form	flat	FP	FP	basin	basin
	Glaciation	Old	Old	New	New	New
			New	dless	dless	dless
Water	Source	prcn	sfce	sfce	prcn	prcn
		sfce	grnd	grnd	sfce	
	Regime	seas	semi	perm	perm	perm
	pH value	5–7	5–7	5–8	3–7	3–5
Soil	Organic %	5–15	5–35	30–90	40–100	50–100
	Clay %	55–95	10–80	5–60	0–50	0–40
Plants	Diversity	modL	modH	high	modL	low
	Trees		few	few	few	
	Shrubs		few	few	few	
	Grasses	many	few	few	few	
	Sedges	few	many	many	few	
	Reeds		many	many	few	
	Mosses				few	few
	Prim. prod. kg/m²/yr	0.3–2.8	1.0–3.5	0.5–1.2	0.3–2.0	0.1–1.9

Source: Mitsch and Gosselink, Wetlands, 1993.

the lower courses of the Mississippi, Missouri, Ohio, and their major tributaries. J. E. Carman has illustrated cartographically the contrast between indeterminate drainage on the Iowan drift-covered plain and a well-developed network of streams in the driftless (dless, in table 2.1) area northeast of Savanna, Illinois, in the upper Mississippi valley.[36] In the former locality, waterlogged soils, sloughs, and lakes abound on a more or less flat surface; in the latter, a dendritic pattern of streams has produced a deeply dissected, well-drained surface. In detail, the obvious difference between wet and well-drained land is that the former occupies only relatively flat surfaces. There is no equivalent in the Midwest of blanket bog, which covers hilly country in Ireland and western Scotland, on slopes occasionally exceeding 15 degrees.[37] Instead, the edge of a level surface is

36. J. Ernest Carman, The Mississippi valley between Savanna and Devonport, Illinois, in Ill. State Geol. Surv. Bull. 13 (1909) 54–55.

37. A. G. Tansley, The British Islands and their vegetation (Cambridge University Press 1939) 714, 718.

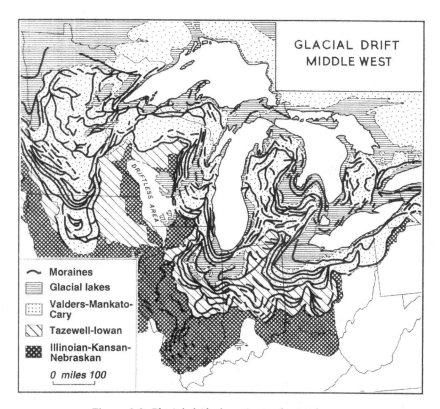

GLACIAL DRIFT
MIDDLE WEST

DRIFTLESS AREA

~ Moraines
Glacial lakes
Valders-Mankato-Cary
Tazewell-Iowan
Illinoian-Kansan-Nebraskan

0 miles 100

Figure 2.2 Glacial drift deposits in the Midwest

Source: F. T. Thwaites, *Outline of glacial geology* (F. T. Thwaites, Madison 1953).

generally marked by an abrupt change from waterlogged to dry conditions. On till plains of the older drift (Old, in table 2.1) in Ohio, Indiana, Illinois, and Iowa the flat, waterlogged surface is a high-level upland (upld) surface. From a field observer's viewpoint, J. T. Curtis identified wet prairie as "lowland prairie," drawing a distinction between wetlands seasonally flooded, supporting hydrophytic plants, and drier tall grasslands occupying somewhat higher ground.[38] In relation to other wetlands, wet prairies clearly occupy upland situations by contrast with low-lying river bottomlands and lakeshore marshes. In the hummocky terrain of the newer drift (New, in table 2.1) in Michigan, Wisconsin, and Minnesota, wet-

38. J. T. Curtis, *The vegetation of Wisconsin* (University of Wisconsin Press, Madison 1959) 284–95.

lands lie at the bottom of low-lying depressions. In Wisconsin, some low-lying basins and river floodplains lie in the driftless area.

Water

Studies of wetlands in a context of general circulation of water and of water as a medium in which wetlands are formed and aquatic plants and animals grow are of recent origin. Most important scientific papers on these subjects have been published in the past thirty years. In the hierarchical ordering of wetlands in the 1979 U.S. Fish and Wildlife classification, the two highest orders, "systems" and "subsystems," are based on wetland hydrology. The first-order systems divide wetlands into marine, estuarine, riverine, lacustrine, and palustrine according to whether water is salt or fresh, and whether it flows in rivers and lakes or simply floods land permanently, seasonally, or intermittently. The second-order subsystems divide wetlands according to depth and duration of flooding. The palustrine system, to which most wetland in the Midwest belongs, is not divided into subsystems but into third-order classes based on characteristics of the substrate, such as "rock bottom," "unconsolidated bottom," or according to vegetation, such as "moss-lichen," "scrub-shrub" or "forested wetland."[39]

The importance of hydrology in the formation of wetlands contrasts starkly with the small amount of research on the topic that has been published. In 1993, Mitsch and Gosselink boldly asserted that "hydrology is probably the single most important determinant of the establishment and maintenance of specific types of wetlands and wetland processes."[40] All wetlands receive some water from precipitation (prcn, in table 2.1) and where precipitation exceeds evapotranspiration the surplus will collect in enclosed basins as lakes or pools. Lakes will be colonized by aquatic plants, and remains of dead plants will accumulate and be laid down on lake bottoms, forming beds of peat. Most fens and almost all bogs and swamps are permanently flooded (perm), peat-accumulating wetlands. Some fens are semipermanently (semi) inundated. Fens and swamps may receive from and discharge into flowing surface (sfce) streams and floods more water than they receive from precipitation. Fens may also be subjected to small inflows and outflows of groundwater

39. Cowardin et al., *Classification of wetlands*, 1993, 5.
40. Mitsch and Gosselink, *Wetlands*, 1993, 68.

(grnd). The flow of groundwater rarely exceeds the net gain from precipitation over evapotranspiration but may be significant in bringing lime-rich water, thereby raising pH values. Occasional stream floods may deluge a swamp with 50 times as much water as it receives from total annual precipitation, and movements of surface water are important in lowering concentrations of acidity. Studies of annual water budgets, showing how amounts of water received from different sources are balanced against amounts discharged, have been conducted in an alluvial swamp in southern Illinois, in a marsh on the shore of Lake Erie in northern Ohio, and in ten prairie potholes in North Dakota.[41]

Marshes and wet prairies are either semipermanently or seasonally (seas, in table 2.1) flooded. In marshes, the water level is raised by frequent surface water floods and to a small extent by variations in the level of groundwater. Wet prairies are waterlogged by snowmelt and heavy downpours of rain in spring and may be flooded occasionally by rainstorms in summer. Because prairie subsoils are generally impermeable or have a low degree of permeability, most waterlogging results from rainstorms or surface flooding. The distinctive pattern of high water levels in spring in the Kankakee River bottomlands in northeastern Illinois has been described and the hydroperiod has been characterized by Mitsch and others as "a hydrologic signature of each wetland type."[42] The hydroperiod defines the frequency and duration of rising and falling levels of surface and subsurface water in a wetland. It expresses the changing balance of all inflows and outflows of water through time. An effect of water standing in a closed basin for a long period is to increase its acidity, to lower its pH value. Northern bogs have some

41. W. J. Mitsch, Interactions between a riparian swamp and a river in southern Illinois, in R. R. Johnson and J. F. McCormick (tech. coords.), *Strategies for the protection and management of floodplain wetlands and other riparian ecosystems,* U.S. Forest Service General Technical Report WO-12 (Washington, D.C. 1979) 63–72; W. J. Mitsch and B. C. Reeder, Nutrient and hydrologic budgets of a Great Lakes coastal freshwater wetland during a drought year, *Wetlands Ecology and Management* 1 (1992) 211–23; J. B. Shjeflo, Evapotranspiration and the water budget of prairie potholes in North Dakota, *U.S. Geol. Surv. Prof. Paper* 585-B. 1968; T. C. Winter, Hydrologic studies of wetlands in the northern prairie, in A. G. van der Valk (ed.), *Northern prairie wetlands* (Iowa State University Press, Ames 1989) 16–54; Mitsch and Gosselink, *Wetlands,* 1993, 77–80.

42. W. J. Mitsch, W. Rust, A. Behnke, and L. Lai, *Environmental observations of a riparian ecosystem during flood season,* University of Illinois Water Resources Center Research Report 142 (Urbana 1979); Mitsch and Gosselink, *Wetlands,* 1993, 72–74.

highly acid waters, whereas fens, marshes, and wet prairies fall
within a normal range from mildly acid to slightly alkaline. High
degrees of acidity inhibit plant growth in bogs, restricting species diversity and also limiting primary productivity.[43] Few detailed studies
have yet been made at wetland sites in the Midwest of water
budgets, hydroperiods, and relations between hydrology, species diversity, and biomass productivity.

Soil

Reviewing evidence afforded by maps and statistics, Leslie Hewes
concluded that soil types provide the clearest and most reliable index of past drainage conditions.[44] This view was subsequently
elaborated by his pupil, Bert Burns, in a detailed study of a wet
prairie area in Minnesota and is one technique currently used to delineate wetlands under guidelines issued by the U.S. Army Corps of
Engineers.[45] The outstanding feature of all poorly drained soils in
the Midwest is their high organic content. The presence of undecomposed organic matter gives them a distinctive black or dark
brown coloration. It is true that some forest soils also have a dark
surface layer, but under conditions of normal drainage it is neither
as deep nor as well preserved. In a poorly drained soil the process
of decomposition is retarded or arrested by a high level of groundwater and imperfect aeration (fig. 2.3). Soils in low-lying basins, subject to permanent waterlogging, are composed almost entirely of
black, undecomposed organic matter. In 1975 the U.S. Soil Conservation Service defined organic soils as having more than 20% to
35% organic matter, and true peat soils as containing not less than
50% organic matter.[46] Soils on upland till plains are largely composed of mineral particles derived from underlying parent material

43. J. G. Gosselink and R. E. Turner, The role of hydrology in freshwater wetland
ecosystems, in R. E. Good, D. F. Whigham, and R. L. Simpson (eds.), *Freshwater wetlands:
Ecological processes and management potential* (Academic Press, New York 1978) 63–78;
M. L. Heinselman, Landscape evolution and peatland types and the Lake Agassiz Peatlands
Natural Area, Minnesota, *Ecol. Monograph* 40 (1970) 235–61; Mitsch and Gosselink,
Wetlands, 1993, 104–9.

44. Leslie Hewes, *Northern wet prairie*, 1951, 313–15.

45. Bert Earl Burns, Artificial drainage in Blue Earth County, Minnesota, Ph.D. diss.,
Geography, University of Nebraska 1954, 3; U.S. Army Corps of Engineers, *Federal manual
for identifying and delineating jurisdictional wetlands* (Washington, D.C. 1989).

46. U.S. Soil Conservation Service, *Soil taxonomy*, 1975.

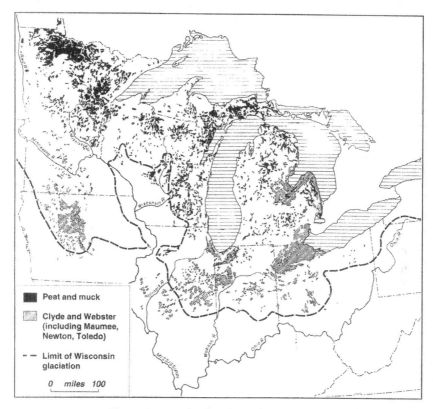

Figure 2.3 Wetland soils in the Midwest

Bog and swamp (Peat and Muck) and wet prairie (Clyde and Webster) soils are differ-
entiated. Compiled from maps at a scale of 1:2,500,000 (pl. 5, sects. 1, 2, 7, 8).
Source: Curtis F. Marbut, Soils of the United States, *Atlas of American agriculture*
(U.S. Department of Agriculture, Washington, D.C. 1931) plate 5.

and tend to be somewhat lighter in color at the surface. A reason
why wet prairie soils are seasonally waterlogged is that they contain
a very high clay or silt fraction and rest on impervious clay or silty
subsoils.

Plants

Fundamental differences in relief and soils between wetlands in the
southern and northern Midwest are, to some extent, obscured by
the use of vegetational terms to indicate the divisions, but to classify
wetlands in the south as wet uplands or claylands and those in the
north as wet lowlands or peatlands would be to ignore the rich di-

TABLE 2.2 Primary productivity of some North American wetlands

	Average kg/m^2/yr
Cattail marshes	2.74
Reed marshes	2.10
Swamp forests	1.05
Sedge fens	1.04
Bogs and fens	0.91
Wet grasslands	0.51

Sources: Michael Williams (ed.), *Wetlands: A threatened landscape* (Blackwell, Oxford 1990)23; after C. J. Richardson, Primary productivity values in freshwater wetlands, in P. E. Greeson, J. R. Clark, and J. E. Clark (eds.), *Wetland functions and values: The state of our understanding* (Amer. Water Resources Assoc., Minneapolis 1979) 131–45.

versity of their flora and the manifold complexity of their ecosystems. Differences in relief and soils are obvious and may be studied in the field in all parts of the Midwest, whereas native plant communities can be observed only in a few unoccupied or rarely visited localities. Plants examined in relatively undisturbed habitats are recognized as belonging to many different ecosystems, and ecologists have continued to argue about the classification and ordering of local variants since the scientific study of ecology began in the early years of the twentieth century. Through time, increasing emphasis has been given to studies of aquatic plants and transitions from open water to emergent vegetation. Also the theory of evolution has been modified in the light of systems theory and quantitative analysis. During the same period, American studies have diverged from European studies, Europeans tending to concentrate more on woody plants and competition between individual species while Americans have become more interested in the behavior and functioning of groups or associations of living organisms.

Species diversity in wet prairie is moderately low (modL, in table 2.1). Growth of trees and shrubs is suppressed by frequent burning and grazing by large mammals. Tallgrasses, including big bluestem (*Andropogon gerardi*) and bluejoint grass (*Calamagrostis canadensis*), are absolutely dominant. Varieties of rushes (*Scirpus*), sedges (*Carex*), cattails (*Typha*), and other herbaceous plants are also present, but the dense mass that forms a prairie sod is almost entirely composed of stems and roots of tallgrasses. Primary productivity (Prim. prod, in table 2.1) of wet prairie is fairly high, averaging less than 2 kg/m^2 per year (table 2.2).

Intermediate in character between wet prairie and marsh is wet meadow, having somewhat higher species diversity. Waterlogging is more persistent and deeper than in wet prairies and the mat of grass roots is less dense. Willow (*Salix*) and a few other woody plants have succeeded in establishing themselves, and reeds (*Phragmites*), rushes, and sedges are more widely represented. Primary productivity is generally somewhat higher than for wet prairie, averaging no more than 2 kg/m^2 per year (table 2.2).

Marsh is semipermanently flooded. It has a much richer assemblage of plants, its species diversity being moderately high (modH, in table 2.1). Characteristic plants include trees and shrubs, many varieties of rushes (*Juncus* and *Scirpus*), smartweeds (*Polygonum*), sedges, cattails, and wild rice (*Zizania aquatica*). Among trees and shrubs, alders (*Alnus*), willows, and occasional pines are present. Primary productivity is generally high and in some places is as high or higher than that of any ecosystem in the world.

Fen has a higher biodiversity but lower productivity than marsh. Fen carr may be dominated by alder or black spruce (*Picea mariana*). Pine, northern white cedar (*Thuja occidentalis*), and bog birch (*Betula pumila*) may also be present. Large numbers of herbaceous plants, reeds, rushes, sedges, and leatherleaf (*Chamaedaphne calyculata*) are likely to form a thick ground layer. The decayed remains of fen plants yield deposits of brushwood peat, reed peat, and sedge peat. Fens, like swamps and bogs, accumulate organic matter.

Swamp has moderately low diversity of species and fairly low primary productivity, averaging about 1 kg/m^2 per year. Swamp water is derived mainly from precipitation running from surrounding slopes, occasionally from floods. Water may be highly acid and soil wholly or largely organic. A layer of deep peat may support trees and shrubs, characteristically tamarack (*Larix*), cedar, less frequently spruce and bog birch. A ground layer is commonly formed by sedges and cotton grasses (*Eriophorum*)[47] (table 2.3).

Bog is the most elemental type of peat-accumulating wetland. It has the lowest diversity of plants, being dominated almost exclusively by mosses, characteristically *Sphagnum*. Where sphagnum covers the surface little else will grow. Patches of cotton grass and

---

47. Curtis, *Vegetation of Wisconsin*, 1959, includes in a swamp category "southern forests—lowland," 156–68, "northern forests—lowland," 221–42, and "alder thicket," 355–57.

TABLE 2.3 Species richness and hydrologic regime in northern Minnesota peatlands

	Tree	Shrub	Field Herb	Grass Fern	Ground layer	Total	Flow conditions
1. Rich swamp forest	6	16	28	11	10	71	Good surface flow, minerotrophic
2. Poor swamp forest	3	14	17	12	5	51	Downstream from 1, sluggish flow
3. Cedar string bog and fen	3	10	10	12	4	39	Better drainage than 2
4. Larch string bog and fen	3	9	9	12	4	37	Similar to 3, sheet flow
5. Black spruce feather moss	2	9	2	2	10	25	Gentle flow, semiconvex template
6. Sphagnum bog	2	8	2	1	7	20	Isolated, little standing water
7. Sphagnum heath	2	6	2	2	5	17	Wet, soggy on convex template

Sources: Mitsch and Gosselink, Wetlands, 1993, 106; after Heinselman, Landscape evolution, 1970, 235–61; Gosselink and Turner, The role of hydrology, 1978, 63–78.

clumps of ling (*Erica*), may establish themselves here and there, but generally grasses and shrubs cannot survive among mosses. Primary productivity measured by dry weight is extremely low, but the volume of organic matter swollen by water is bulky and little or none of the biomass escapes from a bog.

In a context of neighboring plant communities, wetlands are situated in two major regions. Wet prairies belong to a region in which grassland predominates over broad expanses of upland plains and woodland clings to strips of steeply sloping land along valley sides (fig. 2.4). In this region the boundary between woodland and grassland is sharp, but between dry and wet prairie indistinct. Open bogs, swamp forests, and fens, on the other hand, belong to a forested region in which they appear as openings or lightly wooded tracts set among generally close-canopied woodland. Here the boundary between vegetation of dry and wet sites is sharper than the transition either from hardwood to coniferous forest or from swamp forest to open bog. The common characteristic of all wetland vegetation in the Midwest is its paucity of trees. Is it plausible therefore to attribute treelessness to waterlogging? May an absence of trees on the prairies be explained by poor drainage and ought the distinction between so-called "wet" and "dry" prairies to be dismissed?

Wet Prairie on Upland Till Plains

Wet prairies are approached from the south across the broad bottomlands of the Mississippi, the Missouri, the Ohio, or one of their tributaries. These are the lowest lands in the Midwest, lower in places than the beds of the rivers, threatened by floods when rivers overflow their banks. Upstream the valleys narrow, their flanks become steeper, the horizon rises and closes in. Climbing the side of one of these valleys, one reaches the prairie surface. The break of slope is abrupt, much more pronounced than along the front of low hills, the subdued ranges of terminal moraines, which now no longer form a continuous edge to the glaciated area. The southern rim of the prairie surface is deeply bitten into and scalloped by many active streams flowing toward the Mississippi; to the north the view opens out over an upland plain stretching toward lines of fresher, more recent terminal moraines. The surface of the flat prairie has undergone little modification since ice of Illinoian, Iowan, Tazewell, and Cary age laid down sheets of till, or since beds of extinct glacial lakes were exposed. A periglacial landscape is

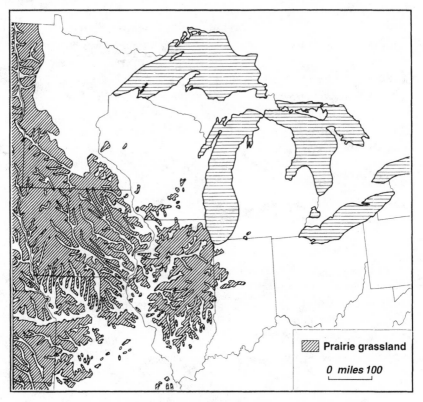

Figure 2.4 Prairie grassland in the Midwest

Source: H. L. Shantz and Raphael Zon, Natural vegetation, in *Atlas of American agriculture* (U.S. Department of Agriculture, Washington, D.C. 1923) figure 2.

buried deeply beneath glacial drift and a new drainage network has begun to develop on the drift surface. In the north, on fresh drift, drainage becomes indeterminate and the ground miry. The level of the land descends step by step to the lakeshore lowlands and bogs of the northern lake country. Westward into Iowa and Minnesota, headstreams of the Mississippi and Missouri have cut less deeply than those in Illinois, and the surface of both old and new drift is well preserved. It is a smooth, flat upland, an unbroken expanse of open prairie.

In Ohio, Indiana, and Illinois wet prairie almost invariably occupies broad, flat uplands; it extends across interfluves but not valleys; above all, it is characteristic of flat upland till plains. One of the most extensive of these plains is the Grand Prairie of Illinois, on whose level upland surface natural drainage is imperfect.[48] Before ditches and drains were provided, water used to stand on the ground until it either evaporated or found its way through the soil to shallow channels.[49] A large part of the prairie upland on either side of the upper Illinois valley was reported by Carl Sauer to have been too poorly drained originally to produce normal crops.[50] The river has cut a deep trench where it flows southward, in its middle course. Some 200 feet above the valley bottom, the monotonously level plains of Bureau and Delavan Prairie were unimproved grazing lands and haunts of wildfowl until better drained slopes had been settled.[51] In 1857 Frederick Gerhard described and mapped a hundred such prairies whose edges were notched with deep valleys but whose flat tops were ill drained.[52]

In north central Iowa the whole surface of the land is appreciably flatter than in Illinois. Because valleys are shallower and interfluves correspondingly lower, the difference between the level of valley bottoms and uplands is very slight. But valley slopes, however gentle, are drier than uplands and the shallowest depressions on the uplands are subject to waterlogging.[53] In Minnesota and north of the Canadian border, the most extensive area of wet prairie lies below the level of the highest land, but above the present floodplain of the Red River. It is situated on a flat expanse once occupied by glacial Lake Agassiz, one of the largest tracts of level land in the world.[54] Its position is intermediate between the summit-level wet prairies of Illinois and marshes and bogs which occupy many former lake beds in Michigan and Wisconsin.

48. Wesley Calef and Robert Newcomb, An average slope map of Illinois, *Annals Assoc. Amer. Geog.* 43 (1953) 314.

49. G. W. Pickels and F. B. Leonard, *Engineering and legal aspects of land drainage in Illinois*, in *Ill. State Geol. Surv. Bull.* 42 (1928) 168.

50. Carl O. Sauer, *Geography of the upper Illinois valley and history of its development*, in *Ill. State Geol. Surv. Bull.* 27 (1916) 87.

51. Harlan H. Barrows, *Geography of the middle Illinois valley*, in *Ill. State Geol. Surv. Bull.* 15 (1910) 68–74; Stanley D. Dodge, Sequent occupance on an Illinois prairie, *Bull. Geog. Soc. Philadelphia* 29 (1931) 209; John K. Rose, Delavan Prairie: An Illinois corn belt community, *Journ. Geog.* 32 (1933) 2.

52. Frederick Gerhard, *Illinois as it is* (Chicago 1857) 200–259.

53. Hewes and Frandson, Occupying the wet prairie, 1952, 24.

54. Arthur Moehlman, The Red River of the north, *Geog. Rev.* 25 (1935) 79; U.S. Department of Agriculture, *Soils and men: Yearbook of agriculture, 1938* (Washington, D.C. 1938) 1077.

No sharp dividing line can be drawn between a broad zone extending from Ohio to Minnesota in which upland wet prairie is the commonest type of wetland and a northern area in which lowland bogs prevail. In the catchment area of the Great Lakes, in lands drained by the Maumee, for instance, the gradient from upland to floodplain is imperceptible and wetland once stretched without interruption from one to the other. To the south of the Maumee River lay 1,500 square miles of the Black Swamp of Ohio. It formerly extended from the treeless wet prairie upland, where grass grew eight feet high on ground that was firm and dry for the greater part of a year, to the banks of the river itself, where the surface was frequently inundated to a depth of three feet and dense swamp forest was the predominant type of vegetation.[55] Wetlands intermediate between wet prairie and open bog once blanketed most of northern Indiana; both the crests of dividing ridges and intervening lowlands were waterlogged and inhospitable.[56] To the west, wet prairies of the southernmost glacial lake beds graded down to marshes, bogs, and sloughs of the Kankakee country, graphically described by Alfred Meyer.[57] Still further west lies the Green River lowland, another waterlogged depression in this borderland between wet prairie and bog country.

Marsh conditions may be found in large depressions and incipient drainage ways in dissected drift in upland prairie regions of Ohio, Indiana, Iowa, and Minnesota, but marshes are extremely rare on uplands.[58] Less than one acre in ten thousand has deep peat deposits on the uplands of ten central Illinois counties.[59] Ribbons of marsh occupy bottomlands of major rivers. In the middle Illinois valley only a fraction of the area was considered cultivable in 1910;

55. Martin R. Kaatz, The Black Swamp: A study in historical geography, *Annals Assoc. Amer. Geog.* 45 (1955) 1–35.

56. Richard Lyle Power, Wet lands and the Hoosier stereotype, *Miss. Valley Hist. Rev.* 22 (1935) 37; Power, *Planting corn belt culture: The impress of the upland Southerner and Yankee in the Old Northwest* (Indiana State Hist. Soc., Indianapolis 1953) 63–91.

57. Alfred H. Meyer, The Kankakee "marsh" of northern Indiana and Illinois, *Papers Mich. Acad. Sci., Arts and Letters* 21 (1935) 359–96; Meyer, Circulation and settlement patterns of the Calumet region of northwest Indiana and northeast Illinois: The first stage of occupance; the Pottawatomie and the fur trader, 1830, *Annals Assoc. Amer. Geog.* 44 (1954) 245–74.

58. Wallace E. McIntyre, Land utilization of three typical upland prairie townships, *Econ. Geog.* 25 (1949) 263; Burns, Artificial drainage, 1954, 49.

59. F. C. Gates, A bog in central Illinois, *Torreya* 11 (1911) 205–11.

the remainder, in the words of Harlan Barrows, was "an unoccupied waste of marshes, sloughs, lakes and ponds," which was considered likely to remain uninhabited.[60] In addition to the difficulty of draining bottomlands, the threat of floods acted as a powerful deterrent to their settlement and reclamation.[61]

While it is true that all wetlands lie on relatively flat surfaces, it is also evident that they may occupy either flat uplands or flat lowlands. The difference between the two, if not explicit in the terms "wet prairie" and "bog," is clearly observable on the ground. Uplands in the southern Midwest are wet because they are flat, because their drainage pattern is poorly developed, because they are underlain by impervious till. Drainage of almost all upland till plains is improved by the provision of tile drains, but the extent of improvement varies from place to place, from swales that in their natural state are more or less permanently waterlogged, incapable of raising a normal crop without ditches and drains, to rising ground that rarely fails to produce a normal crop, but may benefit from supplementary draining.

In contrast to the permanence of high water levels in bogs, the depth and extent of waterlogging in wet prairies change from year to year and season to season, according to the amount and incidence of precipitation and intensity of snowmelt. Large areas may remain completely waterlogged throughout the year, while in other years wet patches may vanish after only a brief spell of flooding. These periodic variations in the degree of waterlogging are a characteristic feature of wet prairie. There is no doubt that the original land surveyors were inconsistent in their identification of wet prairies because their work had to be carried out at different seasons over a number of years.

Silty Clay Loam Soils of Wet Prairies

The existence, or former existence, of wet prairie is indicated by the presence of soils whose surface characteristics appear to differ little, if at all, from other prairie soils. Their surface layer or A. horizon is

60. Barrows, *Middle Illinois valley,* 1910, 6.

61. Pickels and Leonard, *Land drainage in Illinois,* 1928, 24–25; James B. Anderson, White River: Historical influences observed in Morgan County, *Indiana Mag. Hist.* 43 (1947) 249.

at least 8 inches thick, rich in humus, and dark brown or black in color. It is in the subsoil or B. horizon that the effect of imperfect drainage manifests itself. A distinctive mottled gray and yellow or brown layer is indicative of impeded soil drainage. The parent material in the C. horizon ranges widely in character from a highly calcareous, stiff, blue clay, derived from compact impervious glacial till, to silts developed from less well consolidated material that looks like loess.[62]

Poor drainage in prairie soils may result from one or more of three causes. It may be due to soil texture, to the presence of an impermeable hardpan layer in the subsoil, or to topography. Widely differing conditions of drainage may be observed in soils that seem to be identical in other respects. Most prairie soils are derived from glacial drift. Where the parent material is blue clay till, soil drainage is almost invariably poor, but where it is derived from coarse-grained, stony till or outwash sands and gravels, good drainage may be expected, unless a hardpan layer has formed or the soil has developed in an enclosed depression.

In southeast Wisconsin the relationship between poor drainage conditions and compact soils derived from fine-grained glacial till was recognized by T. C. Chamberlin in 1883.[63] In Illinois the term "Black Gumbo" succinctly describes the sticky nature and water-retentive properties of black organic clay loams of the Clyde and Webster series which cover many flat prairie uplands.[64] Clyde soils occur mainly in eastern Iowa, in central Illinois, and on the southern border of Wisconsin, in areas underlain by older drift of Iowan, Illinoian, or Kansan age. Webster soils occur in three major areas: the first in north central Iowa and adjoining parts of Minnesota, the second in east and northeast Illinois with isolated patches extending into southeast Wisconsin, the third in west central Indiana. They are derived from calcareous till laid down during later phases of the Wisconsin glaciation.[65] To the northwest, on either side of the Red

62. C. F. Marbut, Soils of the United States, pt. 3 of O. E. Baker (ed.), Atlas of American Agriculture (U.S. Department of Agriculture, Washington, D.C. 1935) 64.

63. T. C. Chamberlin, Geology of Wisconsin, Survey of 1873–1879 (Madison 1883), vol. 1: 681.

64. Edith Muriel Poggi, The prairie province of Illinois, Ill. Studies in Soc. Sci. 19 (University of Illinois Press, Urbana 1934) 54.

65. USDA, Yearbook of Agriculture, 1938, 1053–54. The soils in a broad belt extending from Minnesota through Iowa, Illinois, and Indiana into the Black Swamp of Ohio and the

River valley in Minnesota, vast expanses of poorly drained Fargo clays and clay loams occupy basins of glacial Lake Agassiz. They have a very heavy black surface layer extending to a depth of over a foot, below which lies a gray or olive gray calcareous clay subsoil.[66]

In some wet prairie uplands it is possible to distinguish between soils which have inherently poor surface drainage because of a high clay fraction in the topsoil and those which have poor under-drainage because of an impervious layer in the subsoil. In southern and western Illinois, the planosol group of soils, having a truncated profile, developed on till of Illinoian age and silts presumed to be loess. Typical of this group are the Putnam series, whose light-colored, leached topsoils rest upon a layer of hardpan.[67] In these soils poor underdrainage is the chief cause of waterlogging. Poor drainage over much of central Ohio, Indiana, and Illinois has been attributed specifically to the effects of a tight clay subsoil.[68] In Illinois and Iowa the amount of artificially drained farmland in 1930 exceeded the amount of land classed as having poor natural drainage. The remainder was land with an impervious subsoil, diffi-cult to work under normal conditions. It was not so ill-drained as to have been incapable of raising a crop without ditches and tile drains but was greatly improved by being provided with adequate under-drainage.[69]

Relief is everywhere a contributory cause of poor drainage, more important in the north than in the south. Leslie Hewes states that "poor drainage in the areas of Wisconsin glaciation is due chiefly to topography, ponding or flatness, whereas southward rela-tively impervious subsoil becomes the chief cause."[70] Where pond-ing occurs and waterlogging is permanent rather than seasonal, or-ganic matter accumulates and remains undecomposed to form a

---

Saginaw lowland of Michigan were classified as Clyde by C. F. Marbut, Soils, 1935, 26–27, but the later classification followed here distinguishes Clyde soils in the west from Newton, Maumee, and Toledo soils in the east.

66. USDA, *Yearbook of Agriculture, 1938,* 1077.

67. Ibid., 1104.

68. P. B. Sear, The natural vegetation of Ohio: II. The prairies, *Ohio Journ. Sci.* 26 (1926) 129; W. LeRoy Perkins, The significance of drain tile in Indiana, *Econ. Geog.* 7 (1931) 381; McIntyre, Land utilization, 1949, 260, 263; Arthur Weldon Watterson, *Economy and land use patterns in McLean County, Illinois,* Department of Geography Research Paper 17 (University of Chicago, Chicago 1950) 21, 25–27.

69. Hewes, Northern wet prairie, 1951, 317.

70. Ibid., 310.

layer of peat. Black prairie soils contain between 5% and 20% vegetable matter in their upper layers; peat soils contain not less than
50%, frequently more than 80%. Prairies are bordered on the north
by a number of lowlands whose soils are derived in part from clays,
silts, sands, and gravels and in part from accumulations of peaty material. Soils grade from silty clay loams on prairie uplands, through
muck and degraded peat into deep peat. Such transitions may be
observed in the Green River lowland, the Kankakee Marsh, the
Black Swamp, the Saginaw lowland, in several other lowlands
around the shores of Lake Huron, and in many intermorainal
troughs.[71] Profiles of Newton, Maumee, and Toledo soils show
some features which may be found in situations transitional between
wet prairies and bogs. A layer of dark gray or black muck forms the
top 8 to 12 inches, below which lies a mottled sand or silt permanently waterlogged under natural conditions. Newton and Maumee
soils in the Green River lowland and Kankakee Marsh are dark gray
silt loams overlying glacial or fluvioglacial gravels, silts, and clays.[72]
Toledo soils in the Black Swamp of Ohio and Saginaw lowland of
Michigan are deep, black silt loams and silty clay loams developed
either on lacustrine silts and clays or on highly calcareous till of
Wisconsin age.[73]

All wet prairie soils are predominantly mineral soils, mostly
clays, silts, and silty clay loams, with an appreciable admixture of
organic matter. They are exceptionally well endowed with plant nutrients in a form which can be assimilated easily. They are rich in
available nitrogen, phosphorus, and potassium, and most of them
derive a high lime content from parent till. Most wet prairies have a
neutral or mildly acid topsoil, their colloids being rich in calcium
even where no free lime is present.[74] They are heavy soils to cultivate but under proper management produce the highest yields of
corn in the world.

71. George J. Miller, Some geographic influences in the settlement of Michigan and in
the distribution of its population, *Bull. Amer. Geog. Soc.* 45 (1913) 344; Meyer, Kankakee
"marsh," 1935, 364; Kaatz, Black Swamp, 1955, 1, 15–18, 26.

72. USDA, *Yearbook of Agriculture, 1938,* 1115.

73. Ibid., 1117.

74. Ibid., 971.

Prairie Vegetation and Poor Drainage

Wet and dry prairies differ very little from one another. The great midcontinental grasslands of North America are characterized by a uniformity of vegetation, while also containing large numbers of nondominant species[75] (fig. 2.4). They are marked by an absence of trees and a paucity of shrubs, and all prairies, whether wet or dry, are dominated by grasses, in particular by big bluestem (*Andropogon gerardi*).[76] Big bluestem constitutes about 78% of the plant cover of a typical prairie studied by J. E. Weaver. It produces over 1 kg/m^2 of dry, underground plant parts in the top 10 centimeters of soil and another 0.3 kg/m^2 in the next 20 centimeters.[77] Every sample of soil is permeated with fine, absorbing rootlets to a depth of over a meter or to the water table. Grass roots form a thick, closely knit, almost solid mat, seldom penetrated by invading plants. The toughness of the sod acted as a serious impediment to its early cultivation. In its virgin state it was nearly impossible to plow and vast areas were first broken with axes. After breaking, it was normal practice to leave it to rot for two or three seasons until it became loose and workable.

Imperfectly drained prairie is distinguished by an exceptionally luxuriant growth of tallgrass, some of which reaches a height of three meters or more. The composition of its flora differs from that of dry prairie in the type of grasses present and also in the variety of other plants, mostly hydrophytes, not frequently found in dry prairie. Bluejoint grass (*Calamagrostis canadensis*), present in 74.3% of dry prairie plots, is represented in 91.0% of the wet prairies examined by J. T. Curtis in Wisconsin, whereas big bluestem, whose presence is recorded in 96.8% of dry prairie plots, is present in only 68.2% of wet prairies.[78] Of other species, slough grass (*Spartina pectinata*) is the most widespread and reliable indicator of wet localities.[79] Rushes, sedges, cattails, reed bent grass, wild rice, Canadian wild rye, and tall panic grass are also prevalent

75. John Ernest Weaver, *North American prairie* (Johnsen, Lincoln, Neb. 1954) 118.

76. Homer C. Sampson, An ecological survey of the prairie vegetation of Illinois, *Ill. Nat. Hist. Surv. Bull.* 13, art. 16 (1921) 523–77.

77. Weaver, *North American prairie*, 1954, 150, measures the productivity as 4.1 tons of dry plant parts per acre in the top four inches of soil and a further 1.3 tons per acre in the next eight inches.

78. Curtis, *Vegetation of Wisconsin*, 1959, 284–85.

79. Hewes, Northern wet prairie, 1951, 307, 322.

although they are by no means as common or as numerous as in wet meadows or open marshes, constituting only a minority of the component plants.

Wet prairie is dominated by a drought-resisting flora. Hence it is argued that the climate must be "more favorable to grass than to trees or shrubs or, indeed, to any other type of vegetation."[80] Specifically, as J. T. Curtis puts it, plant communities dominated by grasses can exist only under a climatic regime "characterized largely by an excess of potential evaporation over available rainfall."[81] Detailed analysis of rainfall data shows that this argument has much to commend it. Although the total precipitation is rarely insufficient for cultivated plants, it is extremely variable and unevenly distributed throughout the year.[82] In summer and autumn wet prairie has fewer rain days and lower relative humidities than forest regions. During July and August the whole grassland zone may be affected occasionally by severe and prolonged droughts accompanied by abnormally hot, scorching conditions.[83] The occurrence of such droughts is held to be a primary cause of the formation of prairies.[84]

An entirely different explanation for the absence of trees has been put forward by several writers who believe it is caused by waterlogging in the soil rather than dryness in the atmosphere. The coincidence of prairie and ill-drained land has been noted in Ohio, Illinois, and Iowa. In three typical upland prairie townships in Illinois, Wallace McIntyre's maps show that areas of grassland recorded by the original land survey correspond exactly with areas of poor natural drainage.[85] In the upper Illinois country Carl Sauer observed that "a large part of the prairie surface is imperfectly drained."[86] The late settlement of prairies in north central Iowa is attributed by W. J. Berry to their waterlogged condition, and in Illinois, Delavan Prairie remained unoccupied until after the Civil

80. Weaver, *North American prairie*, 1954, 3, 7; E. N. Transeau, The prairie peninsula, *Ecology* 16 (1935) 423–37.

81. Curtis, *Vegetation of Wisconsin*, 1959, 302.

82. John Lorence Page, *Climate of Illinois* (University of Illinois Press, Urbana 1949) 43–51; S. S. Visher, *The climate of Indiana* (Bloomington 1944) 403–10.

83. John R. Borchert, The climate of the central North American grassland, *Annals Assoc. Amer. Geog.* 40 (1950) 5, 9, 12, 31, 33.

84. B. Shimek, The prairies, *Lab. Nat. Hist. Univ. Iowa Bull.* 6 (1911) 169–240.

85. McIntyre, Land utilization, 1949, 264–65.

86. Sauer, Upper Illinois valley, 1916, 87.

War, as J. K. Rose suggests, "because of its swampy condition."[87] P. B. Sears roundly declares, "the prairies of Ohio represent areas of inadequate drainage."[88] If throughout the broad zone extending from Ohio to Iowa, prairies represent areas of inadequate drainage, it is not unreasonable, as F. T. Thwaites infers, that "the origin of prairies is clearly related to poor subsoil drainage which inhibited tree growth."[89] From this it is a short step to asserting that "prairies are due to the undrained condition of the flattish interstream areas."[90]

The two explanations offered seem at variance with each other. The climatic explanation accounts for the dominance of drought-resisting grasses; the edaphic explanation accounts for the absence of trees. Neither suffices to explain why a drought-resisting flora is growing on waterlogged soils. The difficulty may be resolved to some extent by considering the seasonal incidence of droughts and floods. The most notable characteristic of the rainfall regime in the prairie region is its variability and marked seasonal variation. Periodic extremes in the distribution of rainfall may be critical for the growth of trees. Severe droughts may occur in summer or early autumn, while extremely heavy downpours may fall in spring and early summer.[91] It is during the spring and early summer that surplus water, unable to find an outlet from flat, retentive clay surfaces, stands in pools and lakes for weeks or months until it evaporates.[92] In Minnesota large floods spread over wet prairies at snowmelt and may persist until midsummer.[93] For part of nearly every year the land is under water, a breeding ground for mosquitoes and a haunt for wildfowl.[94] At a later season the same spot may be desiccated, with cracks and fissures extending three feet or more in the sub-

87. W. J. Berry, The influence of natural environment in north central Iowa, *Iowa Journ. Hist. and Politics* 25 (1927) 280–82; Rose, Delavan Prairie, 1933, 2.

88. Sears, Natural vegetation of Ohio, 1926, 145.

89. F. T. Thwaites, *Outline of glacial geology* (Madison 1953) 63.

90. Barrows, *Middle Illinois valley,* 1910, 77.

91. E. N. Transeau, Precipitation types of the prairie and forested regions of the central states, *Annals Assoc. Amer. Geog.* 20 (1930) 44.

92. Pickels and Leonard, *Land drainage in Illinois,* 1928, 168; L. M. Turner, Grassland in the flood plains of Illinois rivers, *Amer. Midland Nat.* 15 (1934) 770–80; McIntyre, Land utilization, 1949, 263.

93. Moehlman, Red River, 1935, 80.

94. Rose, Delavan Prairie, 1933, 2; Watterson, *McLean County,* 1950, 75; Gerhard, *Illinois as it is,* 1857, 259.

soil.[95] Under such conditions dry prairie may be regarded as "merely the late summer aspect of wet prairie."[96]

The division between prairie and forest is marked by differences in their precipitation regime, but the dryness or wetness of an area cannot be measured by precipitation alone. The incidence of drought or flood is related to moisture present in the soil. In C. W. Thornthwaite's scheme, "moisture in the soil is regarded as being a balance between what enters it as a result of precipitation and what leaves through evaporation and transpiration."[97] A water deficit or water surplus represents the difference between the amount of rain or snow falling on the earth's surface and the amount needed for evaporation and plant transpiration, that is, the amount that would disappear as evapotranspiration under optimum moisture conditions from soil continuously at field capacity. In terms of water balance regimes, only very slight differences are observed between forest and prairie regions, and the border shifts back and forth depending on prevailing conditions subject to appropriate time lags. The incidence of both water deficit and water surplus is almost identical in Madison, Wisconsin, which is located in a forest region, and Ames, Iowa, located in the prairie. A late summer water deficit in a forest region in Michigan, however, is smaller than in prairie regions and is much smaller than at places such as Greenville, Mississippi, in a southern forest. According to these measurements, prairies experience more severe droughts than forest regions. Spring and early summer water surpluses, on the other hand, are more pronounced in prairies than in neighboring forests but somewhat smaller than at Hartford, Connecticut, on the eastern seaboard. Proposed climatic origins for prairie need reconsidering in the light of Thornthwaite's results. If it is claimed that grasses dominate only where potential evapotranspiration exceeds available precipitation, prairies might be expected to extend much further south to the lower Mississippi valley. Alternatively, if their formation is attributed to excessive spring rainfall, they might be expected in New England as well as in the Midwest.

At least two other related factors need to be taken into account:

95. Transeau, Precipitation types, 1935, 423–37.

96. Sears, Natural vegetation of Ohio, 1926, 129.

97. C. W. Thornthwaite, John R. Mather, and Douglas B. Carter, *Three water balance maps of eastern North America* (Resources for the Future, Washington, D.C. 1958) 9.

first, the exposed situation of the upland plains; secondly, the role of fire. Flat upland prairies are exposed to the full force of winds from all quarters. Blizzards, snowstorms, hailstorms, and tornados sweep fiercely across the smooth, unprotected surfaces. Wind checks the free growth of trees and prevents them from reaching their full stature. It increases the rate of evaporation and transpiration and it also drives before it autumnal prairie fires. Fire spreads unchecked across a level plain until it reaches a break of slope at the brink of a valley or rise of a hill.[98] Relief and exposure determine the extent and limits of burning and also the margin of the prairie. Local residents have long recognized this fact by using the word "prairie" in a physiographic rather than a vegetational sense. To them a prairie is a flat treeless expanse, not a particular kind of grassland.[99]

Prairies are also, to a large extent, "ancient cultural features," repeatedly burned over before Europeans entered the region.[100] Native Americans turned fire loose on dry grass to drive game out of cover, to improve and extend the range of grazing, and to make clearings for cultivation.[101] The extent and nature of such burnings may be judged from early-nineteenth-century eyewitness accounts which were almost unanimous in the opinion that fires were responsible for the absence of trees on the prairie.[102] The observed effect of particular fires in recent times supports the view that both wet and dry prairies are perpetuated by fire but does not account for the origin of their vegetation.[103] Many tracts of forest have been reduced to barrens by repeated burning without being replaced by prairies. In Wisconsin J. T. Curtis observed that "oak woods can never be converted to true prairie by the agency of fire."[104] If, as he

98. Carl O. Sauer, *Agricultural origins and dispersals* (Amer. Geog. Soc., New York 1952) 17.

99. Arthur G. Vestal, Preliminary account of the forests in Cumberland County, Illinois, *Trans. Ill. Acad. of Sci.* 12 (1919) 240.

100. Carl O. Sauer, The agency of man on earth, in William L. Thomas (ed.), *Man's role in changing the face of the earth* (University of Chicago Press, Chicago 1956) 55.

101. Omer C. Stewart, Fire as the first great force employed by man, in Thomas, *Man's role*, 1956, 120–21.

102. Carl O. Sauer, Grassland climax, fire, and man, *Journ. Range Management* 3 (1950) 16–21.

103. James C. Malin, The grassland of North America: Its occupance and the challenge of continuous reappraisals, in Thomas, *Man's role*, 1956, 351.

104. Curtis, *Vegetation of Wisconsin*, 1959, 302.

proposed, prairie, and in particular wet prairie, originated on burned-over forest land, its location is related to the nature of the preceding community, which, he suggested, is likely to have been a solid, close-canopy maple forest. Once fire has destroyed the cover of maples, invading prairie plants can compete successfully with the shade-demanding herbs of a maple forest. Regeneration of trees is subsequently repressed by fire. Burning is critical in the formation and maintenance of wet prairies which, "in the absence of fire, are quickly invaded by trees, including aspens, willows, cottonwoods, ashes, elms and oaks."[105]

Assuming that grassland communities prevail only where trees will not grow, no single explanation accounts for the absence of trees on prairies. Drought alone is not sufficient to prevent trees growing at the present time on prairie land provided with drainage and protected from fire; nor, unless the climate was wetter or Curtis's interpretation is wrong, could it have been sufficient to prevent the establishment of a climax forest before the coming of the prairie. Wind and fire both contribute to arresting the regeneration of trees in prairie areas but on many similar sites trees successfully recolonize burned-over land. Of all characteristics, waterlogging in spring and early summer offers the most plausible explanation for the limits of the treeless area. Not only is it a distinctive attribute of many, if not all, prairie surfaces, it also contributes to the openness of wet meadows, marshes, and bogs. The only extensive areas of forested wetlands in the Midwest are swamps.

The boundary between open grassland and forest has a knife-edged sharpness. Woodland ascends the slopes of the southern valleys; it also hems in grassy plains on the north where the flat till surface gives way to hummocky drift. Broadest in the west, the prairie forms a treeless peninsula, narrowing toward its eastern extremity in Ohio. Its limits are certain, by contrast with the ambiguous division between wet and dry prairie. Little prairie is dry all the time, little is wet all the time, but most is waterlogged for some time almost every year. Prairie vegetation is maintained as long as tree growth is repressed by intermittent incidence of drought, wind, fire, grazing, and, above all, waterlogging. Almost all prairie is wet prairie for a time.

105. Ibid., 305.

Low-lying Bogs on Newer Drift

In a northerly zone extending from Michigan through Wisconsin to northeastern Minnesota the drift-covered surface is pitted with tens of thousands of low-lying shallow depressions occupied by lakes and bogs. Apart from 800 square miles of marsh in central Wisconsin lying beyond the edge of the older drift, practically all wetlands in this northerly zone lie within the area of the most recent glacial deposition, in the province of the Cary and Mankato stages of the Wisconsin glaciation.[106] To the north and east of the Cary end moraine, which is the Kettle moraine in southeast Wisconsin, over two thousand lakes and many thousand marshes were known in 1879.[107] It has since been estimated that over one-tenth of the glaciated area of Wisconsin has characteristic wetland soils.[108] The driftless area, on the other hand, has barely 1%. The proportion of the total area of Wisconsin occupied by naturally ill-drained soils is 8.3%. This is lower than the 11.6% for Michigan which lies almost wholly within the province of the newer drift.[109]

Unlike the broad, flat till plains upon which upland wet prairies rest, the surface of the newer drift is relatively uneven, hummocky, and in places stony. It is a fresh, rough, and irregular landscape whose largest continuous tracts of flat, miry land lie on the beds of former glacial lakes. Wetlands in central Wisconsin occupy the bed of former glacial Lake Wisconsin; 50 square miles of Horicon Marsh and 16 square miles of Sheboygan Marsh have similar origins.[110] Michigan also has large bogs of this type.[111]

Most bogs are located in innumerable small hollows which are less than a square mile in extent. They have formed in pits left by the melting of buried blocks of ice, in valleys plugged by drift, in troughs between morainic ridges, or in basins within glacial lobes,

106. Thwaites, *Glacial geology,* 1953, 80–88.

107. Chamberlin, *Geology of Wisconsin,* 1883, 283.

108. A. R. Albert and O. R. Zeasman, *Farming muck and peat in Wisconsin,* Univ. Wisc. Coll. Agric. Circular 456 (Madison 1953) 2, 3.

109. Frank Leverett, *Surface geology and agricultural conditions of the southern peninsula of Michigan,* in Mich. Geol. and Biol. Surv. Bull. 9 (Lansing 1912) 85; Towar, *Peat deposits of Michigan,* 1900, 157.

110. Lawrence Martin, *The physical geography of Wisconsin,* in *Wisc. Geol. Nat. Hist. Surv. Bull* 36 (State of Wisconsin, Madison 1932) 279–81, 289–91, 362–64; J. Riley Staats, The geography of the central sand plain of Wisconsin, Ph.D. diss., Geography, University of Wisconsin 1933, 46, 132–40.

111. Davis, *Peat in Michigan,* 1907, 114–20.

but most of all in shallow depressions and extinct channels in ground moraine.[112] Most wetlands in Michigan are produced by glacial modification of drainage.[113] As in Wisconsin, waterlogged conditions prevail in low-lying sites, in numerous small hollows.

With the exception of a few relatively small tracts of wet prairie in southern Wisconsin and southwest Michigan, the only wet uplands are situated in the far north, in the highland lake country of northern Wisconsin, in the upper peninsula and High Plains of Michigan. They are similar to wet prairies of the south only in their elevation; in every other respect they resemble adjoining tracts of lowland bog occupying depressions and valley bottoms. They are characteristically numerous, small in size, and unevenly scattered. As Charles M. Davis has observed in the High Plains of Michigan, they are found "along river channels, but not along all river channels. Their place in the major lineaments of the area is sporadic."[114]

Bogs, like lakes, are permanently waterlogged, enclosed depressions. The level of groundwater is rarely more than a few inches below the surface and they are frequently inundated. In course of time, shallow stretches of open water are filled by an accumulation of sediments and a growth of vegetation, colonized in turn by open bogs, by brushwood, and eventually by forest trees. If, as less frequently happens, the water level rises, forest will give way to bog and ultimately to a lake.[115] Such changes generally take place very slowly; their duration may be measured in centuries rather than in years. On the other hand, very rapid changes can be brought about either by damming outlet channels or by cutting drainage ditches.

Because of a bog's low-lying position, the work of draining it differs greatly from draining a tract of upland wet prairie. Before draining, the level of groundwater in a low-lying basin will be determined by the height of water in its outlet channel. Where a stream is ponded back by a rock outcrop or other obstruction the level of groundwater may fall only exceptionally below that of water passing over the threshold of the obstructed channel. The first steps

112. Nevin M. Fenneman, *The lakes of southeastern Wisconsin,* in *Wisc. Geol. Nat. Hist. Surv. Bull.* 8 (Madison 1902) 4–13.

113. Davis, *Peat in Michigan,* 1907, 114.

114. Charles M. Davis, The High Plains of Michigan, *Papers Mich. Acad. Sci. Arts and Letters* 21 (1936) 339.

115. Davis, *Peat in Michigan,* 1907, 135–37.

in artificial drainage are taken by lowering the height of the outlet passage, by removing obstructions, by deepening, dredging, and straightening the course of the main channel through the bog. But these works alone may not secure adequate drainage, since a bog is waterlogged not only by rain falling directly upon it but also by water collected from its catchment area. It may, therefore, need both draining and protecting from flood. Water flowing from surrounding uplands may have to be diverted before it enters a basin and may then be discharged below the main outlet. The low-lying situation of a bog is also important for its climate. Downdrafts of cold air drain from surrounding uplands and the floor of a basin is likely to experience killing frosts late in spring and early in autumn.

Peat and Muck Soils of Bogs

In discussing drainage conditions in the southern peninsula of Michigan, Frank Leverett expressly excluded till plains from the area of waterlogged soil in that region because they generally had "sufficient slope to permit of easy drainage."[116] Wet spots in Michigan and other northern lakes states are low-lying depressions in which water collects and organic matter accumulates to form beds of peat or muck. A peat soil is composed almost entirely of organic matter in slightly decomposed form and carries little mineral material. The accumulation of peat may vary in depth from a few inches in wet meadow soils to as much as 40 feet in a Michigan bog.[117] A muck or marsh border soil consists of moderately or thoroughly decomposed organic matter and contains more than 50% mineral material. It may be formed on sites with a high but fluctuating water table or it may be developed from peat under conditions of improved drainage.

The composition of peat varies from place to place according to the type of plants from which it is formed. In the formation of bog peat, mosses are the most important constituent, occasionally supplemented by a small admixture of grasses and sedges. Their partly decayed remains are so loosely consolidated and contain so much water that the peat shakes and trembles under foot. Fen peat is generally formed from brushwood, reeds, and sedges; it is normally firm but tends to be soft and spongy when very wet. Swamp peat,

116. Leverett, *Southern peninsula of Michigan*, 1912, 63.
117. Davis, *Peat in Michigan*, 1907, 110.

derived in part from trees and shrubs, is usually firm enough to walk on.[118]

One important distinction between peat soil and wet prairie soil is that peat occurs on many types of bedrock and varies greatly in acidity. In southeast Wisconsin where it overlies limestone and glacial drift derived from limestone it is neutral or only slightly acid in reaction. But in most of northern Wisconsin, particularly in the central marshes, it has a much higher degree of acidity than any wet prairie soil.[119] Again, peat soils are darker and richer in organic matter than prairie soils. After drainage and aeration they will yield abundant supplies of nitrogen but insufficient phosphates or potash to promote and sustain normal plant growth.

Farmers who came from successfully reclaimed wet prairies in Illinois and Iowa were unaware of the problems of managing peat and muck soils in Michigan, Wisconsin, and northeast Minnesota. Wet prairie uplands are almost entirely devoid of deep peat deposits. In the early years of the twentieth century, the Illinois Soil Survey recorded only 1,986 acres of deep peat in ten central Illinois counties, representing 0.04% of their total area. Of this small amount, 1,779 acres were situated along the floodplain of the middle Illinois River. In three north and north central counties peat accounted for only 0.15% of the area; in the Kankakee valley the proportion was still as low as 0.83%; and in three northeastern counties it reached no more than 4.67%.[120] To farmers familiar with wet prairie soils, peatland might at first have appeared to offer unexpectedly attractive opportunities for agriculture. T. C. Chamberlin called peat the most abundant of "natural fertilizers" in Wisconsin.[121] It is not liable to bake in summer or to clod in winter. It is looser and more easily worked than the intractable clays of the wet prairies. But early settlers were soon disappointed. They found it cold, sour, infertile, and even its looseness proved a disadvantage. In 1859 James Caird observed that it would not support long-strawed or heavy-eared varieties of wheat.[122] Unless peat was compacted by rolling or cattle treading, crops were likely to lodge

118. Ibid., 109.
119. Whitson, *Soils of Wisconsin,* 1927, 84.
120. Gates, A bog in central Illinois, 1911, 205–11.
121. Chamberlin, *Geology of Wisconsin,* 1883, 685.
122. James Caird, *Prairie farming in America* (New York 1859) 82.

and soil to be blown away. Draining brought fresh problems. Peat shrank like a sponge as it dried out and wasted away as bacteria broke down its fibrous structure. In its dry state it readily ignited and wind rapidly eroded it.

Open Bog, Swamp Forest, and Marsh Vegetation

A gradual transition from wet prairie through marsh to open bog may be observed in many localities on the northern border of the prairie zone.[123] Upland wet prairie is dominated by one or two species of tallgrass, open bogs by one or two species of moss. Marsh is dominated by reeds and sedges, but on sites above the water table it contains a high proportion of prairie plants such as bluejoint grass, and on sites permanently at or near the water table it contains fen or swamp plants such as shrubs and reeds. An abundance and diversity of grasses provided pioneer farmers with valuable supplies of wild hay, which in Wisconsin in 1846 were regarded as "highly important aids in the settlement of a new country."[124] When mowing was discontinued much artificially maintained marsh and wet meadow was invaded by alder thicket. Marshes in general are much less homogeneous in their composition, and form much less stable communities than either wet prairies or open bogs. Slight changes in the height of the water table, in the frequency of burning, grazing, or mowing quickly transform them into open bogs or swamp forests.[125]

In permanently waterlogged depressions, where the water table is only a few inches below the surface, open bog vegetation prevails. Bog plants are shallow-rooted with weak, short-lived stems and long, narrow leaves. Few plants attain heights of more than five feet and the ground is continuously carpeted with moss, generally

123. A. B. Stout, A biological and statistical analysis of the vegetation of a typical wild hay meadow, *Wisc. Acad. Sci., Arts and Letters* 17 (1914) 438; Sears, Natural vegetation of Ohio, 1926, 129; A. Hayden, A botanical survey in the Iowa lake region of Clay and Palo Alto Counties, *Iowa State Coll. Journ. Sci.* 17 (1943) 277–416; Dean Finley and J. E. Potzger, Characteristics of the original vegetation in some prairie counties of Indiana, *Butler Univ. Bot. Studies* 10 (1952) 117, points out that large areas reported by the original land survey as wet prairie "were very likely chiefly sedge meadow"; Burns, Artificial drainage in Blue Earth County, 1954, 49; Weaver, *North American prairie*, 1954, 195.

124. Increase A. Lapham, *Wisconsin: Its geography and topography* (Milwaukee 1846) 74.

125. Curtis, *Vegetation of Wisconsin*, 1959, 365–77.

Sphagnum. A gradient from open water to moss to reeds to sedges to rushes to cattails to grass may be observed at the edges of many thousand lakes and ponds.[126]

In forested regions of Michigan, Wisconsin, and Minnesota open bogs grade into swamp forests on sites that are a little above the water table, the wettest land occupied by open bogs, the borders by swamp forests. Many trees present in deciduous swamp forests, including alder, black willow, cottonwood, green ash, American elm, river birch, and silver maple, are also present in upland forests.[127] Coniferous swamp forests are largely dominated by tamarack (*Larix laricina*) and black spruce (*Picea mariana*), with tamarack dominating the wettest localities, and jack pine, white cedar, balsam fir, yellow birch, and American elm selecting firmer peat.[128] Tamarack swamps are lightly timbered, often with crooked, hollow, diminutive trees whose crowns rarely reach 50 feet. Trees generally cover less than 60% of the surface and are associated with a dense growth of minor vegetation, the ground itself being thickly carpeted with moss.

Conditions promoting the growth of bogs and swamp forests differ in degree but are similar in kind to those promoting prairies. Waterlogging of the ground is the fundamental cause of their formation. They have originated mostly from infilling of lakes by the building of plant material from lake bottoms or by the encroachment of vegetation from their sides or by the drifting of plant remains carried into them by streams.[129]

Climatic conditions in the northern zone generally are not inimical to forest growth. Losses of soil moisture by evaporation and transpiration are lower and precipitation is slightly more evenly dis-

126. Davis, *Peat in Michigan,* 1907, 109–10, 135–37; Curtis, *Vegetation of Wisconsin,* 1959, 378–84.

127. Leslie A. Kenoyer, Forest distribution in southwest Michigan as interpreted from the original land survey, 1826–1832, *Papers Mich. Acad. Sci., Arts and Letters* 19 (1933) 110; Meyer, Kankakee "marsh," 1935, 365; Kaatz, Black Swamp, 1955, 18; Curtis, *Vegetation of Wisconsin,* 1959, 156–68.

128. Chamberlin, *Geology of Wisconsin,* 1877, 240; Leslie A. Kenoyer, Ecological notes on Kalamazoo County, Michigan, based on the original land survey, *Papers Mich. Acad. Sci., Arts and Letters* 11 (1929) 215; Curtis, *Vegetation of Wisconsin,* 1959, 221–42.

129. N. S. Shaler, Fresh water morasses of the United States, *United States Geological Survey Tenth Annual Report,* pt. 2 (Washington, D.C. 1890) 263; Davis, *Peat in Michigan,* 1907, 120; Samuel Weidman and Alfred R. Schultz, *The underground and surface water supplies of Wisconsin,* in *Wisc. Geol. Nat. Hist. Surv. Bull.* 35 (1915) 18.

tributed throughout the year than in prairies. Considerable quantities of water are also gained from condensation on cold, foggy nights when lowland plants may be bathed in a saturated atmosphere for periods of eight to ten hours. An abundance of lichens and bryophytes is attributed by J. T. Curtis to this heavy nocturnal condensation.[130] Some localities, however, receive too little rain in late summer to produce normal crop yields. In central Wisconsin, where rainfall is extremely irregular in July and August, the growth of crops is affected adversely by droughts eight years in every nine.[131] The water deficiency for plants is further increased by differences of as much as 40°F or 50°F between air temperatures and soil temperatures. The amount of water transpired by leaves of plants subject to air temperatures of 90°F or 100°F exceeds the amount of water absorbed by their roots in soil, insulated by a mat of moss and peat, at temperatures of about 50°F. In early spring, when the ground is frozen, plants flowering in warm air cannot replenish any water lost because their roots cannot absorb water in its solid state; like Coleridge's Ancient Mariner they have "water, water everywhere, nor any drop to drink." Drought-resisting features of bog plants are probably related to temperature differences between their transpiring and absorbing surfaces as well as to irregularity in summer rainfall.[132]

As in the prairie zone, destructiveness of fire is increased by drought, by wind, and by flatness of terrain. Almost half the forest land in northern Wisconsin was burned over at least once before 1898 and it was reported that the worst fires "started in the dense tamarack and cedar swamps of the sandy areas where the most complete destruction has taken place."[133] Repeated burning, by repressing trees and stripping the peat surface down to the water table, prepared sites for invasion by bog or marsh vegetation. In peatlands containing remains of tamarack logs and cones, the presence of layers of charcoal is clear evidence that they have been converted from swamps by this means.[134]

130. Curtis, *Vegetation of Wisconsin,* 1959, 238–39.

131. P. E. McNall, H. O. Henderson, A. R. Albert, and W. R. Abbott, Farming in the central sandy area of Wisconsin, *Univ. Wisc. Agric. Expt. Sta. Bull.* 497 (1952) 23.

132. Curtis, *Vegetation of Wisconsin,* 1959, 382.

133. Filibert Roth, *On the forestry conditions of northern Wisconsin,* in *Wisc. Geol. Nat. Hist. Surv. Bull.* 1 (Madison 1898) 13.

134. Curtis, *Vegetation of Wisconsin,* 1959, 374–76.

As plant communities, open bog and swamp forest are closely related, waterlogging exerting a constant controlling influence on them from the time of their formation while fire, wind, drought, and other incidents also affect their development. Plant communities of northern wetlands and also factors promoting their growth more closely resemble those in the southern Midwest than those in neighboring upland forests. The vegetational boundary between wet prairie and bog is less clearly defined than the boundary between treeless or lightly wooded wetland and dense forest.

On the basis of terrain characteristics and degree of waterlogging, it is easy, without drawing a precise line between them, to identify two zones of wetland. A southerly zone, extending from Ohio through Indiana, Illinois, and Iowa into Minnesota, is characterized by its broad expanses of upland wet prairies with narrow ribbons of marsh following valley bottoms. A northerly zone embraces the shores of the Great Lakes including most of Michigan, Wisconsin, and northeast Minnesota. Here, thousands of low-lying bogs account for much of the wetland. Upland wet prairie is negligible. Between these two major divisions lies a transition zone in which both wet prairie and bog occur, the one grading imperceptibly into the other. The greatest contrast between wet prairie and bog is a function of relief, wet prairie situated on upland till plains, bog occupying low-lying basins in newer drift.

Soils of wet prairie and bog are clearly differentiated. Wet prairie soils are silts, clays, silty clay loams, or other mineral soils containing appreciable amounts of vegetable matter, whereas peat, muck, and marsh border soils are composed wholly or largely of organic matter. Wet prairie is stiff, heavy land; bog is light and puffy. In these two contrasting physical settings draining and the success of human settlement have been very different.

Biological Productivity of Midwest Wetlands

The biological productivity of some wetlands in the Midwest is fairly low but far above that of tundra scrub or dry deserts (table 2.4). On the other hand, some marshes are among the most productive ecosystems in the world, ranking alongside tropical rainforests or highly cultivated Midwestern cornfields.[135] Some of the highest

135. Williams, *Wetlands,* 1990, 23, based on H. Lieth, Primary production of the major vegetation units of the world, in H. Lieth and R. H. Whitaker (eds.), *Primary productivity of*

measures of primary productivity are obtained for individual plants in freshwater marshes. In a prairie pothole in Iowa, for example, a yield of 2.8 kg/m^2 per year has been recorded for slough sedge (*Carex atherodes*), 2.3 kg for cattail (*Typha glauca*), and 1.1 kg for bur reed (*Sparganium eurycarpum*).[136] Between sedge, cattail, and reed marshes whose annual production may exceed 2 kg/m^2 and wet grasslands producing around 0.5 kg/m^2, lie fens, swamp forests, and open bogs.[137] Productivity in fens falls at the lower end of the scale while swamp forests and open bogs register at the upper end. Generally, where species diversity is rich, primary productivity is low; and where diversity is poor, productivity is high. In mixed alder and ash (*Alnus-Fraxinus*) swamp forests in Michigan, G. R. Parker and G. Schneider report yields of about 0.6 kg/m^2 per year.[138] At the upper end of the range, in swamps in Minnesota containing northern white cedar (*Thuja occidentalis*) and bog birch (*Betula pumila*), W. A. Reiners reports values over 1 kg/m^2 per year.[139] Relatively low productivity in mixed swamp forests and fen carrs in Michigan is associated with species richness and higher inflows and outflows of surface water than observed in boreal swamps, where flows of water are sluggish and only one or two trees dominate plant communities.

By comparison with woodlands and grasslands on well-drained sites, all wetland ecosystems are highly productive, yielding between two and five times as much as their dryland counterparts. Most wetland plants produce more leaf than woody tissue; most are perennials, growing whenever the sun shines and water is not frozen. Plants standing in water have to keep growing in order to raise their leaves

the biosphere (Springer-Verlag, New York 1975) 203–15; also I. K. Bradbury and J. Grace, Primary production in wetlands, in A. J. P. Gore (ed.), *Ecosystems of the world, 4A; Mires: Swamp, bog, fen, and moor, general studies* (Elsevier, Amsterdam 1983) 285–310.

136. A. G. van der Valk and C. B. Davis, Primary production of prairie glacial marshes, in R. E. Good, D. F. Whigham, and R. L. Simpson (eds.), *Freshwater wetlands: Ecological processes and management potential* (Academic Press, New York 1978) 21–37; Mitsch and Gosselink, *Wetlands*, 1993, 352–54.

137. C. J. Richardson, Primary productivity values in freshwater wetlands, in P. E. Greeson, J. R. Clark, and J. E. Clark (eds.), *Wetland functions and values: The state of our understanding* (Amer. Water Resources Assoc., Minneapolis 1979) 131–45.

138. G. R. Parker and G. Schneider, Biomass and productivity of an alder swamp in northern Michigan, *Canadian Journ. Forest Research* 5 (1975) 403–9, cited by Peter D. Moore, Soils and ecology: Temperate wetlands, in Williams, *Wetlands*, 1990, 102.

139. W. A. Reiners, Structure and energetics of three Minnesota forests, *Ecological Monographs* 42 (1972) 71–94.

TABLE 2.4 Net primary productivity of world vegetation types

	Range (kg/m^2/yr)	Approximate mean
Tropical rain forest	1.0–3.5	2.0
Freshwater swamp and marsh	0.8–4.0	2.0
Warm temperate mixed forest	0.6–2.5	1.0
Cultivated land	0.1–4.0	0.7
Boreal forest	0.2–1.5	0.5
Freshwater lake and stream	0.1–1.5	0.5
Temperate grassland	0.1–1.5	0.5
Tundra dwarf scrub	0.1–0.4	0.14
Dry desert	0.0–0.01	0.003

Sources: Williams, Wetlands, 1990, 23; after H. Lieth, Primary production of the major vegetation units of the world, in H. Lieth and R. H. Whitaker (eds.), Primary productivity of the biosphere (Springer-Verlag, New York 1975) 205.

and flowers above the surface. Below the water surface, the quantity of root and fibrous structures honeycombed with air spaces is at least as great as the quantity of emergent plant material. In northern bogs much of the primary production takes place below water level or below ground.[140] In prairie potholes in Iowa, perennial grasses produce more biomass below ground than above.[141]

Wetland Habitats in the Midwest

Little of the vast production of biomass is grazed by herbivores. Most dies and is decomposed very slowly through the agency of fungi, larvae, protozoa, and bacteria. Fungi are more active and bacteria less active than in dry environments. Much has still to be learned about processes of decomposition in water that is very acid, without oxygen, and covered by ice for long periods. Acidity, anoxia, and low temperatures are more conducive to pickling organic matter than promoting decay; indeed, pollen grains, diatoms, molluscs, trunks of trees, and bodies of mammals have been preserved for thousands of years in northern peat bogs. Some decaying remains are consumed by worms, shellfish, crustaceans, and salmon, but most are deposited underwater as layers of peat. Living

140. R. J. Reader, Primary production in northern bog marshes, in Good et al., Freshwater wetlands, 1978, 53–62, cited in Mitsch and Gosselink, Wetlands, 1993, 398–99; Moore, Soils and ecology, in Williams, Wetlands, 1990, 103.

141. Van der Valk and Davis, in Good et al., Freshwater wetlands, 1978, 21–37.

or dead plants and the soil and water in which they grow or accumulate provide habitats for a variety of insects, fishes, birds, and mammals.

Swarms of insects are the most profuse and characteristic forms of life in all Midwestern wetlands. Mosquitoes, midges, black flies, crane flies, aphids, and moths abound in wet prairies and northern bogs. Many flies are herbivorous, feeding on grasses, reeds, and sedges during their adult stages, while in their larval stages they live in water, feeding on dead plant remains. In turn, insects and their larvae provide food for fishes, frogs, and waterfowl.[142] In northern bogs, carnivorous plants are able to overcome a nitrogen deficiency by trapping and digesting insects. These specialized bog plants include the pitcher plant (*Sarracenia*) and sundew (*Drosera*).[143]

It is not clear what contribution freshwater marshes and swamps make to maintaining fish life in the Midwest. Northern pike, common carp, and other species frequent marshes for reproduction but spend most of their lives in fast-moving streams. Detritus carried by surface outflows from wetlands adds to food supplies for fish in rivers and lakes.

Wetlands play an essential role in sustaining bird life, serving as feeding grounds, breeding places, and resting places for seasonal migrants. The Mississippi flyway is followed by over 30% of North America's migratory birds on their way from the Canadian tundra to wintering areas in the lower Mississippi valley and Gulf of Mexico. Migrating flocks find temporary accommodation on thousands of lakes and wetlands in the upper Midwest, especially in Minnesota and the Dakotas.[144] An overwhelming majority of birds that visit northern wetlands are not wholly dependent on wetland habitats, nor are they game birds.[145] Some seek shelter, water, and food in wet potholes situated in the midst of almost completely cultivated upland prairies. Others make nests in trees and shrubs in swamp

142. Mitsch and Gosselink, *Wetlands*, 1993, 349.

143. A. W. H. Damman and T. W. French, *The ecology of peat bogs of the glaciated northeastern United States: A community profile*, U.S. Fish and Wildlife Service Biol. Report 85 (7.16) (Washington, D.C. 1987) 100; Mitsch and Gosselink, *Wetlands*, 1993, 394.

144. F. C. Bellrose and N. M. Trudeau, Wetlands and their relationship to migrating and winter populations of waterfowl, in D. D. Hook et al. (eds.), *The ecology and management of wetlands* (Croom Helm, London 1988), vol. 1: 183–94, cited in Williams, *Wetlands*, 1990, 25–27.

145. D. E. Kroodsma, Habitat values for nongame wetland birds, in Greeson et al., *Wetland values and functions*, 1979, 320–29.

forests and fens or at edges of marshes and bogs. Game birds, particularly ducks, geese, and other waterfowl, can survive only in wetland habitats or on open water.[146] They require fish and aquatic plants for food as well as reeds, rushes, and sedges as cover for nests. Different sites are occupied by different species: loons and diving ducks fish in deeper waters, wading birds together with geese, swans, and dabbling ducks occupy shallow waters and graze on marsh plants. Terns, teal, and some ducks nest in upland sites and feed along the margin of marsh and open water. Rails, owls, and other solitary birds range widely over many types of wetlands.[147]

Large numbers of mammals visit wetlands in search of food and shelter, including moose, bear, deer, hares, lynx, otter, mink, and many small shrews, voles, mice, moles, lemmings, squirrels, and weasels. Some of these are more or less permanent refugees from vanishing forested areas but none is entirely dependent on wetlands. Muskrats, nutria, beavers, and marsh rats, on the other hand, are among wetlands' fully dependent residents. Large mammals have been preyed upon by humans for fur and flesh and their populations are now closely controlled by human action.[148]

Wetlands in the Midwest exhibit the furthest extremes of plenty and want, abundance and scarcity, luxuriance and privation. Southern wet prairies have soils that are rich in plant nutrients, supplying plenty of nitrogen, phosphorus, and potassium for plant growth. They contain neither trees nor shrubs and very few hydrophytic plants. They are dominated by tallgrasses that produce enormous quantities of biomass. At present, hardly any patches remain unmodified by artificial draining. They have been written off as nonwetlands, but widespread summer floods in 1973 and 1993 demonstrate that they are liable to revert periodically to a waterlogged state.

In the northern lakes region, bogs, swamps, and marshes teem with insect life, bird life, and furbearing mammals, yet their plant growth displays extremes of richness and poverty. Some marshes are

146. L. D. Flake, Wetland diversity and waterfowl, in Greeson et al., *Wetland values and functions,* 1979, 312–19; Weller, *Freshwater Marshes,* 1987.

147. M. W. Weller and C. S. Spatcher, *Role of habitat in the distribution and abundance of marsh birds,* in *Iowa State Univ. Agric. and Home Econ. Expt. Sta. Spec. Report* 43 (Ames, Iowa 1965); Mitsch and Gosselink, *Wetlands,* 1993, 350–51, 396–97.

148. Glaser, *Patterned boreal peatlands,* 1987.

more productive than any other plant communities in the world while some bogs are less productive than neighboring dryland vegetation associations. Wetlands in the north suffer from extremely short growing seasons, much water is highly acid, and plants have difficulty in adapting to anaerobic conditions. When drained, peat soils are easy to cultivate and rich in nitrogen once exposed to air. Draining causes peat to shrink, exposure to air activates processes leading to rapid decomposition, and dry peat is combustible. One way or another, the valuable substance is soon dissipated. Northern wetlands' functions as wildlife habitats are extremely sensitive to human interference.

Many questions about the physical geography of wetlands in the Midwest remain unanswered. Two such questions are important for this study. First, it is important to know more about the history of origins and changing forms of wetlands. Very slight changes in water level, in length and intensity of hydroperiods, in the balance of precipitation, and in the inflow and outflow of surface water produce very large changes in the extent and character of wetlands. How are these delicately balanced regimes started and how are they maintained in a continually changing regional setting? A second question concerns the introduction and extinction of individual plants and animals, constituent elements within ecosystems. At what date did Kentucky bluegrass enter Midwestern prairies? What is the history of wild rice (*Zizania aquatica*)? When did beavers arrive and depart? What are the dates of origin and reasons for dispersal of different species of insects, fishes, and birds? Behind these questions lies a question about possible roles played by human agents.

3

Native American Occupation

Before the coming of Europeans, wetlands in the upper Mississippi valley and Great Lakes region were fringed by native American settlements. Wetlands were highly productive ecosystems. Hunters, fishers, and gatherers were drawn across miry land to the edge of open water. They came in pursuit of deer, moose, waterfowl, sturgeon, whitefish. They gathered berries, fruit, nuts, wild rice, moss, medicinal plants. Wetlands, lakes, and rivers supplied them with water, food, and clothing. Waterways were major channels of communication and canoes the principal means of transport. On their travels by river and lake, Europeans first encountered natives in watery habitats.

Without evidence from written records, geographies of early native American cultures, economies, and societies are difficult to reconstruct. The lack of documentary evidence is matched by a deficiency of archaeological material. No traces of stone buildings, few pieces of metalwork, and no rich variety of decorated pottery fragments remain.[1] Interpretation of ethnographic information is complicated because of the displacement and migration of native American people in the 500 years following the landing of Europeans on the Atlantic coast. First, native populations were devastated by diseases brought by Europeans and their livestock. Then, surviving natives were driven westward from the St. Lawrence lowlands, New England, and the Appalachians. People from woodlands and mountains were pushed onto grassland plains and marshes and former marsh-dwellers were dislodged to drier lands further west. Descriptions of native American life by European explorers in the upper Mississippi valley began in the seventeenth century after some

---

1. Jesse D. Jennings, *Prehistory of North America* (McGraw Hill, New York 1968) 220–21.

movement had taken place.[2] By the time the U.S. government came to survey and sell lands to settlers in the nineteenth century, most of the original inhabitants had been removed. Population in the Midwest suffered a catastrophic fall in the fifteenth and sixteenth centuries followed by gradual recovery in the seventeenth and eighteenth centuries as migrants from the East entered the region and European fur traders created a new economy. When the fur trade declined, white Americans expelled natives and built their own new settlements.

This chapter discusses four questions but offers no conclusive answer to any of them. First, how did native Americans in the precontact period modify their environments, particularly wetlands? Second, what effects on the environment followed the steep fall in native American populations in the century after 1492? Third, to what extent did fur traders induce native Americans to change their environments, especially wetlands? Fourth, in what ways did removals contribute to the disappearance of traces of native American occupation? Debate over these questions will continue but the questions themselves call for a fuller examination of environmental history in the period before the arrival of white Americans. The investigation challenges an assumption that the plats and notes of the U.S. original land survey recorded "original" landscape and depicted "natural" vegetation.[3] An article published as recently as 1993 falls into the trap of treating observations by about two dozen early-nineteenth-century scientific expeditions, all conducted by white men, as "the only real basis for knowing about the presettlement grasslands" of the midcontinent plains.[4] Clearly, the grasslands were not "presettlement." All the scientific observers referred to in that article confirmed that Indians were present in the areas visited and described their settlements. Nineteenth-century observers were also aware that Indians had modified their environment.

---

2. Carl O. Sauer, *Seventeenth century North America* (Turtle Island, Berkeley 1980) 138.

3. Evidence of the federal land survey is reviewed in E. A. Bourdo, A review of the General Land Office survey and its use in quantitative studies of former forests, *Ecology* 37 (1956) 754–68.

4. Dwight A. Brown, Early nineteenth-century grasslands of the midcontinent plains, *Annals Assoc. Amer. Geog.* 83 (1993) 591.

Environmental Changes in the Pre-Columbian Period

Returning to the time of the earliest contacts between Europeans and native Americans, William Denevan asked whether "the landscape encountered in the sixteenth century [was] primarily pristine, virgin, a wilderness, nearly empty of people, or was it a humanized landscape, with the imprint of native Americans being dramatic and persistent?"[5] He was inclined to think that the humanized view "may be more accurate" and the pristine view was "to a large extent an invention of nineteenth-century romanticist and primitivist writers."[6] The myth of a virgin land continued to flourish. A popular book, Seeds of change: Christopher Columbus and the Columbian legacy, published by the Smithsonian Institution in 1991, declared that "pre-Columbian America was still the First Eden, a pristine natural kingdom. Native people were transparent in the landscape, living as natural elements in the ecosphere."[7] The notion that disturbance of the environment by early inhabitants was barely perceptible was reinforced by a stereotype of native Americans as passive people.

European observers in seventeenth-century New England marveled at the "natural" fecundity and prodigious size of wild animals, birds, fishes, and plants, offering bountiful supplies of food just waiting to be collected.[8] These early colonists' accounts also noted that native Americans did not domesticate animals and made little effort to store crops they gathered. When stocks ran out they went hungry. In winter their villages broke up into small family hunting bands and they were reported to hunt whatever game might sustain them until the following spring.[9] It was also surmised that they pre-

5. William M. Denevan, The pristine myth: The landscape of the Americas in 1492, Annals Assoc. Amer. Geog. 82 (1992) 369.

6. Ibid.

7. S. Shetler, Three faces of Eden, in H. J. Viola and C. Margolis (eds.), Seeds of change: A quincentennial commemoration (Smithsonian Institution, Washington, D.C. 1991) 226, cited in Denevan, Pristine myth, 1992, 370.

8. William Cronon, Changes in the land: Indians, colonists, and the ecology of New England (Hill and Wang, New York 1983) 22–24.

9. Cronon, Changes in the land, 1983, 41, 47; the question of hunger is discussed in Ronald Fritz, Roger Suffling, and Thomas Ajit Younger, Influence of the fur trade, famine, and forest fires on moose and woodland caribou populations in northwestern Ontario from 1786 to 1911, Environmental Management 17 (1993) 477–89; references to starvation 478, 480, 483–88.

ferred passively to lie in wait or set traps for their prey rather than
actively chase or stalk.[10] It was convenient for European commentators to represent native Americans as feckless savages who did
little to modify their surroundings apart from waiting for game and
fish to fall into their traps and collecting fruit and nuts as they
ripened. White settlers' takeover of native American hunting
grounds was held to be justified because the land was unclaimed,
unfenced, and unimproved.

After living among white Americans for several generations,
some native Americans began to accept and repeat an image of
themselves as inactive. In deciding the future of their reservation in
Wisconsin in the 1950s, some Menominee complained of the apa-
thy of their compatriots. One described a majority as "dependents.
They gripe continually about problems and conditions but do absolutely nothing about them. Most of them do nothing."[11] Another
said: "They're just damn lazy fishermen. That one don't care to fish
at all and the other can hardly hold up the pole. They better be
careful they don't fall asleep and fall in the river. Those fishing
should get up and go to work with the others. Being lazy will never
help."[12] Stereotypes of pristine land and passive people continue to
be uttered, but a wider debate over contributions made by native
Americans to changing the landscape is advancing slowly.

10. Jennings, *Prehistory of North America,* 1968, 104–7; George Spindler and Louise
Spindler, *Dreamers without power: The Menomini Indians* (Holt, Rinehart and Winston, New
York 1971) 34.

11. Spindler and Spindler, *Dreamers without power,* 1971, 180.

12. Ibid., 181, 197–98. At the end of local discussions, the Menominee reservation was
terminated in 1961 and tracts of tribal land were sold to a developer who formed an artificial
lake and built homes for people from outside. A private corporation bought the sawmill and
embarked on a program of forest clearance. Angered by the dissipation of cherished tribal
resources, a majority of Menominee voted to revoke the earlier decision and regain their
reservation. In 1973, after a hard-fought campaign, treaty rights were confirmed and tribal
property was restored to a trust. A newly incorporated Menominee Enterprises started to buy
back swamp and forest lands funded partly by profits from a casino. In 1993, forestry operations began to yield a profit. Progress toward sustained development in a modern economy is
reported in Deborah Shames (ed.), *Freedom with reservation: The Menominee struggle to
save their land and people* (National Committee to Save the Menominee People and
Forests, Washington, D.C. 1972); Duncan A. Harkin, The significance of the Menominee
experience in the forest history of the Great Lakes region, in Susan L. Flader (ed.), *The
Great Lakes forest: An environmental and social history* (University of Minnesota Press,
Minneapolis 1983) 96–112; Robert E. Deer, A Menominee perspective, in Flader, *Great
Lakes forest,* 1983, 113–18; Caspar Henderson, Chainsaw massacre? Not in Wisconsin,
Independent (London 4 March 1996) 18.

Fire and Tallgrass Prairies

Explanations for openings in hardwood forests in the East no longer ignore actions by native Americans. Seventeenth- and eighteenth-century writers noted that native people cleared land for cultivation. They taught French and English settlers how to grow corn, beans, and squashes. They chopped down wood for winter fuel and burned brush to rouse game from coverts and also, as early-nineteenth-century traveler Timothy Dwight remarked: "to produce fresh and sweet pasture for the purpose of alluring deer to the spots on which they had been kindled."[13] All these activities implied direct intervention in shaping the natural environment and also implied making provision for the future. Few native Americans spent all their time sitting waiting for fish to bite or wild rice to ripen; many more managed plants and animals and cultivated the soil.

Explanations for changes in pre-Columbian environments in the Midwest have shifted from almost exclusive attention to natural processes toward a fuller consideration of human activity. The extension of the tallgrass prairie peninsula to Ohio, deep in the humid East, was attributed by Frederic Clements to the prevalence of a late summer dry season, a feature of the climate he thought had remained constant since Tertiary times.[14] Analysis of pollen deposits by H. E. Wright and others led to the conclusion that climatic changes during the Pleistocene era were sufficient to account for an origin of tallgrass prairies about 10,000 years before the present.[15] Studies of the spread of grassland at the expense of forests during the Holocene raised the question of fire.[16] John Borchert consid-

13. Cronon, *Changes in the land*, 1983, 51.

14. F. E. Clements, *Plant succession: An analysis of the development of vegetation*, Carnegie Institution Publ. 242 (Washington, D.C. 1916), cited in Brown, Nineteenth-century grasslands, 1993, 591; an extended examination of Clements's views is in Donald Worster, *Nature's economy: A history of ecological ideas* (Cambridge University Press, Cambridge 1977) 205–20.

15. Edward J. Cushing, Problems in the Quaternary phytogeography of the Great Lakes region, in H. E. Wright and David G. Frey (eds.), *The Quaternary of the United States* (Princeton University Press, Princeton 1965) 403–16; J. E. Kutzbach and H. E. Wright, Simulation of the climate of 18,000 yr BP: Results for North America/North Atlantic/European sector, *Quaternary Sci. Rev.* 4 (1986) 147–87; Thompson Webb III, Patrick J. Bartlein, and John E. Kutzbach, Climate change in eastern North America during the past 18,000 years: Comparisons of pollen data with model results, in W. F. Ruddiman and H. E. Wright (eds.), *The geology of North America: North America and adjacent oceans during the last deglaciation* (Geol. Soc. of America, Boulder 1987) 447–62.

16. J. C. Bernabo and T. Webb III, Changing patterns in the Holocene pollen record of

ered that the present climatic regime, especially seasonal distribution of precipitation, was conducive to the persistence of tallgrass prairies.[17] He also thought that fires ignited by lightning and fanned by strong winds in hot, dry spells in July and August, or less frequently in late spring, contributed to the suppression of tree growth. John Weaver dismissed fire as "damaging and destructive," a rare and detrimental disturbance in an ecosystem dominated by climate.[18]

Later writers have come to recognize that recurrent fires were essential in maintaining tallgrass prairies: when burning stopped, woody plants invaded the grassland.[19] It has also come to be acknowledged that many if not most fires were either started or broadcast by humans, above all to drive or attract wild game. As far back as 1926, Walter Hough discussed the role of fire in early cultures, but a deeper understanding of fire-making in sustaining the dominance of grasses had to await progress in archaeological, ethnographic, and demographic studies.[20] Deposits of charcoal are closely associated with native American trails, campsites, and villages: they increase in frequency through time, occurring more or less every year toward the end of the period of native American oc-

north-eastern North America: A mapped summary, *Quaternary Research* 8 (1977) 64–96.

17. J. R. Borchert, The climate of the central North American grassland, *Annals Assoc. Amer. Geog.* 40 (1950) 1–39.

18. J. E. Weaver, *North American prairie* (Johnsen, Lincoln 1954); Weaver, *Prairie plants and their environment* (University of Nebraska Press, Lincoln 1968), cited in Scott L. Collins, Introduction: Fire as a natural disturbance in tallgrass prairie ecosystems, in Scott L. Collins and Linda L. Wallace (eds.), *Fire in North American tallgrass prairies* (University of Oklahoma Press, Norman 1990) 3.

19. Stephen J. Pyne, *Fire in America: A cultural history of wildland and rural fire* (Princeton University Press, Princeton 1982) xvi, 75–76, 96–99; Grant Cottam, The phytosociology of an oak wood in southern Wisconsin, *Ecology* 30 (1949) 171–287; H. A. Gleason, The vegetational history of the Middle West, *Annals Assoc. Amer. Geog.* 12 (1932) 39–85, cited in J. T. Curtis, The modification of mid-latitude grasslands and forests by man, in W. L. Thomas (ed.), *Man's role in changing the face of the earth* (University of Chicago Press, Chicago 1956) 735.

20. Walter Hough, *Fire as an agent in human culture*, U.S. National Museum Bulletin 139 (Washington, D.C. 1926); E. V. Komarek, Fire ecology: Grasslands and man, *Tall Timbers Fire Ecology Conference Proceedings* 4 (1965) 169–220; Komarek, Fire: And the ecology of man, *Tall Timbers Fire Ecology Conference Proceedings* 6 (1967) 143–70; Carl O. Sauer, Grassland climax, fire, and man, *Journ. Range Management* 3 (1950) 16–21; Omer C. Stewart, Burning and natural vegetation in the United States, *Geog. Rev.* 41 (1951) 317–20; Stewart, Fire as the first great force employed by man, in Thomas, *Man's role*, 1956, 115–33, esp. 127.

cupation. They are also present in greatest numbers near well-trodden trails and much-visited campsites and are fewest where remains of human occupation are thinnest.[21] The relations between recurrent burning and grazing of large mammals, especially bison and deer, are beginning to be investigated.[22] Leaving aside Stephen Pyne's sweeping claim that nearly all midcontinental grasslands in North America "were created by man, the product of deliberate, routine firing," most writers now subscribe to two views: that fire is second only to climate as a factor in the formation of prairies, and that fire, in conjunction with grazing, is responsible for maintaining tallgrasses wherever they survive in the humid east.[23]

Hunting, Gathering, and Wetlands

In hunting bison and deer, fire was directed to driving them toward open water or wetland, generally toward habitual watering places. There hunters could wait to slaughter the fleeing animals. Trees and brushwood were able to grow at the water's edge, protected from burning at ground level. Indians kept grazing areas open for bison and deer, regulated the movement of herds, and controlled their numbers. It is uncertain whether they also modified areas of water and wetland or how they influenced aquatic flora and fauna.

By trapping beaver, native Americans must have exerted an indirect effect upon wetlands. Beavers built dams on streams, "backing up water across great expanses, creating wetlands where none existed before, and possibly even altering global carbon biogeochemistry."[24] Beaver ponds have much higher rates of methane production than other boreal wetlands. Beavers also have had a significant effect on the flooding of northern peatlands.[25] In the seven-

21. J. T. Curtis, *The vegetation of Wisconsin* (University of Wisconsin Press, Madison 1959); J. E. King, Late quaternary vegetational history of Illinois, *Ecol. Monogr.* 51 (1981) 43–62; D. I. Axelrod, Rise of the grassland biome, central North America, *Bot. Rev.* 51 (1985) 163–202.

22. Paul G. Risser, Landscape processes and the vegetation of the North American grassland, in Collins and Wallace, *Fire in tallgrass prairies,* 1990, 133–46, esp. 135–36.

23. Pyne, *Fire in America,* 1982, 84; Collins and Wallace, *Fire in tallgrass prairies,* 1990, 3.

24. William J. Mitsch and James G. Gosselink, *Wetlands,* 2d ed. (Van Nostrand Reinhold, New York 1993) 70; R. J. Naiman, T. Manning, and C. A. Johnston, Beaver population fluctuations and tropospheric methane emissions in boreal wetlands, *Biogeochemistry* 12 (1991) 1–15.

25. Naiman et al., Beaver population, 1991, 1–15, based on observations in northern

teenth century, a French governor of Acadia, Nicolas Denys, thought that the greater part of ponds and lakes in the interior of the province had been made by beavers.[26] Deep in the heart of the continent, seventeenth-century traveler Peter Radisson described a plain on the south shore of Lake Superior changed into a bog by the industry of beavers who had cut down trees to the extent that no wood was left for kindling a fire.[27] In the mid-nineteenth century, George Perkins Marsh was disposed to think that "more bogs in the Northern States owe their origin to beavers than to accidental ob-structions of rivulets by wind-fallen or naturally decayed trees; for there are few swamps in those States, at the outlets of which we may not by careful search, find the remains of a beaver dam."[28] Beaver was highly valued by native Americans both for its fur and its flesh. Before rivalry between French, English, and American fur traders hastened the destructive exploitation of the species, population numbers in the Laurentian Shield and northern lakes seem to have been maintained by judicious culling. At the end of the seventeenth century early signs of overkill may be inferred from traders' com-plaints that beaver pelts supplied by Illinois were poor quality. Prairie streams had smaller colonies of smaller-sized beavers than those in the lakes and marshes around the northern Great Lakes.[29] At that time, Miamis and Illinois were reported to be raiding Iroquois territory to the north and indiscriminately killing "all [beavers], male and female, against all native custom."[30]

There is no clear evidence that Indians "carefully watched and nurtured" beavers, nor is there any reason why they should have discouraged beavers from building dams and creating ponds.[31] On

---

Minnesota.

26. Nicolas Denys, *The description and natural history of the coasts of North America (Acadia)*, ed. and trans. W. F. Ganong, Publication of the Champlain Society 2 (Toronto 1908), cited in Sauer, *Seventeenth century North America*, 1980, 93, 221.

27. Peter Esprit Radisson, *Voyages*, ed. Gideon D. Scull, Publication of the Prince Society (Boston 1885), cited in Sauer, *Seventeenth century North America*, 1980, 123.

28. George Perkins Marsh, *Man and nature*, ed. David Lowenthal ([1864]; Belknap, Harvard University Press, Cambridge 1965) 32.

29. Sauer, *Seventeenth century North America*, 1980, 191.

30. Baron de Lahontan, *Nouveaux voyages* [1703], cited in Sauer, *Seventeenth century North America*, 1980, 215.

31. Wilbur R. Jacobs, The Indian and the frontier in American history: A need for revi-sion, *Western Historical Quarterly* 4 (1973) 43–56, ref. on 50, cited in David Ward (ed.), *Geographic perspectives on America's past* (Oxford University Press, New York 1979) 70–76; W. R. Jacobs, *Dispossessing the American Indian: Indians and whites on the colonial fron -*

the contrary, there was much to be gained from extending the area of ponds, lakes, and wetlands. Water surfaces were highly productive breeding and feeding places for fish, waterfowl, and many different furbearing animals. The question whether native Americans assisted beavers or independently constructed dams has not, as far as I know, been examined. A chronology of stratified lake deposits might throw light on the possibility or impossibility of human interference in processes leading to the formation of lakes and wetlands. If lakes and wetlands were formed before the arrival of humans, there can be no question of human agency, but if they originated when humans were present, the possibility of human intervention is worth considering.

Selection and Protection of Useful Plants

Whether or not native Americans had a hand in making or enlarging wetlands, there is little doubt that they modified plant life both by fostering native plants and introducing new species. In the river basins of the Tennessee, Ohio, and upper Mississippi, Karl Butzer concludes that "manipulations of weedy seeds gradually led to domestication of marsh elder (sumpweed, *Iva*) and maygrass, *Phelaris*, by 4000 BP."[32] During the next 2,000 years native squash (*Cucurbita*)was domesticated, and bottle gourd (*Lagenaria*) of Mexican origin and cultivated maize (*Zea*) from central America were introduced into the Midwest. The range of cultivated plants was enlarged in the period of the Woodland culture (3000—1000 BP) by the introduction of sunflower (an oil plant, *Helianthus*), goosefoot (a starchy seed, *Chenopodium*), eight-rowed "flint," and twelve-rowed "dent" maize and tobacco from Mesoamerica.[33] In addition, wetlands supported a great variety of useful plants that needed only a small amount of attention to yield valuable crops.

A staple food of several northern tribes was wild rice (*Zizania aquatica*). In their language, the Menominee people were named after the plant: they were wild rice people. When the grain ripened, large flocks of waterfowl came to feed and could be netted for their meat. The birds also disseminated seed, and native Americans pro-

tier (Scribner, New York 1972).

32. Karl W. Butzer, The Indian legacy in the American landscape, in Michael Conzen (ed.), *The making of the American landscape* (Unwin Hyman, Boston 1990) 30.

33. Ibid., 32.

Figure 3.1 Chippewa harvesting wild rice in northern Minnesota

Source: United States Information Service. Photo: Tom Hollyman, Photo Researchers Inc.

moted the spread of the plant by periodic weeding. Seventeenth-century travelers observed the method of gathering the harvest, a particularly detailed account by Father Jacques Marquette describing a Menominee harvest. A birchbark canoe was paddled among the plants and the ripe ears were shaken into the hull of the canoe (fig. 3.1). The ears were dried on wooden griddles, put into leather bags, trodden to separate grain from husks, then winnowed.[34] Other

34. Sauer, *Seventeenth century North America,* 1980, 123, 130, 131, 140.

plants that supplemented largely meat and fish diets of wetland villages were plums, cherries, grapes, peaches, pawpaw, persimmon, crab apples, pecan, hickory nuts, hazelnuts, butternuts, blueberries, cranberries, raspberries, and wild strawberries. Among the Menominee in 1923, Huron Smith listed 227 plants that had medicinal value, although not all of them were being used at that time. In addition, he listed 45 food plants, 17 that could be used for making twine or thread, none of which were then used for that purpose, and 13 plants that served as charms or love potions.[35]

Seed Planting and Gardening

The gathering of fruit, nuts, wild leeks, and wild rice was not entirely superseded by the selection and propagation of useful plants. Agriculture developed alongside the establishment of more or less permanent villages, some securely palisaded.[36] In the upper lakes region villages were sited close to streams or wetlands and hunting, fishing, and gathering remained more important than cultivation.

In the Mississippian age from about 2000 BC to AD 1000, virtually all known settlement sites were located on slopes of low, sandy ridges above the level of annual floods because "cultivating this wetland was too difficult without the benefit of an elaborate drainage or levee system."[37] On the other hand, prairies, although subject to seasonal waterlogging, are thought to have been "well suited to Mississippian agriculture."[38] Cultivators possessed neither plows nor draft animals. Indeed, the sole contribution that native Americans made to animal domestication was to keep turkeys and feed dogs that scavenged near their dwellings.[39] Without dung, soils were occasionally fertilized with fish and ashes.[40]

In the most recent pre-Columbian period, advanced cultures in the Midwest developed a highly efficient system of cultivation. The

35. Huron H. Smith, *Ethnobotany of the Menomini Indians,* Bulletin of the Public Museum of the City of Milwaukee 1923 (reprinted by Greenwood Press, Westport 1970).

36. Sauer, *Seventeenth century North America,* 1980, 228.

37. Gregory Waselkov, Prehistoric agriculture in the central Mississippi valley, *Agric. Hist.* 51 (1977) 518.

38. Ibid., 519.

39. Carl O. Sauer, *Agricultural origins and dispersals* (American Geog. Soc., New York 1952) 73.

40. Erhard Rostlund, The evidence for the use of fish as fertilizer in aboriginal North America, *Journ. Geog.* 56 (1957) 222–28; Rostlund, *Freshwater fish and fishing in native North America,* University of California Publ. Geography 9 (Berkeley 1952).

surface of the land was broken up with digging sticks and heaped
into hillocks or ridges. Into these hills or ridges seeds of maize,
beans, and squash were sown. The corn grew upward toward the
light; tendrils of the beans clung to the tall corn stalk; foliage and
gourds of the squash covered the ground. Roots of the corn reached
down for anchorage and groundwater; bean roots fixed nitrogen in
the soil; and squash shaded the hill and retained soil moisture. It
was a unique symbiotic complex.[41] Besides carbohydrates all these
plants provided vegetable oils, and beans and squash were fairly
rich in protein. Tomatoes, tobacco, and sunflowers were sown close
to or in the same hills as maize, beans, and squash. They are mentioned in seventeenth-century travelers' accounts.[42]

Many ridged gardens were sited on seasonally waterlogged land
or close to permanently flooded land. The raising of earth into hills
or ridges ensured that roots of young seedlings could grow above
the level of summer floods. Traces of ridges are common in Wisconsin, where such features have enabled no fewer than 175 native
American gardens to be located.[43] In Michigan, Ohio, Indiana, and
Illinois many ridges and hills "were constructed in order to expand
cultivation into wetlands by elevating planting surfaces on a waterlogged subsoil above the water surface."[44] Other explanations for
the practice include avoidance of ground frost in low-lying depressions and enrichment of the soil by piling up nutrient-rich prairie
sod for a seed bed.[45] The technique was adapted, above all, to the
utilization of seasonally waterlogged soils. During the 1970s and
1980s, further research by geographers and others has confirmed
that the pre-Columbian ridged fields, raised fields, ditched fields,

41. Sauer, *Agricultural origins,* 1952, 64.

42. Sauer, *Seventeenth century North America,* 1980, 132, 229, 235, 236.

43. Denevan, The pristine myth, 1992, 375, refers to W. G. Gartner, The Hulbert Creek
ridged fields: Pre-Columbian agriculture near the Dells, Wisconsin, M.A. thesis, Geography,
University of Wisconsin, Madison 1992.

44. William E. Doolittle, Agriculture in North America on the eve of contact: A reassessment, *Annals Assoc. Amer. Geog.* 82 (1992) 394, cites Melvin Fowler, Middle
Mississippian agricultural fields, *American Antiquity* 34 (1969) 365–75.

45. Thomas J. Riley and Glen Freimuth, Fields systems and frost drainage in prehistoric
agriculture of the upper Great Lakes, *American Antiquity* 44 (1979) 27–85; James P.
Gallagher and Robert F. Sasso, Investigations into Oneota ridged field agriculture on the
northern margin of the prairie peninsula, *Plains Anthropologist* 32 (1987) 141–51; Gallagher,
Prehistoric field systems in the upper Midwest, in William I. Woods (ed.), *Late prehistoric
agriculture: Observations from the Midwest,* Studies in Illinois Archaeology 7 (Illinois
Historic Preservation Agency, Springfield 1993).

and island fields situated alongside streams, in lake basins, and at springs at many places in Mesoamerica were constructed as a means of cultivating wetlands. They were associated with the growth of dense concentrations of population in the Preclassic period from 1500 BC to AD 1, and the system endured until the arrival of Spanish colonists in the sixteenth century.[46] In the Midwest, similar methods of ridging originated at a later date and continued in places into the seventeenth century.

A new picture is emerging from close reading of early explorers' journals and reports as well as from archaeological research showing native Americans in the precontact period actively engaged in exploiting wetland resources and modifying habitats to promote the growth of useful plants, fish, birds, and mammals. While no evidence for animal domestication, apart from turkeys and dogs, has yet come to light, it is now clear that native Americans used fire to improve grazing for bison and deer and to drive herds of animals toward wet places for ambush. They also managed stocks of fish, birds, and beaver by encouraging reproduction and by killing selectively. Knowledge of plant domestication is sketchy. It is not clear whether Indians propagated wild rice by pulling out weeds and scattering seed in appropriate places in bogs. Nor is it known how Kentucky bluegrass invaded and established itself throughout the prairies. Alfred Crosby thinks it was "probably brought to the Appalachian area by the French," but how did it spread from there to grasslands beyond the Mississippi?[47] Did native Americans assist its dispersal? Mounding and ridging of land liable to seasonal floods was an ingenious device for cultivating poorly drained sites, as was the interplanting of corn, beans, and squash. The notion that native Americans made no provision for the future is not true of Midwestern people. They employed elaborate techniques for curing and drying fish and meat. They dried and winnowed wild rice, dried

46. Andrew Sluyter, Intensive wetland agriculture in Mesoamerica: Space, time, and form, *Annals Assoc. Amer. Geog.* 84 (1994) 557–84.

47. Alfred W. Crosby, Ecological imperialism: The overseas migration of western Europeans as a biological phenomenon, in Donald Worster (ed.), *The ends of the earth: Perspectives on modern environmental history* (Cambridge University Press, Cambridge 1988) 114; Lyman Carrier and Katherine S. Bort, The history of Kentucky bluegrass and white clover in the United States, *Journ. Amer. Soc. Agronomy* 8 (1916) 256–66; Robert W. Shery, The migration of a plant: Kentucky bluegrass followed settlers to the New World, *Natural History* 74 (1965) 43–44; G. W. Dunbar, Henry Clay on Kentucky bluegrass, 1838, *Agric. Hist.* 51 (1977) 522.

squashes, and stored corn, beans, and other seeds. Recent research has dispelled an Edenic myth that native Americans inhabited an environment without laboring to change it. Richard White has concluded that "academic historians have produced a respectable body of work on humans and the environment that concentrates on how Indian peoples shaped the natural world they lived in."[48]

Effects of Sixteenth-Century Depopulation

When French and English explorers entered the Midwest early in the seventeenth century the land was very different from that known to its inhabitants in the pre-Columbian period. The most important difference was that there were many fewer people, many fewer settlements, and many fewer signs of human occupation. The first tentative estimates of population were made by seventeenth-century French Jesuit missionaries. James Mooney's conservative evaluation of data drawn from the best available primary sources put the total population of Indians in the central states, roughly equivalent to the area of the Midwest, in the second half of the seventeenth century at 75,800.[49] A more generous interpretation of the same evidence might increase the total to 106,070[50] (table 3.1; figure 3.2).

There are no comparable estimates for the population of the Midwest at the end of the fifteenth century. No written records appear to have survived and archaeological evidence is very scanty. If the total was four times as large as Mooney's estimate for the late seventeenth century it would still be very small and very thinly spread over a vast area; a population of a city the size of present-day Oshkosh was scattered over a territory covering nearly half a million square miles. It would take an enormous inflation of figures to raise an overall population density to one person per square mile. In the

48. Richard White, "Are you an environmentalist or do you work for a living?": Work and nature, in William Cronon (ed.), *Uncommon ground: Toward reinventing nature* (W. W. Norton, New York 1995) 175; a bibliographical essay is presented in Richard White, Native Americans and the environment, in W. R. Swagerty (ed.), *Scholars and the Indian experience: Critical reviews of recent writing in the social sciences* (Indiana University Press, Bloomington 1984) 179–204.

49. James Mooney, *The aboriginal population of America north of Mexico*, ed. J. R. Swanton, Smithsonian Institution, Miscellaneous Collection 80 (GPO, Washington, D.C. 1928).

50. Douglas H. Ubelaker, The sources and methodology for Mooney's estimates of North American Indian populations, in William M. Denevan (ed.), *The native population of the Americas in 1492* (University of Wisconsin Press, Madison 1992) 243–88.

TABLE 3.1 Estimated populations of native American tribes
in the Midwest in the late seventeenth century

	Mooney 1928	Ubelaker 1992
Erie	4,000	12,000
Fox	3,000	4,000
Illinois confederation	8,000	8,000
Kickapoo	2,000	3,000
Mascouten	1,500	1,600
Menominee	3,000	4,170
Miami	4,500	24,000
Ojibwa (Chippewa)	35,000	35,000
Potawatomi	4,000	4,000
Sauk	3,500	3,500
Shawnee	3,500	3,000
Winnebago	3,800	3,800
Total	75,800	106,070

Sources: James Mooney, The aboriginal population of America north of Mexico, ed. J. R. Swanton, Smithsonian Miscellaneous Collection 80 (GPO, Washington, D.C. 1928); Douglas H. Ubelaker, The sources and methodology for Mooney's estimates of North American Indian populations, in William M. Denevan (ed.), The native population of the Americas in 1492 (University of Wisconsin Press, Madison 1992) 243–88.

whole region only one town, Cahokia, near St. Louis, had as many as 30,000 inhabitants.[51] It was a culture that had no urban institutions, no organized commerce, no manufacturing industries. Individual tribes might muster two or three thousand warriors but political and administrative organizations were rudimentary.[52] By no stretch of the imagination could the pre-Columbian Midwest be considered a densely peopled country.

The loss of population during the first generation following contact was grievous even though no estimate of the death toll can be made. Native people had little or no immunity to diseases brought by Europeans, and infections transmitted by domestic animals were almost invariably fatal. Influenza, smallpox, measles, mumps, diphtheria, and pneumonic plague were the first mass killers.[53] No

51. Denevan, Pristine myth, 1992, 377; Melvin Fowler, The Cahokia atlas: A historical atlas of Cahokia archaeology, Studies in Illinois Archaeology 6 (Ill. Hist. Preservation Agency, Springfield 1989).
52. Ubelaker, Methodology for Mooney's estimates, 1992, 243–88.
53. Karl W. Butzer, The Americas before and after 1492: An introduction to current geographical research, Annals Assoc. Amer. Geog. 82 (1992) 351 and 364 n. 3; also W. George Lovell, Heavy shadows and black night: Disease and depopulation in colonial Spanish

record was kept of the numbers who died following encounters
with white people or their pigs, cattle, horses, sheep, and poultry.
Neither native Americans nor Europeans recognized until too late
that newcomers and their livestock at large in woods and pastures
were carriers of deadly diseases.

By the time the first white explorers, missionaries, soldiers, and
traders reached the interior of the continent the initial, most virulent
epidemics had passed and populations were beginning to recover.
In the Midwest, Europeans did not witness the highest mortality.
They observed an empty land, remarked on the absence of villages,
noted traces of deserted fields, and above all were curious about the
very small groups of people they met. There were not enough able-
bodied people to carry out elaborate forms of cultivation, nor could
the survivors defer their immediate want of food until crops
ripened. They were forced to rely for daily subsistence on hunting,
fishing, and gathering.

As the total number of hunters was reduced, stocks of game
increased. Seventeenth-century travelers were amazed at the size of
herds of deer, moose, and bison. Bison, also known as buffaloes or
wild cattle, formed most spectacular herds of several hundred animals. In 1669, south of Green Bay, Wisconsin, Father Allouez
observed "in every direction prairies as far as the eye can see," over
which buffalo roamed in herds of four and five hundred.[54] In
1688, a young soldier, Pierre Deliette, accompanied a band of
Illinois on a summer hunt. In the first drive, 120 buffaloes were
killed. The carcasses were dried slowly over small fires. The party
then followed a buffalo trail from marsh to marsh to water for about
60 miles, taking a further 1,200 buffaloes as well as deer, bear,
turkeys, and other game. On their way they gathered nuts, medlars,
grapes, plums, and crab apples.[55] Normally, relations between
neighboring tribes were friendly, but hunting parties occasionally
pursued game onto hunting grounds of others and quarrels led to
war. Hunters and gatherers on the move were more likely to clash
than settled cultivators. Some bands that lost young men in war
resorted to polygamy; others initiated into their tribes white people

---

America, *Annals Assoc. Amer. Geog.* 82 (1992) 426–63; Alfred W. Crosby, *The Columbian
exchange: Biological and cultural consequences of 1492* (Greenwood, Westport 1972).

54. Sauer, *Seventeenth century North America,* 1980, 130.

55. Ibid., 195.

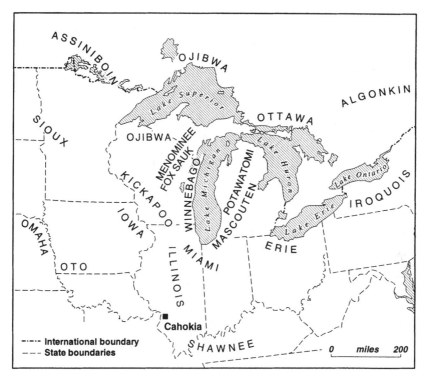

Figure 3.2 Native American tribal areas, circa 1680

Source: Douglas H. Ubelaker, The sources and methodology for Mooney's estimates of North American native populations, in William Denevan (ed.), *The native population of the Americas in 1492* (University of Wisconsin Press, Madison 1992) 243–88.

taken prisoner in raids; others adopted offspring of mixed parentage.[56]

Detailed local studies of sixteenth-century vegetation changes are needed to show what effects, if any, reductions in human population had on the extent of grasslands, forests, and wetlands. In the absence of such studies it might be presumed that a few survivors would continue to burn prairies periodically and that pressure from grazing animals would repress tree growth. In the absence of firm evidence it is not possible to know how wetlands might have responded to fewer human occupants.

56. Ubelaker, Methodology for Mooney's estimates, 1992, 268; Gary B. Nash, *Red, white and black: The peoples of early America* (Prentice Hall, Englewood Cliffs 1982) 275–80.

Impact of the Fur Trade on Native Americans

While missionaries and soldiers kept journals and wrote reports on their travels into the interior of North America, most important relations with natives were conducted by fur traders. From the mid–seventeenth to the mid–nineteenth century a few hundred French, English, and American traders maintained contacts with all nations and tribes in the Great Lakes country and upper Mississippi basin. A very small number of white visitors exerted a profound influence on the native population, transforming their economy, society, and culture, causing them to modify their environment.

In the first quarter of the seventeenth century, before Jean Nicollet reached Green Bay, Wisconsin, in 1634, Huron acted as intermediaries collecting furs on the west shore of Lake Michigan from Potawatomi, Ottawa, and others, selling them to traders in Ontario. In 1660 Sieur de Groseilliers and Pierre Radisson ventured deep into Wisconsin and Minnesota and returned to Montreal with a fleet of canoes heavily laden with valuable pelts. Contacts were made with Winnebago, Menominee, and eastern Sioux.[57] In 1670, the English founded the Hudson Bay Company and began trading with Cree north of Lake Superior and Ojibwa, also known as Chippewa, south of the lake.

In the beginning, natives were acquiescent but not eager partners in the trade. They hunted for subsistence and to some extent for prestige. Normally, they suffered no shortage of food. The Great Lakes provided an unimaginable abundance of fish, including sturgeon, whitefish, and lake herring. In the spawning season, rivers teemed with sturgeon and salmon.[58] Woods yielded nuts, fruit, and maple syrup. From wetlands copious harvests of wild rice, berries, and many varieties of wildfowl were gathered. As population recovered from devastating losses inflicted by European diseases, ridging and garden cultivation were resumed. Native people were not dependent on fur traders for food; indeed, traders were largely fed by natives.[59]

Traders did not find it easy to persuade trappers to bring them

57. Sauer, *Seventeenth century North America,* 1980, 121–25.

58. Jeanne Kay, Wisconsin Indian hunting patterns, 1634–1836, *Annals Assoc. Amer. Geog.* 69 (1979) 402–18, esp. 403, 412.

59. Thomas R. Wessel, Agriculture, Indians, and American history, *Agric. Hist.* 50 (1976) 13.

precious furs of beaver, otter, and mink. A taste for iron kettles, pans, knives, axes, and woolen blankets was promoted but there was a limit to the number of such articles natives wanted to buy. It was not difficult to generate a demand for firearms, but trading in weapons was fraught with danger. Guns were used not only to shoot deer and bison but to attack other natives and, when opportunity arose, were turned against traders and European soldiers. Even more harmful was the traffic in alcohol. A little wine or brandy helped traders obtain favorable bargains in their negotiations, but when natives sought to exchange furs for brandy they were likely to damage their health and efficiency as hunters. Many native people lacked enzymes to metabolize alcohol, so effects of intoxication were injurious to liver, heart, and brain. On the Atlantic seaboard, where they could obtain unrestricted supplies of spirits from fishermen not only from France, but also from Britain, the Netherlands, and northern Europe, native Americans in the second quarter of the seventeenth century at last "learned that the source and origin of all their trouble is intoxication. . . . They gorge themselves for several days at a time each year on the arrival of ships from France; they have no restraint in what they eat or drink, and in the absence of self control, why seek any further explanation of the prevailing Mortality."[60] In the interior of the continent, Jesuits and monopolistic trading companies attempted to regulate the sale of liquor. In practice, traders usually offered a pint of brandy for a pelt, which kept rates of consumption below the level of incapacitating hunters. A number of people moved to the upper lakes country to get away from the debauchery of maritime Canada and New England. Many Huron, Ottawa, and other natives took neither wine nor brandy. In the late seventeenth century drunkenness and disorder near trading posts in the Midwest were less frequent than in small market towns in Brittany.[61]

Traders clearly understood that they could not gain more furs by offering higher prices. If they dispensed more brandy, hunters were rendered incapable of catching more game; if they supplied more kettles and blankets they flooded the market and exhausted

60. François du Creux, *The history of Canada or New France, 1625–1658,* ed. James B. Conacher, Publ. Champlain Society 30 (Toronto 1951), vol. 1: 88–90, cited in Sauer, *Seventeenth century North America,* 1980, 116.

61. Sauer, *Seventeenth century North America,* 1980, 148.

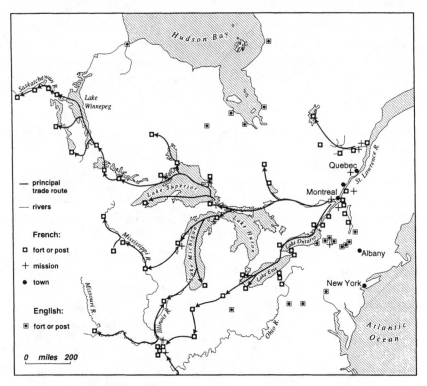

Figure 3.3 French and English fur trade, circa 1750

Source: Cole Harris, France in North America, in Robert D. Mitchell and Paul A. Groves (eds.), *North America: The historical geography of a changing continent* (Hutchinson, London 1987) 88.

demand. For their part, hunters did not wish to deplete stocks of furbearing animals, nor were they motivated to amass profits by raising output. Traders and white observers misinterpreted both these attitudes. Native Americans' reluctance to increase production was put down to lack of effort and enterprise; their later destructive exploitation of renewable resources was attributed to lack of foresight and ignorance of a conservation ethic. A myth of native idleness and fecklessness was constructed out of disinclination to respond to inducements offered by European traders. Traders themselves were not discouraged by the myth they had invented and propagated. They continued to expand their activities, setting up new posts further west in the Mississippi valley and beyond, offering a wider range of goods and allowing credit facilities (fig. 3.3). During the eighteenth century, enough native Americans had succumbed to European sales talk to start taking more animals than could be replaced by natural reproduction. Inevitably populations declined. Jeanne Kay traces the gradual reduction in numbers of beaver and otter collected at trading posts on the shores of Lake Michigan and later from Lac du Flambeau in the interior. Throughout Wisconsin, elk and black bear became scarce in 1800 and bison were extinct by 1832. Sauk and Fox acquired domesticated horses and moved to Iowa in pursuit of remaining herds of bison. In Wisconsin the fur trade was sustained by diminishing numbers of muskrat, deer, and raccoon. Fish, corn, and maple syrup were sold in place of furs. During the Napoleonic wars French traders were displaced by British, and after the War of 1812 British companies were replaced in the Midwest by Jacob Astor's American Fur Company. Some local agents remained at their posts and switched allegiance to new companies. The American Fur Company was unable to retain its monopoly against fierce competition from many independent traders and shopkeepers.[62]

While control of trading companies passed from one nation to another, all their representatives in the Midwest kept up relentless pressure on hunters to collect more and more furs. They devoted great ingenuity and incessant efforts to make hunters dependent on goods supplied by trading posts. Lewis B. Cass described the dilemma facing a native American hunter in 1830:

The rifle was found a more efficient instrument than the bow and arrow; blankets were more comfortable than buffalo robes; and cloth than dressed skins. The exchange was altogether unfavorable to them. The goods they received were dear and the peltry they furnished was cheap. A greater number of animals was necessary to the support of each family, and increased exertion was required to procure them. We need not pursue this subject further. It is easy to see the consequences both to the Indians and their game.[63]

Discussion of native American resource use and environmental beliefs focuses on two questions: whether native people abandoned a precontact conservation ethic or whether at any time in the past

62. Kay, Wisconsin Indian hunting, 1979, 403–14.
63. Lewis B. Cass, Considerations on the present state of the Indians and their removal to the west of the Mississippi, *North American Review* 66 (1830) 5, cited in Kay, Wisconsin Indian hunting, 1979, 417.

they deliberately practiced methods of conserving game populations
at levels that habitats could sustain. Jeanne Kay thinks that "many
tribes did not change their beliefs and historical evidence indicates
that they considered they were taking from nature only enough to
satisfy their essential wants." In effect, European traders substituted
costly manufactured products for natives' homemade utensils and
garments. This change in material consumption required more animal pelts to supply equivalent goods and led to a depletion of
wildlife without a shift in basic environmental attitudes.[64]

Another reason why native Americans became more deeply dependent on the fur trade was that their population was increasing at
a faster rate than traditional means of subsistence would support.
The fur trade created new work for natives: growing food for trading post personnel and their livestock, acting as guides, gathering
firewood, performing many kinds of domestic service. Craftsmen
found new employment building and repairing canoes, mending
nets, curing skins, tanning, making leather goods. Native American
settlements grew up alongside trading posts and forts. Many children were born out of unions between European men and native
women, and these children "remained in almost all cases within
Indian society."[65] In addition, "throughout the colonial period,
much to the horror of the leaders of white society, colonists ran
away to Indian settlements, or, when they were captured in war and
had lived with a tribe for a few years, often refused to return to
white society."[66] The movement was invariably toward increasing
native American populations; no native American or child of mixed
blood was ever recorded as coming to join white society.
Throughout the seventeenth century a stream of displaced tribes
from the Atlantic colonies swelled the native American population
in the Midwest.

---

64. Jeanne Kay, Preconditions of natural resource conservation, *Agric. Hist.* 59 (1985)
124–35, quotation from 131; alternative views are put forward in Calom Martin, *Keepers of
the game: Indian-animal relationships and the fur trade* (University of California Press,
Berkeley 1978); Shepard Krech III (ed.), *Indians, animals, and the fur trade: A critique of
"Keepers of the game"* (University of Georgia Press, Athens 1981).
65. Nash, *Red, white and black,* 278.
66. Ibid., 279.

Hunting for Furs and the Ecology of Wetlands

The fur trade was articulated along waterways of the upper lakes country. The earliest trading posts were situated on the shores of the Great Lakes close to native villages near Detroit, St. Ignace, Sault Ste. Marie, Green Bay, Milwaukee, Chicago, and other lake ports. Traders carried their canoes inland over portages between the Fox and Wisconsin rivers and from Chicago to the headwaters of the Illinois River and so reached the upper Mississippi. French place-names along the banks of the great river, from St. Cloud, St. Anthony, St. Paul, St. Croix, La Crosse, Prairie du Chien, Dubuque, and Moline down to St. Louis, Baton Rouge, and New Orleans, mark the consolidation of French power over the principal waterway through the interior of the continent from the late seventeenth to the eighteenth century. British and Americans also entered the region by water. Native settlements likewise were aligned along waterways, and winter hunting expeditions followed rivers and streams, parties of hunters carrying their equipment to hunting grounds by canoe.[67] The game itself was pursued across frozen ground. Hunters on snowshoes stalked moose and deer; beaver were speared through holes in the ice, and deer were driven over the snow toward prepared positions for ambush.[68] From March or April, when fish began to run up rivers to spawn and ducks and geese arrived from the south to begin nesting, until September, when eels were at their best, hunters lived mainly on fish and fowl[69] (fig. 3.4).

Open water and wetlands were habitats not only for wild rice, berries, fish, and wildfowl but for a great majority of game animals. Open water was an essential element in the habitat for beaver and muskrat and was of some importance for white-tailed deer.[70] Beavers were more abundant in headwater streams and muskrat in lowland swamps and marshes, the richest feeding grounds being provided by large river basins. In addition, mink, otter, and raccoon were either semiaquatic or frequent foragers on lakeshores and in wetlands. Deer and elk had to have regular access to lakes and

67. Kay, Wisconsin Indian hunting, 1979, 410.
68. Sauer, *Seventeenth century North America*, 1980, 80, 86, 94, 101, 105, 110.
69. Ibid., 89–90.
70. Kay, Wisconsin Indian hunting, 1979, 406–7.

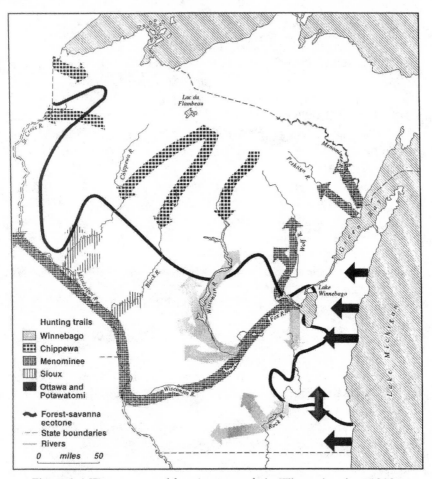

Figure 3.4 Waterways and hunting grounds in Wisconsin, circa 1810

Flow bands indicate routes frequently used by different groups of hunters to reach their winter hunting grounds. *Source:* Jeanne Kay, Wisconsin Indian hunting patterns, 1634–1836, *Annals Assoc. Amer. Geog.* 69 (1979) 411.

streams for water and cover.[71] Very few of the highly valued fur-bearing animals were more numerous outside wetlands; marten, fisher, and lynx lived in coniferous forests, bison on the prairies, but they all had to visit watering places (table 3.2). The most prolific species and also one of the most valuable furbearers was the musk-

71. Ibid., 410.

TABLE 3.2 Values of habitats for game animals in Wisconsin

	Shrubs	Hardwood trees	Conifer trees	Herbaceous wetland	Open water
Beaver	4	5	–	4	5
Muskrat	1	–	–	4	5
Raccoon	2	4	–	1	5
White-tail deer	4	4	4	3	2
Black bear	4	4	3	1	1
Marten	–	1	5	–	1

Value of habitat: 5 essential; 4 very important; 3 important; 2 of some importance; 1 of slight importance; – of no importance.
Source: Jeanne Kay, Wisconsin Indian hunting patterns, 1634–1836, *Annals Assoc. Amer. Geog.* 69 (1979) 406.

rat, which stocked wetlands from Ohio to northern Minnesota and survived the most intensive hunting in the early nineteenth century. Jeanne Kay emphasizes the crucial importance of wetlands in the geography of hunting. The most productive hunting grounds and the most favored destinations for hunters from neighboring tribes were wetlands lying on the "ecotone between northern forest and prairie, extending from Illinois to Alberta."[72]

Increasing native American populations, decreasing catches of furbearers, widespread burning of forests and prairies, wet summers followed by deep winter snows, and a return of epidemic diseases presented the fur trade with a series of intractable problems in the first forty years of the nineteenth century. Supplies of muskets, powder and shot, steel knives, iron cooking pots, and blankets made it easier for native American hunters to pursue game and lighten their packs. Hunting parties returned with more pelts, bought more goods, and raised more children. Some children were fathered by white men, some were fed and clothed from company stores and traders' potato patches. Native gardens furnished trading posts with "country provisions" including corn, beans, squashes, wild rice, fish, wildfowl, and meat.[73] The growth of population and rise in food production beyond the bag of game resulted in more hunters chasing more furs to trade. This set in motion a spiraling predation crisis in which diminishing stocks of furbearers were hunted by increasing numbers of hunters. The most highly prized beaver and ot-

72. Ibid., 412.
73. Fritz et al., The influence of the fur trade, 1993, 477–89.

ter first declined in Ohio and Michigan, then in Wisconsin and
Minnesota, and finally in the Canadian muskeg. A reduction in
numbers of beaver "must have initiated extensive environmental
changes, for thriving beaver populations had created frequent forest
clearings, elevated water tables and created ponds."[74] Fewer beaver
would create fewer ponds, and stocks would continue to shrink.
Unprofitable trading posts closed and native Americans were left
without the supplies of powder and shot they required to continue
hunting.

Extensive fires in the period from 1790 to 1805 reduced
browse and cover for deer but increased grazing for bison and
moose. Herds of bison were exterminated on open plains by
hunters on horseback. Year by year buffalo drives and the ensuing
slaughter moved further and further west from tall grasslands to
shortgrass prairies. native people lamented their passing. When "the
buffalo went away, the hearts of my people fell to the ground and
they could not lift them again," declared a Crow chief, adding,
"After this nothing happened."[75] The history of his tribe ended
when hunting ceased and life no longer had meaning for him when
the bison had gone.

Recurrent cold, wet summers and deep snowfalls in winter in
1816–17, again in 1824–25, and once more in 1830–31 were accompanied by widespread starvation. Wet summers caused failures
of fish, wild rice, and cultivated crops; and deep winter snows impeded travel and stalking of big game. A few people were saved by
supplies of potatoes from trading posts; others succeeded in trapping hares and small animals. Cold and famished, many succumbed
to outbreaks of measles and influenza in 1819. Woolen clothing
and blankets gave them less warmth and protection than native buffalo robes.[76]

74. David M. Gates, C. H. D. Clark, and James T. Harris, Wildlife in a changing envi-
ronment, in Flader (ed.), *Great Lakes forest,* 1983, 52–80, quotation on 62; G. J. Knudsen,
History of beaver in Wisconsin (Wisconsin Conservation Department, Madison 1963); A. W.
Schorger, The beaver in early Wisconsin, *Trans. Wisconsin Acad.* 54 (1965) 147–79.

75. Frank Linderman, *Plenty-coups: Chief of the Crows* ([1930]; reprint Lincoln 1962),
cited in William Cronon, A place for stories: Nature, history, and narrative, *Journ. Amer.
Hist.* 78 (1992) 1366.

76. Roger Suffling, Catastrophic disturbance and landscape diversity: The implications
of fire control and climate change in subarctic forests, in Michael R. Moss (ed.), *Landscape
ecology and management,* Proceedings of the First Symposium of the Canadian Society for
Landscape Ecology and Management (Polyscience Publications, Montreal 1988) 111–20;

Native Americans had no way of recovering from the effects of falling fur production. Traders were neither willing nor strong enough to resist the incursion of white American settlers who sought only vacant land. In this situation, native Americans were at best a nuisance and were perceived by whites as a threat to peaceful settlement.

Loss of Native American Independence

The insidious effects arising from overkill of furbearing animals coincided with a period of struggle between France and Britain for control of the St. Lawrence, Great Lakes, and Mississippi waterways. Up to the middle years of the eighteenth century the French maintained friendly relations with many Indian tribes, based on an enduring military alliance with the Algonkin and stable trading connections with Chippewa, Potawatomi, Menominee, Winnebago, Ottawa, Fox, Sauk, Illinois, and others. The French offered natives a share in their culture and some protection in return for trading privileges, knowledge of native ways of life, and some military assistance. The British held sway over the Mohawk in the Adirondack Mountains and the Iroquois nation, including Oneida, Onondaga, Cayuga, and Seneca south of Lake Ontario. Unlike the French, they tried to avoid entering into partnerships with natives; there was no attempt to reciprocate benefits or accommodate native American aspirations.[77] The British incited the Iroquois to wage war on tribes trading with the French and claimed sovereignty over their lands, but Iroquois chiefs did not recognize these claims. They were not conquered by the British and they wanted to continue governing themselves.[78]

The Seven Years War, 1756–63, between Britain and France ended in the defeat of France and the loss of her colonies in Canada as well as the Midwest. Following the Treaty of Paris in 1763, a British royal proclamation drew a line at "the heads or sources of

---

Roger Suffling and Ron Fritz, The ecology of a famine: Northwestern Ontario in 1815–17, in C. R. Harington (ed.), *The year without a summer? World climate in 1816* (Canadian Museum of Nature, Ottawa 1992) 203–17.

77. D. W. Meinig, *The shaping of America: A geographical perspective on 500 years of history,* vol. 1, *Atlantic America 1492–1800* (Yale University Press, New Haven 1986) 112–13, 212.

78. Ibid., 262.

any rivers which fall into the Atlantic Ocean," south from the Gulf of St. Lawrence to the mouth of St. Mary's River on the northern border of Florida. To the west of that line, corresponding roughly to the Allegheny Mountain watershed, land was reserved for native American Indian occupation. By 1763, several thousand Europeans had already invaded tracts further west between the Mononga-hela and Kanawha rivers. In 1771, a new Proclamation Line was drawn to open for European settlement a great wedge of territory south of the Ohio River from the Great Forks at Fort Pitt to the Kentucky River. Beyond the Ohio River, Britain expressly granted land "under our sovereignty, protection and dominion, for the use of said Indians." On the ground, sovereign power could not be exercised and there was no way a distant colonial government could protect Indian rights. The fur trade was rapidly destroyed by uncontrolled competition while aggressive speculators and land-grabbers swarmed in from New York, Pennsylvania, and Virginia. In a belated and forlorn attempt to restore order in the Midwest, the British government, through the Quebec Act in 1774, handed back power to administer native American Indian lands to Montreal and Detroit.[79] Two years later the whole of the West was thrown into turmoil by the outbreak of the American War of Independence.

During the third quarter of the eighteenth century, the popula-tion of colonial North America, excluding native American Indians but including French and Spanish inhabitants, doubled in size from about 1,300,000 in 1750 to somewhere between 2,325,000 and 2,600,000 in 1775. Natural increase accounted for most of the growth, but numbers of immigrants arriving from Ulster, Scotland, and Germany rose sharply.[80] Many young men and women were ready to seek their fortunes in the bluegrass country beyond the Appalachians, following a trail through the Cumberland Gap opened by Daniel Boone in 1775.[81] Almost all natives in the inte-rior were opposed to expansive American nationalism although they were lukewarm in their loyalty to the British crown.

The American War of Independence was a conflict for control of sealanes and ports on the Atlantic coast, in which the British were

79. Ibid., 284–88.
80. Ibid., 288–89.
81. Ibid., 293.

beaten on land and only narrowly succeeded in keeping command of the seas. Native Americans who had professed support for Britain were presumed by white Americans on the western frontier to have forfeited their lands and were expected to be removed. In 1783, in a letter to a congressional committee set up to deal with Indian affairs, George Washington sought to moderate demands for immediate expropriation of native American lands. He hoped Indians would acknowledge that "their true Interest and safety must now depend on *our* friendship," and he wished to reassure them that the American government would "establish a boundary line between them and us beyond which we will *endeavor* to restrain our People from Hunting or Settling, and within which they shall not come, but for the purposes of Trading, Treating, or other business unexceptionable in its nature."[82] These hopes and promises had no prospect of being fulfilled. Native Americans had not recognized the sovereignty of France or Britain, nor were they willing to surrender to the United States. On the other hand, white American settlers could not be held back by a remote power in Washington. On the frontier, native Americans were enemies who had to be killed or driven away.

It was urgently necessary for the U.S. government to impose some order on the furious, rough-and-tumble scramble for land. Claims had to be legalized and registered, property boundaries had to be drawn, new counties created, new territories and states admitted to the federal union. The Land Ordinance of 1785 established a system for surveying lands in the public domain in regular square-mile sections and 36-square-mile townships. Federal Land Offices were set up to sell by auction at not less than one dollar per acre, freehold to the first purchaser, sections, quarter-sections, and forty-acre lots.[83] The ordinance failed to stop unlawful invasions of American Indian territory, nor did it enable the government to evict squatters from Miami and Shawnee homelands. While confusion reigned on the frontier, Congress, in dire need of money, ignored its own enactment and sold at far below the minimum price extensive tracts of land to eastern speculators. In 1787, an association of New

82. Ibid., 408, cites George Washington, *The writings of George Washington,* vol. 27, ed. John C. Fitzpatrick (GPO, Washington, D.C. 1938) 134–36, 138.

83. Roy M. Robbins, *Our landed heritage: The public domain, 1776–1936* (Princeton University Press, Princeton 1942) 8–10.

England war veterans, incorporated as the Ohio Company, acquired one and a half million acres of land on the Ohio and Muskingum rivers at just over 8 cents per acre. At about the same time, a group of congressmen organized the Scioto Company and helped themselves to five million acres on the Ohio east of the Scioto River. In 1788, a third contract at a similar bargain price was obtained by another congressman from New Jersey for a tract of land between the Great and Little Miami rivers. All these large-scale ventures were dissolved after a few months and they benefited no one on the frontier.[84] In 1787, Congress made a fresh attempt at regulating sales of public lands. The so-called Northwest Ordinance placed the territory northwest of the Ohio River under direct jurisdiction of the federal government and laid down stages by which districts within the territory might progress to statehood and full membership of the union.

The first condition for an orderly entry of white settlers into this vast northwestern region was to make treaties with American Indians to gain cessions of land and define areas for reservations. The Iroquois who had actively supported the British found themselves harried as defeated enemies by hostile American officials and speculators. Some fled to Canada; the rest ceded their western lands to the United States. Algonkin who occupied lands in the Ohio valley faced the main advance of white American settlers. The tribes united to resist the onslaught. In 1790 and again in 1791 they drove back American militias in the Miami valley. In 1794, a force of regular soldiers under the command of General Anthony Wayne defeated a large party of American Indians at Fallen Timbers on the lower Maumee. Effectively, this ended all prospect of concerted resistance and by the Treaty of Greenville in 1795 American Indians ceded to the United States a large portion of the Northwest Territory.[85] The signatories to the treaty came as free people, recognized as rightful owners of their lands. They went away unfree, their leader having declared: "We do now, and will henceforth, acknowledge the fifteen United States of America to be our father."[86] A decisive shift in the relationship had taken place. The federal government threw its weight behind the settlers, American Indian peo-

---

84. Ibid., 10–11.
85. Meinig, *Atlantic America,* 1986, 353–54.
86. Ibid., 409.

TABLE 3.3 Population of the United States 1790–1820

	1790	1800	1810	1820
Ohio	–	45,365	230,760	581,434
Indiana	–	5,641	24,520	147,178
Illinois	–	–	12,282	55,211
Rest of USA	3,929,214	5,257,477	6,972,319	8,854,630
Total	3,929,214	5,308,483	7,239,881	9,638,453

Source: U.S. Bureau of Census, Seventeenth Census of Population, 1950, vol. 1: 1–8, 1–9.

ple were treated as subordinates, and their lands were placed in trusteeship. It is ironic that the republic that had so recently won its independence from colonial rule should impose colonial status on native people living within its borders, in the Northwest Territory.

Removal of Native Americans

In the first forty years of its existence as a nation, the population of the United States more than doubled in size. It increased by 35% in the last ten years of the eighteenth century, from 3,929,214 in 1790 to 5,308,483 in 1800; it increased again by 36% in the first decade of the nineteenth century to 7,239,881 in 1810 and by a further 32% in the second decade to 9,638,453 in 1820 (table 3.3). Pressure on the western frontier was increasing inexorably. At the end of the eighteenth century about 45% of the surface area of the United States lay in unceded American Indian territories. At a stroke, in July 1803, the Louisiana Purchase added over a million square miles, thereby doubling the national area. The vast extent of newly acquired land, mostly lying in the upper Mississippi valley, was perceived by Thomas Jefferson as a means of both satisfying American land hunger and providing adequate reserves for American Indians (fig. 3.5).

From 1795 onward, native Americans everywhere were forced to retreat before armed gangs of pioneers organizing themselves into local militias to guard lands seized in territories legally closed to them. Again and again, against ever-increasing numbers of well-armed invaders, natives fought desperately to defend their homelands, save their reservations, and avoid forcible expulsion. At the end of each war, the survivors had to sign a new treaty, cede more land, and suffer further removal. After a century of repeated re-

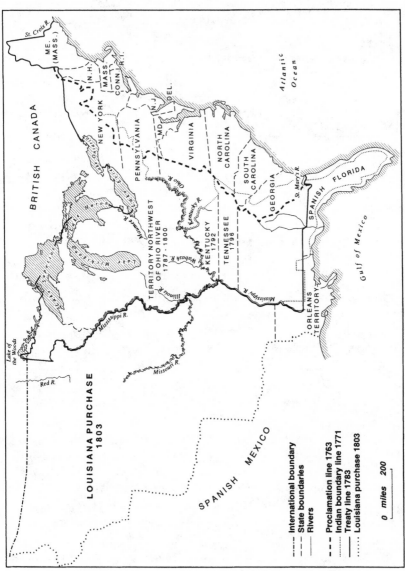

Figure 3.5 Boundaries of the United States to 1803

Sources: Robert D. Mitchell and Paul A. Groves (eds.), *North America: The historical geography of a changing continent* (Hutchinson, London 1987) 151; Charles O. Paullin and John K. Wright, *Atlas of the historical geography of the United States* (Carnegie Institute/American Geographical Society, Washington, D.C. 1932) plates 46, 95.

movals, almost all American Indian lands had been taken by white Americans; almost all native people had been expelled or had perished. A few who remained were confined to small reservations in Michigan, Wisconsin, and Minnesota. As white Americans moved further west their methods of acquiring American Indian land became less honest and more brutal, while natives' fears and resentments deepened and their hostility grew fiercer. In early encounters between individual whites and natives, whites plied their victims with liquor and bartered manufactured goods for possession of land. Some bargains were made under duress, the victim's land being claimed as settlement for unpaid debts or as penalties for alleged offenses, or insolently demanded with menaces at the point of a gun. For their part, native Americans sometimes signed away land belonging to others or land which their tribe had not authorized them to part with. On both sides, ambiguities about definitions of property gave rise to future disputes.

The federal government attempted to keep the opposing factions apart. Their aims were to separate native Americans from whites, negotiate cessions of Indian land by treaties with accredited tribal chiefs, carry out a cadastral survey of the public domain, and hold periodic sales of land that would transfer to incoming settlers legal title to registered purchases. The acquisition of the trans-Mississippi West in 1803 and the redrawing of the Canadian border after the War of 1812 provided space for resettling all who wished to move. Thomas Jefferson envisaged these "illimitable regions" as offering "the means of tempting all our Indians on the East side of the Mississippi to remove to the West."[87] At the negotiating table, few tribes were tempted to take up offers of resettlement on lands beyond the Mississippi. In the first decade of the nineteenth century, treaties with the Wyandot, Ottawa, Munsee, Delaware, Shawnee, Potawatomi, Illinois, and other tribes ceded further tracts in northern Ohio, eastern Michigan, southern Indiana, and Illinois.[88] By way of compensation for these cessions natives were given presents and annuities and a few were granted small areas to be reserved per-

87. D. W. Meinig, *The shaping of America: A geographical perspective on 500 years of history*, vol. 2, *Continental America 1800–1867* (Yale University Press, New Haven 1993) 78–79.

88. Sam B. Hilliard, A robust new nation, 1783–1820, in Robert D. Mitchell and Paul A. Groves (eds.), *North America: The historical geography of a changing continent* (Hutchinson, London 1987) 163.

manently and exclusively for their occupation. Federal agents and missionaries exhorted those who stayed on their homelands to become Christians, learn English, practice agriculture, raise livestock, and abandon hunting. In practice, few were persuaded to adopt white American ways of life.

In Wisconsin, northern Indiana, Illinois, and Missouri, the conclusion of treaties was held up by conflicting Indian land claims. Fox, Sauk, Kickapoo, and Winnebago would not agree to the partition proposed in 1815. In response, the federal government embarked on a plan to build or rebuild forts in the upper lakes country, ostensibly to protect American Indian territory against intrusions by British agents and warmongers, but, on the ground, new garrisons were deployed to extend and consolidate American power over more and more tribes.[89] In the territory effectively dominated by forts, federal commissioners set about confronting natives with harsh directness. In 1817, representatives of six tribes still owning lands in Ohio were called upon to cede their remaining holdings, accept annuities, and either move beyond the Mississippi or be confined together on much smaller reservations. That procedure was repeated all over Indiana, Illinois, and Michigan. By 1821 almost the entire region had passed into American hands but still natives refused to leave (fig. 3.6).

In December 1824 President Monroe introduced a formal proposal to compel American Indians to move to a western sanctuary together with a program for educating them in the arts of American civilization. Congress did not approve the measure. Meanwhile, natives were being expelled from upper New York State, Ohio, and Illinois, and whites were trespassing widely on lands possessed by native American Indians, wantonly destroying fields and villages in Illinois and Wisconsin. These attacks provoked the Sauk to retaliate; the ensuing Black Hawk War, 1829–32, ended in a massacre of the tribe on the banks of the Mississippi. A new president, Andrew

---

89. Ray Allen Billington, *Westward expansion: A history of the American frontier* (Macmillan, New York 1949) 291. Forts restored in 1816 and 1817 included Ft. Wayne and Ft. Harrison in Indiana, Ft. Shelby in Detroit, Fts. Gratiot and Mackinac on the eastern edge of Michigan, and Fts. Dearborn and Clark protecting settlements in Illinois. New fortifications were built in 1816 along the Mississippi at Fts. Edward and Armstrong in western Illinois, Ft. Crawford at Prairie du Chien, and Ft. Howard near the mouth of the Fox River in Wisconsin. Between 1819 and 1822 new forts were built at Ft. Snelling in St. Paul, Minnesota; Ft. Saginaw in Michigan; and Ft. Brady at Sault Ste. Marie.

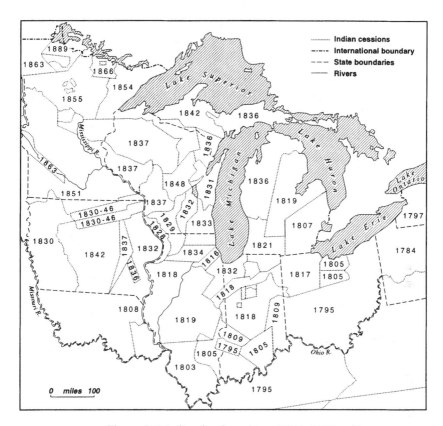

Figure 3.6 Indian land cessions, 1784–1889

Sources: Hildegard B. Johnson, Order upon the land: The U.S. rectangular land survey and the upper Mississippi country (Oxford University Press, New York 1976)
123; Robert D. Mitchell and Paul A. Groves (eds.), North America: The historical
geography of a changing continent (Hutchinson, London 1987) 164; C. C. Royce,
Indian land cessions in the United States, Eighteenth Annual Report of the Bureau of
American Ethnology, pt. 2, 1896–97 (Washington, D.C. 1899).

Jackson, strongly supported by his buckskin-clad admireis, openly
advocated the abolition of treaty-making and the removal, by force
if necessary, of all American Indians. In May 1830, after long and
heated debate, Congress passed by a narrow majority the Indian
Removal Bill. Tribes in the northwest were now pressed to make
large concessions. Remnants of defeated Sauk, Fox, and neighboring Potawatomi were unceremoniously ousted from Illinois and
Wisconsin. After holding out for many years, the Winnebago were
rapidly cleared from central Wisconsin. That left only small reserves

held by Oneida, Munsee, and scattered bands of Chippewa in the northern forests and swamps.

In 1831, a few Shawnee, Seneca, and Ottawa remained in Ohio, all of whom were scheduled for removal to Kansas in the following spring. They tarried until September 1832 before setting off. Indian agent James Gardiner, in charge of the column, described the slow progress of the march:

The Shawnees . . . have almost exhausted our patience. They forfeited their promises and abused every kindness. It seemed impossible to get them to make the least movement . . . Nature has sunk under their beastly intemperance . . . Many are sick; some are wounded . . . and all that can are still drinking whiskey, women as well as men, half-crazy and enfuriated . . . The whites beset us with their barrels and kegs of whiskey, hide out in the woods, and three days were consumed in almost fruitless efforts to remedy the serious evil inflicted by our own citizens . . . We fail if we *stop three days between this and the Mississippi.*[90]

In his next letter, Gardiner reported the first deaths, "mostly children," and by the time they reached the Wabash more had died from bad colds and cholera. As winter approached, they passed into Missouri, still followed by whites who "continue to sell whiskey to our Indians. About twenty of our Ottaways were as drunk as David's sow yesterday. When sober, they are by far the most orderly and manageable of the whole detachment. But drunk, sober or sick, we will move them along."[91] At the beginning of December the party reached its destination, weakened by sickness and fatigue; they then had to start building cabins before all were killed by fever and roaring blizzards.

The permanent new homes promised by the government turned out to be no more than temporary resting places. In 1842, the Sauk and Fox, recently removed to Iowa, were forced to cede their new lands to incoming whites. Winnebago and other tribes relocated from the East became "a source of great annoyance and dissatisfaction" to white settlers who complained that the area was "inhabited by savages and wild beasts."[92] It was demanded that they be re-

90. Gloria Jahoda, *The trail of tears: The American Indian removals, 1813–1855* (Allen and Unwin, London 1976) 70–71.

91. Ibid., 72.

92. R. N. Satz, *American Indian policy in the Jacksonian era* (University of Nebraska Press, Lincoln 1975) 115.

moved again further west. Surveyor General George Jones described them as "one of the lowest and most abandoned of Indian tribes." In 1846, they were forced to give up their lands in Iowa and move to Minnesota.[93] In 1849, an unarmed team of federal surveyors working in northern Iowa was held up by a band of Sioux who stole their horses, tents, blankets, and survey instruments. It was alleged that they also terrorized local squatters, and the army was called in to prevent further disturbances.[94] Again, demands were made that these Indians be dispatched elsewhere.

By the late 1840s, the population of American Indian tribes had been greatly reduced and almost all had been removed to small reserves west of the 95th meridian, in what are now Oklahoma, Kansas, and Nebraska. There were fewer than 400 Shawnee and Seneca from New York, another 900 Shawnee from Ohio, only 400 Sauk and Fox from Missouri, 500 Kickapoo from Illinois, nearly 500 Iowa displaced from their home state, a mixed group of Chippewa, Ottawa, and Potawatomi numbering about 4,400, and about 2,000 Delaware, Wyandot, and others from Indiana and Illinois, but by far the largest contingents, totaling about 74,000, had come from Southern states.[95] Small remnants of native populations clung to some of the least attractive parts of the Old Northwest and some Winnebago had found refuge in Minnesota.

The last and most cruel war leading to removals of almost all American Indians from Minnesota broke out in 1862. The conflict was caused by flagrant derelictions of duty by officials in the federal Bureau of Indian Affairs. The price paid for land cessions and the Removal Act was heavy recurrent expenditure on annuity payments, grants for resettlement, and soaring administrative costs. For many years dealings with American Indians had been riddled with fraud and theft. Political patronage exercised by incoming presidents in nominating commissioners, assistant commissioners, superintendents, and agents to the Indian bureau fomented institutional corruption, venality, and neglect. In Minnesota, an outspoken religious leader, Bishop Henry Benjamin Whipple, declared that the Indian

93. Roscoe L. Lokken, *Iowa: Public land disposal* (State Hist. Soc. of Iowa, Iowa City 1942) 35–36.

94. Ibid., 57–60.

95. Meinig, *Continental America*, 1993, 98, citing the *Report of the Commissioner of Indian Affairs 1845*.

system "commences in discontent and ends in blood."[96] In 1861, it was alleged that senior officials in the Indian bureau in Minnesota had committed many serious frauds, including extensive use of blank vouchers, embezzlement of annuities, and falsification of accounts. These allegations were referred to President Lincoln's special investigator, who collected much evidence, but the head of the bureau, himself implicated in some of the alleged misconduct, suspended the inquiry before incriminating reports were published. No action was taken to stop further abuses. A blind eye was turned to white lumbermen stealing Indian timber and traders cheating Indian customers. In 1862, payments of government annuities were delayed and relief for hungry Indians was callously withheld.

On 17 August 1862, at a farmhouse near Acton in western Minnesota, four young Sioux broke in, demanded food, and killed three men and two women. Fearing instant reprisals, the whole tribe launched a preventive war, looting food stores and killing white settlers over a wide area. The state governor, Alexander Ramsey, called upon President Lincoln to send an army. At this time, Lincoln was preoccupied with the war in the South, where his troops were suffering terrible losses. After some hesitation, he decided to relieve an incompetent, troublesome general, John Pope, of his command in the South and send him north to attend to the Indians. Pope arrived in Minnesota on 16 September and spoke to Colonel Henry Sibley, commander of Minnesota's military forces, a former governor and Indian trader, who regarded Sioux as "devils in human shape." Pope's view was even more extreme and he promptly announced: "It is my purpose utterly to exterminate the Sioux if I have the power to do so and even if it requires a campaign lasting the whole of next year."[97] He ordered Colonel Sibley to set out to destroy Indian farms and stocks of food. Minnesotans were seized by panic. Alarming stories circulated that Indians were perpetrating atrocities: raping, scalping, mutilating, disemboweling, burning farms. False rumors that Winnebago and Chippewa were about to join the insurrection were concocted by people who were casting covetous eyes on valuable Indian lands. Politicians and speculators

96. David A. Nichols, *Lincoln and the Indians: Civil War policy and politics* (University of Missouri Press, Columbia 1978) 75, citing a letter dated 2 November 1863 to the Bureau of Indian Affairs in Washington.

97. Ibid., 84–87, quotation on 87.

deliberately exaggerated the scale of the crisis, grossly enlarging numbers of casualties, numbers of Indians engaged in fighting, and numbers of white settlers fleeing from their homes. Large profits were to be made from supplying munitions and provisions for the army, and huge payments were anticipated in reparations at the end of the conflict.

The uprising collapsed before Sibley had time to mount an offensive. As soon as he called upon the Sioux to surrender they came in droves. By 3 October, about a thousand women and children and 200 men had been taken into captivity. Sibley set up a military commission to try them for war crimes. On 9 October, Pope notified the War Department that the war was at an end and it was intended to execute all responsible for "late horrible outrages." The president was informed and immediately directed that "no executions be made without his sanction." On 5 November, the trials were concluded and a list of 303 Sioux condemned to death was forwarded to Washington. Lincoln refused to authorize this mass slaughter and had his own lawyers reexamine the records of each case. A month later, Sibley was sent a list of 39 men who were to be executed. They were hanged on 26 December.[98] To appease the vengeful feelings of white Minnesotans and retain their support for the Republican cause, the government agreed to the removal forthwith of all Sioux and also innocent Winnebago who were to be dispatched, for a fifth time, to South Dakota. The federal government bore the cost of military operations and promised to pay compensation to those who had suffered injury or loss of property as a result of the war. A large military force stayed in Minnesota and early in 1863 a punitive expedition ranged across the Dakotas searching for and killing Sioux and destroying their belongings. Further expeditions were sent out in 1864, 1865, and 1866, killing natives as far west as the Yellowstone River.[99]

The Sioux uprising and subsequent removals brought to an end a long period of native American occupation in the Midwest. At the beginning of the nineteenth century native Americans regarded themselves as "original inhabitants of America," still living and standing "on the soil of their own territory."[100] The United States,

98. Ibid., 94–118.
99. Ibid., 119–28.
100. Meinig, *Continental America,* 1993, 86.

for their part, officially recognized Indians as "resident foreign nations."[101] From 1830, President Jackson spoke of them as "conquered and dependent people" and the Indian Removal Act redefined their status as "domestic dependent nations."[102] In their modes of address, most Americans at that time showed little respect for natives; some spoke of them as "wild savages," others said they ought to be "exterminated." After 1862, they became stateless people. An act of 1871 declared that "no Indian nation or tribe . . . shall be acknowledged or recognized as an independent nation, tribe or power with whom the United States may contract by treaty."[103] The effect of these measures was to extinguish native American title to land, dispossess native peoples, and, finally, deprive them of the right to reside in areas taken over by whites. They were not even allowed to buy tracts of land in the U.S. public domain that once belonged to their own nations.[104] Not the least remarkable consequence of the policy of removal was the erasure of native Americans from peoples' memories and from recorded history. Frederick Jackson Turner's assertion that American development might be explained by "the existence of an area of free land" fails to acknowledge that "free land" was not a gift of nature but was forcibly taken from its earliest inhabitants. The relegation of native Americans to "prehistory," and the myth that they were a "dying race" doomed to extinction because they rejected the blessings of American civilization, stem from an attitude that wanted them eliminated.

Environment and Native American Removals

When Americans were at war with natives, they were at a disadvantage in wetlands. Natives could disperse, take cover, and lie low among reeds and bushes. American cavalry could not ride across bogs; packhorses, wagons, and artillery could pass only in winter where ice was very thick. In the War of 1812, General Hull's army on its way to Detroit had great difficulty in crossing the Black Swamp in Ohio in mid-June. It rained, mosquitoes were "disagreeable," "man and horse had to travel mid leg deep in mud" and "the

101. Ibid., 179.
102. Ibid., 88, 180.
103. Hilliard, Robust new nation, 1987, 163.
104. Robbins, *Our landed heritage*, 1942, 226–29.

mud was ankle deep in our tents."[105] In 1832, Sauk chief Black Hawk, pursued by American cavalry, retreated to swamps around Lake Koshkonong in southwest Wisconsin. Between mid-May and the end of July, warriors sallying forth from that hiding place killed 200 whites, suffering equal losses themselves. An American force of 4,500 advanced on the swamp but moved so slowly that Black Hawk was able to escape unobserved and run to the Mississippi, where he was trapped by a small party led by General James Henry.

As long as whites avoided wetlands natives were able to find refuge there, subsisting on fish, wildfowl, wild rice, nuts, berries, and many other plants and animals. In Wisconsin a small group of Menominee and in Minnesota groups of Chippewa succeeded in holding on to small tracts of tribal wetlands. Lands that were not permanently waterlogged were completely cleared of native people. Villages were pulled down or burned, ridged fields leveled, new roads aligned on a rectilinear grid. Bison and moose were exterminated, deer and other game greatly depleted. With the removal of native Americans, annual burning of prairies ceased and trees began to spring up. For a brief period, forests encroached upon the edge of grasslands in Wisconsin, Illinois, and elsewhere.[106] Between 1829 and 1854, in prairies and oak openings in southwest Wisconsin, the area of grass is estimated to have decreased by nearly 60% following the cessation of fires. In adjoining parts of northwest Illinois where, in 1823, grass covered 80% to 90% of the surface, dense forests have grown. H. A. Gleason has calculated that after burning ended on prairies in Illinois, forest advanced onto grasslands at a rate of between one and two miles in 30 years.[107] A wilderness was taking over empty spaces, but most of the spaces were not to remain empty for long.

105. Martin R. Kaatz, The Black Swamp: A study in historical geography, *Annals Assoc. Amer. Geog.* 45 (1955) 1–35, quotations on 7.

106. Denevan, Pristine myth, 1992, 373; Michael Williams, *Americans and their forests: A historical geography* (Cambridge University Press, Cambridge 1989) 46.

107. Gleason, The vegetational history, 1922, 39–85; Williams, *Americans and their forests,* 1989, 46.

4

Early Nineteenth-century Views of Wetlands

When almost all native Americans had been removed from their homelands, memories of conflict faded and white Americans quietly forgot the atrocities committed in the course of wars and removals. Academics consigned knowledge of native American occupation to a limbo of "prehistory" and scientific observers began to regard so-called "presettlement" landscapes as largely natural. The frontier of settlement, which had begun as a line contested by tribes of native Americans, advanced by winning cessions from reluctant, confused, often badly deceived chiefs, and ended as a line drawn by the U.S. Bureau of Census separating areas whose density of population exceeded two persons per square mile from areas inhabited by few or no people or where none had been counted by census enumerators.[1] From being a live frontier it changed into a wild frontier where pioneers struggled against bears, forests, tallgrasses, and deep swamps.

The solitude was so vast it seemed to stretch to infinity. Visitors were overwhelmed by a sense of their own isolation and feelings of utter loneliness. They remarked on the boundless emptiness of the prairies, likening the rippling motion of grass to waves on an ocean surface, and they perceived solitary trees as distant sailing ships.[2]

---

1. U.S. Bureau of Census, *Eleventh census of population 1890: Extra Census Bulletin,* vol. 2 (Washington 20 April 1891); Frederick Jackson Turner, The significance of the frontier in American history, reprinted from the *American Historical Association Annual Report* for 1893, 199, states: "In a recent bulletin of the Superintendent of the Census for 1890 appear these significant words: 'Up to and including 1880 the country had a frontier of settlement, but at present the unsettled area has been so broken into isolated bodies of settlement that there can hardly be said to be a frontier line. In the discussion of its extent, its westward move-ment, etc., it can not, therefore, any longer have a place in the census reports.' This brief official statement marks the closing of a great historic movement."

2. Ralph H. Brown, *Historical geography of the United States* (Harcourt, Brace, New York 1948) 210, quotes a typical description of the feeling of solitude induced by a prairie

They dreaded the eerie stillness and deathly silence of empty spaces, although thunder sounded fortissimo, howling winds blew, swarms of insects hummed, and choruses of birdsong were deafening.[3] After many weeks of travel from Ohio, Major Stephen Long's expedition reached southern Minnesota in July 1823. William Keating, geologist to the expedition, observed: "The monotony of a prairie country always impresses the traveler with a melancholy, which the sight of water, woods, & c. cannot fail to remove."[4] A few days later, Keating came to the wooded valley of the Redwood River, which "formed a pleasing contrast with the burned and blasted appearance of the prairie." The charm of the valley "arose from the contrast which it presented with the wearisome views of the boundless prairies."[5] Visiting the same locality in 1848, another geologist, David Dale Owen, wrote: "A remarkable feature of this country consists in the small lakes and ponds scattered over it. Many of these are beautiful sheets of water, having the appearance of artificial basins, which greatly enhance the beauty of the country especially when skirted, as they sometimes are, by groves of trees and frequented by a variety of water fowl which tends to animate and relieve the otherwise almost deathlike silence which so often pervades the prairie."[6] In this desolate space, far from madding crowds of big cities, naturalists marveled at the riotous profusion of teeming organisms. Wilderness and wildlife inspired both deep awe and high exultation.

Wetlands repelled most newcomers. Whereas native Americans had been attracted by an abundance of big game, wildfowl, fish, and a great variety of useful plants, Europeans and white Americans found little they wanted apart from wild hay. Natives were able to paddle small canoes along narrow channels through reeds and

on the borders of Indiana and Illinois near Vincennes, written in 1824 by William Newnham Blane.

3. John A. Jakle, *Images of the Ohio valley: A historical geography of travel, 1740 to 1860* (Oxford University Press, New York 1977) 19, 56, 65.

4. William H. Keating, *Narrative of an expedition to the source of St. Peter's River, Lake Winnepeek, Lake of the Woods, & c. performed in the year 1823. . . . under the command of Stephen H. Long* ([1824]; Ross and Haines, Minneapolis 1959), vol. 1: 360.

5. Ibid., 362.

6. David Dale Owen, *Report of a geological survey of Wisconsin, Iowa, and Minnesota: And incidentally a portion of Nebraska territory* (Lippincott, Grambo, Philadelphia 1852) 492.

rushes, or leap fleet-footed from tussock to tussock. White travelers and their horses had to wallow in mud and water.

White Americans Repelled by First Encounters with Wetlands

To strangers entering the Midwest in the early nineteenth century, wetlands appeared desolate and frightening, not only empty, but well-nigh impassable. David Zeisberger, a Moravian missionary who traveled extensively through northwestern Ohio and southeastern Michigan, recorded in his diary in October 1781 an account of crossing the Black Swamp of Ohio. After leaving the Sandusky River, following a trail to Detroit, he passed into "deep swamps and troublesome marshes" that stretched for many miles "where no bit of dry land was to be seen, and the horses at every step [were] wading in the marsh up to their knees."[7] It took two and a half days to cover 30 miles from the Sandusky to the Maumee River. The country was absolutely flat, "not so much flooded . . . yet wet and swampy." He observed that the soil was clayey, "which is one reason why water remains standing" and the vegetation consisted of "beech-swamp or ash, linden, elm and other trees such as grow in wet places," interspersed with groves of oak.[8] Later travelers, moving west onto prairie uplands, encountered "slues" or depressions which were permanently marshy. Horses might sink up to their chests in mire and insects, particularly mosquitoes and gadflies, were an inescapable torment.[9] "With our horses wading through water, sometimes to the girth," Major Stephen Long's expedition made its way into northern Indiana in 1823. "The country is so wet," the report continues, "that we scarcely saw an acre of land upon which a settlement could be made."[10] Richard Lyle Power argued that "a condition of forbidding wetness" deterred many pioneers from settling or even traveling through northern Indiana. Wetlands covered much of the surface between the Black Swamp in Ohio and the Kankakee Pond in Illinois. The whole of this district

7. Martin R. Kaatz, The Black Swamp: A study in historical geography, *Annals Assoc. Amer. Geog.* 45 (1955) 3–4, quoting from Eugene F. Bliss (trans. and ed.), *Diary of David Zeisberger, a Moravian missionary among the Indians of Ohio* (R. Clarke, Cincinnati 1885), vol. 1: 30.

8. Bliss, *Diary of David Zeisberger*, 1885, vol. 1: 45.

9. Jakle, *Images of Ohio valley*, 1977, 56.

10. Keating, *Narrative of an expedition*, 1959, vol. 1: 73.

shared with Toledo "a widespread and almost universally believed character for insalubrity." The legendary Hoosier came "fresh from the swamps and bogs of Indiana."[11] Further west, in rolling prairie country south of the Minnesota River, Major Long's expedition was again bogged down. "These lakes are more properly marshes, the quantity of water in them varying according to the seasons. We had passed several of them during the day; in one of these marshes our pack horses were several times exposed to much difficulty; and the mule that carried the biscuit having stumbled, part of our provisions were wet and damaged."[12] William Keating, who kept the journal, was sorely afflicted by mosquitoes that "arose in such swarms as to prove a more serious evil than can be imagined by those who have not experienced it."[13]

Many other travelers found that attacks by myriads of green-headed flies, horseflies, deerflies, gnats, and mosquitoes made journeys across wetlands in summer almost unbearable.[14] Injuries inflicted on horses by prairie flies are described by Samuel Burton: "The smallest kind are a beautiful green about twice the size of a common house fly, another kind is about twice as large as these, of a slate color, these this season in riding in the prairies would entirely cover a horse and when fastened they remain until killed either by smoke, or being skinned off with a knife and then the horse will be covered with blood. The only way of riding Prairie by day is to cover horses completely."[15] In northeast Illinois in the early 1830s, John Peck, a missionary who settled at Rock Spring, re-

11. Richard Lyle Power, Wet lands and the Hoosier stereotype, *Miss. Valley Hist. Rev.* 22 (1935) 37, 39, 44.

12. Keating, *Narrative of an expedition,* 1959, vol. 1: 360.

13. Ibid. The entry was dated 18 July 1823.

14. Gordon G. Whitney, *From coastal wilderness to fruited plain: A history of environmental change in temperate North America, from 1500 to the present* (Cambridge University Press, Cambridge 1994) 88, cites Lewis C. Beck, *A gazetteer of the states of Illinois and Missouri* (Charles and George Webster, Albany 1823) 81; W. Oliver, *Eight months in Illinois: With information to emigrants,* March of America facsimile series 81 ([1843]; University Microfilms, Ann Arbor, Michigan 1966) 97; C. W. Short, Observations on the botany of Illinois, more especially in reference to the autumnal flora of the prairies, *Western Journ. of Medicine and Surgery* n.s. 3 (1845) 185–98; A. G. Vestal, Why the Illinois settlers chose forest lands, *Trans. Illinois State Academy of Science* 32 (1939) 85–87.

15. Douglas R. McManis, *The initial evaluation and utilization of the Illinois prairies, 1815–1840,* Department of Geography Research Paper 94 (University of Chicago, Chicago 1964) 59, cites letter of Samuel Burton to Ashsah Burton, 15 September 1821 [Chicago Historical Society Manuscript Collection].

marked that "level prairie is often wet."[16] Throughout Illinois many ponds, marshes, and other wet places were treacherous for unwary travelers. Illnesses popularly called swamp fever and ague were believed to be caused by a miasma or gaseous emission arising from stagnant water, from ground soaked by water and wet, decaying organic matter.[17] Throughout the Midwest "miasmic waters" were far more widespread than in the northeast of the United States. Insects and insect-borne fevers presented serious health hazards, threatening lives of newcomers and their domestic animals. On open prairies in Illinois, George Flower recorded in his diary that early settlers were frequently "destroyed by the ague."[18] Lorin Blodget, an early climatologist, expressed a prevailing fear of river bottomlands: "India itself has not been more certain to break the health of the emigrant than the Mississippi valley."[19]

Charles Dickens wrote a lurid description of wetlands in southern Illinois, drawn from observations on a journey by steamboat from Cincinnati to St. Louis in April 1842 (fig. 4.1). At the confluence of the Ohio River and the Mississippi he looked upon:

A dismal swamp, on which the half-built houses rot away: cleared here and there for the space of a few yards; and teeming, then, with rank unwholesome vegetation, in whose baleful shade the wretched wanderers who are tempted hither, droop, and die, and lay their bones; the hateful Mississippi circling and eddying before it, and turning off upon its southern course a slimy monster hideous to behold; a hotbed of disease, an ugly sepulchre, a grave uncheered by any gleam of promise: a place without one single quality, in earth or air or water, to commend it: such is this dismal Cairo.[20]

From St. Louis, which was the most westerly point of his wander-

16. John M. Peck, *A gazetteer of Illinois in three parts* (R. Goudy, Jacksonville 1834) 9–10, cited in McManis, *Initial evaluation*, 1964, 10.

17. McManis, *Initial evaluation*, 1964, 41 refers to Daniel Drake, *A systematic treatise, historical, etiological, and practical on the principal diseases of the interior valley of North America, as they appear in Caucasian, African, Indian, and Esquimaux varieties of its population* (W. B. Smith, Cincinnati 1850) 709.

18. McManis, *Initial evaluation*, 1964, 42.

19. Lorin Blodget, *Climatology of the United States* (J. B. Lippincott, Philadelphia 1857) 455, quoted in D. W. Meinig, *The shaping of America*, vol. 2, *Continental America 1800–1867* (Yale University Press, New Haven 1993) 240.

20. Charles Dickens, *American notes for general circulation*, ed. John S. Whitley and Arnold Goldman ([1842]; Penguin Classics, Harmondsworth 1985) 215–16. Cairo becomes the site of Eden in *Martin Chuzzlewit*, ed. P. N. Furbank ([1843–44]; Penguin Classics, Harmondsworth 1986) 442–48.

Figure 4.1 Location of settlements in the Midwest in mid-nineteenth century

Dates of admission of states to the United States and places mentioned in the text are indicated. *Source:* John Calvin Smith, *Map of the United States including Canada and a large portion of Texas* (Sherman & Smith, New York 1845).

ings, Dickens made a trip 30 miles east to Looking-Glass Prairie in Illinois.

We had a pair of very strong horses, but travelled at the rate of little more than a couple of miles an hour, through one unbroken slough of black mud and water. It had no variety but in depth. Now it was only half over the wheels, now it hid the axletree, and now the coach sank down in it almost to the windows. The air resounded in all directions with the loud chirping of the frogs, who, with the pigs (a coarse, ugly breed, as unwholesome-looking as though they were the spontaneous growth of the country), had the whole scene to themselves. Here and there we passed a log hut: but the wretched cabins were wide apart and thinly scattered, for though the soil is very rich in this place, few people can exist in such a deadly atmosphere. On either side of the track, if it deserve the name, was the thick 'bush'; and everywhere was stagnant, slimy, rotten, filthy water.[21]

On the way back from the prairie, he noted:

The track of to-day had the same features as the track of yesterday. There was the swamp, the bush, the perpetual chorus of frogs, the rank unseemly growth, the unwholesome steaming earth. Here and there, and frequently too, we encountered a solitary broken-down waggon, full of some new settler's goods. It was a pitiful sight to see one of these vehicles deep in the mire; the axletree broken; a wheel lying idly by its side; the man gone miles away, to look for assistance; the woman seated among their wandering household gods with a baby at her breast, a picture of forlorn, dejected patience; the team of oxen crouching down mournfully in the mud, and breathing forth such clouds of vapour from their mouths and nostrils, that all the damp mist and fog around seemed to have come direct from them.[22]

A further ordeal awaited him on a coach journey from Columbus to Sandusky in Ohio: "A great portion of the way was over what is called a corduroy road, which is made by throwing trunks of trees into a marsh, and leaving them to settle there. The very slightest of the jolts with which the ponderous carriage fell from log to log, was enough, it seemed, to have dislocated all the bones in the human body."[23] Ten years later, over a thousand miles west of Sandusky, there still seemed no end to the mire and water. William Cowan, wading up to his knees at a flooded outpost in the Red River valley, saw around him "the plains like a sea."[24]

On the basis of these first impressions it might be inferred that contemporary observers regarded wetlands as unsafe and unfit for habitation, but few expressed that opinion openly and directly. When they wrote their diaries they were intent on their own immediate safety and survival. At the same time, speculators were promoting land sales by singing the praises of vacant properties, embellishing descriptions with plans of imaginary towns and smiling cornfields. Dickens was roused to anger by the mischief he perceived be-

21. Dickens, *American notes,* 1985, 221–22.

22. Ibid., 227.

23. Ibid., 236–37.

24. William Cowan, Fort Garry Journal, 1852, Ms. notebook, cited in Arthur Henry Moehlman, The Red River of the north, *Geog. Rev.* 25 (1935) 81. The entry is for 13 May 1852.

ing inflicted on innocent newcomers by misleading prospectuses and he wanted intending immigrants to learn the harsh truth about the environment before parting with their precious savings or perhaps losing their lives. No other writer drew such gloomy pictures of swamps, describing the deadly effects of ague and the penalties of remoteness. Roger Winsor has reassessed nineteenth-century travelers' descriptions of wetlands on the Grand Prairie in Illinois. The experience of attempting to cross wet prairies was novel and unpleasant. Later visitors repeated a negative stereotype and were inclined to exaggerate the extent of waterlogging. Winsor concluded that "the attention wet prairies received in contemporary accounts clearly was disproportionate to the area they occupied. Wetness was perceived as the overriding property of the area."[25]

Initial Doubts about the Value of Prairies

Wetlands were unattractive to most pioneer settlers and extensive prairies were regarded as less promising for settlement than woodlands. Among the first to condemn the grasslands was James Monroe. In 1786, in a letter to Thomas Jefferson, he concluded:

A great part of the territory is miserably poor, especially that near Lakes Michigan and Erie, and that upon the Mississippi and the Illinois consists of extensive plains which have not had, from appearances, and will not have, a single bush on them for ages. The districts, therefore, within which these fall will never contain sufficient number of inhabitants to entitle them to membership in the confederacy.[26]

Even a naturalist took little pleasure in the vast expanse of the Grand Prairie of Illinois. In 1823, Lewis Beck called it "a dreary uninhabited waste."[27] In April 1842, Charles Dickens expressed disappointment at Looking-Glass Prairie:

Looking toward the setting sun, there lay, stretched out before my view, a vast expanse of level ground; unbroken, save by one thin line of trees, which scarcely amounted to a scratch upon the great blank; until it met the glowing

25. Roger A. Winsor, Environmental imagery of the wet prairie of east central Illinois, 1820–1920, *Journ. Hist. Geog.* 13 (1987) 394.
26. Arthur Clinton Boggess, *The settlement of Illinois, 1778–1830* (Chicago Historical Society, Chicago 1908) 97. The original is in Stanislaus M. Hamilton (ed.), *The writings of James Monroe* (G. P. Putnam, New York 1898) vol. 1: 117.
27. Beck, *Gazetteer of Illinois and Missouri,* 1823, 9.

sky, wherein it seemed to dip: mingling with its rich colours, and mellowing in its distant blue. There it lay, a tranquil sea or lake without water, if such a simile be admissible, with the day going down upon it: a few birds wheeling here and there: and solitude and silence reigning paramount around. But the grass was not yet high; there were bare black patches on the ground; and the few wild flowers that the eye could see, were poor and scanty. Great as the picture was, its very flatness and extent, which left nothing to the imagination, tamed it down and cramped its interest.[28]

Prejudices against prairies were of two kinds: materialist and mental. Materially, prairies appeared to be deficient in resources essential for pioneer survival and subsistence. Mentally, individuals felt seriously threatened by environmental hazards. Pioneers preferred woodlands to prairies because they needed lumber to build log cabins and barns and required wood for fencing and fuel. Few pioneers were able to afford a six-horse team or three pairs of oxen to draw a steel plow to break tough prairie sod.[29] Alternatively, the task would be accomplished by breaking matted roots laboriously with an ax. Even though many prairies were seasonally flooded and some had water standing on them more or less permanently, it was necessary to dig wells to obtain drinking water. Water standing on the surface was recognized to be a source of dysentery and other maladies. Prairies were wide open and exposed to the dangers of blizzards, flash floods, tornadoes, hailstorms, and fires, which swept unimpeded across flat surfaces. Nothing and no one would escape their sudden onslaught. People also feared prying eyes of Indians and wolves. Smoke from chimneys and candlelight could be seen miles away. Upland prairies were also distant from navigable waterways, which, before the coming of railroads, were the principal channels for transporting heavy goods. All these objections counted most strongly against large tracts of upland wet prairies in northwest Indiana, northeast Illinois, Iowa, and southwest Minnesota. They did not carry much weight against small prairies or savannahs. Indeed, the most favored locations were at the edges of wooded valleys.

While strangers continued to be gravely apprehensive about venturing out of woodlands, those who came to stay gradually

28. Dickens, *American notes,* 1985, 225–26. The bare black patches presumably had been burned the previous autumn.
29. Jakle, *Images of the Ohio valley,* 1977, 100.

learned to appreciate the advantages of occupying grasslands. Among established settlers, prairies were sought as valuable adjuncts to homesteads sited at the margins of woods. In the first quarter of the nineteenth century, in southern and central Illinois, local residents had no doubt in their minds that they wanted at least as much grass as wood on their farms. Behind them, on valley slopes, they obtained drinking water from springs, timber for building, wood for fencing and fuel. Ahead of them lay prairies, where they pastured livestock, cut wild hay, and planted corn in holes dug or cut with an ax in the unturned sod. After one or two years, they carved out and fenced fields in which oats, wheat, and other crops were sown. Once broken, the soil proved to be exceedingly fertile.[30] In letters to relatives and friends in Virginia, Pennsylvania, New York, and Europe, these settlers advised intending immigrants to ignore warnings against prairies.[31]

Small Prairies Gain Favor

A revision of the general view that early settlers shunned all prairies everywhere, without distinction, was proposed by George Fuller in 1916 in discussing the beginnings of American settlement in southern Michigan. There, pioneers clearly preferred areas of mixed forest and prairies.[32] In southeastern Wisconsin, Joseph Schafer documented a strong preference for properties at the edges of prairies.[33] In central Ohio, Ralph Brown contrasted opinions about the Pickaway Plains of the Scioto based on transient excursions with

30. Terry G. Jordan, Between the forest and the prairie, *Agric. Hist.* 38 (1964) 205–16; 208 refers to John Bradbury, *Travels in the interior of America in the years 1809, 1810, and 1811* (Liverpool 1817) 308; Samuel R. Brown, *The western gazetteer; or emigrant's directory* (Auburn, N.Y. 1817) 48; David Thomas, *Travels through the western country in the summer of 1816* (Auburn, N.Y. 1819) 185; Benjamin Harding, *A tour through the western country, AD 1818 & 1819* (New London 1819) 9; S. Augustus Mitchell, *Illinois in 1837* (Mitchell, Philadelphia 1837) 14. McManis, *Initial evaluation,* 1964, 36, refers to Bradbury, *Travels,* 1817, 308–9; Robert Baird, *View of the valley of the Mississippi* (N. S. Tanner, Philadelphia 1832) 230.

31. Jordan, *Between forest and prairie,* 1964, 208, cites a letter from Morris Birkbeck in January 1819 declaring "An erroneous opinion has generally prevailed both in England and the Eastern States, that all prairies partake more or less of the nature of swamps: That they are, in fact, morasses, too wet for the growth of timber."

32. George N. Fuller, *Economic and social beginnings of Michigan* (Michigan Historical Publications, Lansing 1916) 318.

33. Joseph Schafer, *Four Wisconsin counties: Prairie and forest,* Wisconsin Domesday Book 2 (State Historical Society, Madison 1927) 121.

opinions drawn from knowledge acquired by temporary residence. Excursionists viewed the plains as beautiful scenery; residents were more interested in "the Soil, Neighbourhoods, Conveniences & Inconveniences of particular places."[34] Those with knowledge favored mixtures of small prairies and open woodlands.

In 1964, Terry Jordan abstracted and summarized a large amount of literary evidence, indicating conclusively that early settlers selected land which contained both wood and grass.[35] The earliest pioneers, most of whom originally came from Southern states, including Kentucky, Tennessee, and Virginia, almost invariably chose lightly timbered sections on the edge of small prairies in southern and central Illinois. Northerners, who came later, mainly from Pennsylvania, Ohio, and New York, settled in northern Indiana, northern Illinois, southern Michigan, and southeastern Wisconsin. They also sought prairie-forest mixtures, but some ventured on to the edges of large prairies. Migrants from Britain and Germany chose, first of all, mixed prairie and forest lands, preferring, wherever obtainable, land on the edge of small prairies. Table 4.1 indicates that latecomers chose a wider range of vegetation types, but a preference for lightly wooded fringes of small prairies persisted down to 1850. Another study that appeared in 1964, by Douglas McManis, not only reviewed travel and topographical literature but also analyzed original entries in the tract books for the land office districts of Edwardsville, Shawneetown, Springfield, and Vandalia in southern and central Illinois from 1815 to 1850.[36] The study confirmed that settlers perceived small prairies to be valuable acquisitions and selectively purchased sections and parts of sections that were lightly timbered or open grassland. A move toward open prairies after 1830 is much more pronounced than in Jordan's survey, because McManis examined districts where almost all timbered lands were sold before 1840 (table 4.2). People seeking woodlots had to go to Wisconsin or elsewhere to find vacant forested lands.

Both Jordan and McManis focused on the diverse character of

34. Brown, *Historical geography of the United States,* 1948, 205, quoting from observations by R. Ricks, an immigrant from Virginia, on his stay in Chillicothe in September 1807.

35. Jordan, *Between forest and prairie,* 1964, 205–16.

36. McManis, *Initial evaluation,* 1964, 62. Records of entries are kept in the Tract Book Room, Department of the Interior, Washington, D.C. Copies of the federal land survey plats and notes are held in the General Land Office Records, National Archives, Washington, D.C.

TABLE 4.1 Early American and European settlers' choices of vegetation types, 1816–49

Date	Location	Origin	Forest		Prairie-timbered			Open Prairie		
			1	2	3	4	5	6	7	8
1816	Lawrence Co., Ill.	South				X				
1817	White Co., Ill.	South				X				
1818	Edwards Co., Ill.	South				X				
1818	Edwards Co., Ill.	English				X				
1818	Madison Co., Ill.	North				X				
1827	St. Joseph Co., Mich.	North				X				
1829	Kalamazoo Co., Mich.	South				X				
1829	Bureau Co., Ill.	North						X		
1830	Kalamazoo Co., Mich.	North				X				
1831	Madison Co., Ill.	South				X				
1831	Madison Co., Ill.	German				X				
1831	Peoria Co., Ill.	South						X		
1832	Montgomery Co., Ill.	South				X				
1832	Northwestern Ind.	North							X	X
1833	St. Clair Co., Ill.	German				X				
1834	Lee Co., Ill.	North						X		
1834	St. Joseph Co., Mich.	English				X				
1835	Lake Co., Ind.	North						X		
1835	Will Co., Ill.	North						X		
1836	Knox Co., Ill.	North						X		
1836	Milwaukee Co., Wisc.	North	X							
1837	Southeastern Wisc.	North			X	X	X			
1842	Racine Co., Wisc.	English			X	X	X			
1845	Green Co., Wisc.	German						X		

(Continued)

(TABLE 4.1, *continued*)

Date	Location	Origin	Forest		Prairie-timbered			Open Prairie		
			1	2	3	4	5	6	7	8
1847	Calumet Co., Wisc.	German	X		X		X			
1848	Saginaw Co., Mich.	German					X			
1849	Jefferson Co., Wisc.	North				X				
1849	Winnebago Co., Wisc.	German			X	X				
Total			1	–	4	17	5	7	1	1

Key to vegetation types: 1. high forest 2. floodplain forest 3. openings 4. fringe of small, dry prairie 5. fringe of small wet prairie 6. fringe of large prairie 7. large, dry prairie 8. large, wet prairie

Key to origin: South, people from Southern states; North, people from Northern states; English, from England; German, from Germany and German-Swiss.

Source: Terry G. Jordan, Between the forest and the prairie, *Agric. Hist.* 38 (1964) 216.

TABLE 4.2 Tract Book entries by decade and vegetation type
for four southern Illinois districts

	District		Number of land entries by decade					Sections studied
			Before 1820	1820–30	1831–40	1841–50	After 1850	
P-T	Edwardsville		29	3	19	2	0	109
P			17	1	96	43	10	
P-T	Shawneetown		0	2	28	10	7	58
P			0	1	17	15	27	
P-T	Springfield		17	35	26	0	0	127
P			13	17	76	47	27	
P-T	Vandalia		0	0	13	4	4	54
P			0	1	18	17	27	
P-T	Total		46	40	86	16	11	
P			30	20	207	122	91	

Key to vegetation types: P-T, prairie-timbered; P, open prairie

Sections studied: Number of square-mile (640 acre) land survey sections included in study. Number of land entries includes sections (640 acres), half-sections (320 acres), quarter-sections (160 acres), and forties (40 acres). Total number of entries is greater than number of sections studied.

Source: Douglas McManis, *The initial evaluation and utilization of the Illinois prairies, 1815–1840,* Department of Geography Research Paper 94 (University of Chicago, Chicago 1964) 65.

Midwest grasslands, ranging from openings within forests to treeless plains, and from well-drained rolling terrain to flat waterlogged surfaces. Both studies concluded that timbered prairies and small prairies close to woodlands were keenly sought by the earliest pioneers. John Peck described 90 settlements that were established and flourishing on prairies in Illinois, mostly in the south and center of the state, as early as 1834.[37] In northern Illinois, almost 80% of the earliest fields enclosed by squatters were taken wholly or in part from prairies on the edges of woodland.[38] Few early settlers ventured far into vast expanses of wet prairies in northwest Indiana, northern Illinois, Iowa, and southern Minnesota. Swamp fever was endemic, large prairies were infested with prairie flies and mosquitoes, their soils were too wet to raise normal crops every year, and roads were churned to mud in winter and spring. Because of their great extent it was difficult to reach supplies of drinking water and fuel and it was also difficult to carry produce to markets.[39] Large prairies were to remain remote until railroads opened them to commerce and communication. In McLean County in north central Illinois, large, treeless plains continued to be regarded as inferior to wooded land, and a stigma attached to people who settled there. Timber people, aristocrats in the early community, called them "prairie rats."[40] Long after small prairies were regarded with favor and successfully settled, large prairies, particularly wet prairies, were considered unfit for cultivation.

Differing attitudes toward small prairies accessible to timber and large, open, upland prairies are reflected in differing chronologies of land entries in two counties in south central Illinois. Sangamon County, within which the Springfield land office was situated, lies less than 100 miles south of McLean County, which was also served by the Springfield land office. All land in Springfield township was quickly disposed of (fig. 4.2). The first entries in 1823 were in timbered sections; the last entries in 1837 were in floodplain swamps. Flat, upland prairies were occupied about ten years after the first en-

37. McManis, *Initial evaluation,* 1964, 83–85, abstracts a list of prairie settlements from John Peck, *Gazetteer of Illinois,* 1834.

38. William D. Walters Jr. and Floyd Mansberger, Initial field location in Illinois, *Agric. Hist.* 57 (1983) 292.

39. McManis, *Initial evaluation,* 1964, 47, 89, 91–92.

40. Arthur Weldon Watterson, *Economy and land use patterns of McLean County, Illinois,* Department of Geography Research Paper 17 (University of Chicago, Chicago 1950) 45.

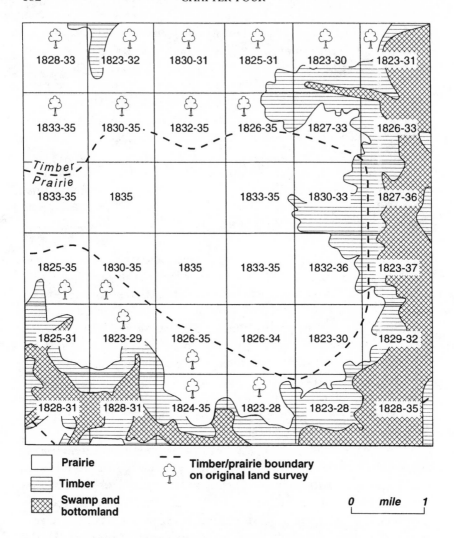

Figure 4.2 Entry dates in Springfield Township, Illinois

The dates in each square-mile section indicate the range of years during which the first land entries were made in the section. The boundaries of vegetation types are outlined in *Sangamon County Soil Report*, Univ. Illinois Agric. Expt. Sta. (Urbana 1912). The timber/prairie boundary is based on plats and notes of the original land survey carried out in the 1820s. *Source:* Douglas R. McManis, *The initial evaluation and utilization of the Illinois prairies, 1815–1840*, Department of Geography Research Paper 94 (University of Chicago, Chicago 1964) 67.

tries on the timbered fringe and all prairie sections had been sold by 1835. It is not clear exactly where the edge of the timbered land lay in 1820. The federal survey mapped a much smaller area of open grassland than that indicated by later soil surveys. The discrepancy may be attributed to errors or omissions in the federal survey, but it may be explained partly by encroachment of tree growth on surfaces that had ceased to be burned over repeatedly and were no longer grazed by bison and deer. Certainly, no part of the prairie was more than two miles distant from timber. Many early settlers in Springfield came from Southern states, but speculators, including some Yankees, were extremely active in purchasing prairies after 1830.[41]

In McLean County in 1822, the first two families occupied a wooded tract later known as Blooming Grove. In the 1830s other Northerners from Ohio, New York, and New England purchased timberland. Extensive upland prairies were used freely as open range grazing. Figure 4.3 indicates the original timbered areas, mapped in 1856 by Peter Folsom, the county surveyor. Arthur Watterson has shown that there is "a striking correlation between the forested areas and the lands entered by early settlers."[42] By 1850, hardly any land more than a mile from the edge of woods had been occupied. Large areas of open prairie were acquired by large landowners. Among the largest was the holding of Isaac Funk, who bought 27,000 acres between 1841 and 1861.[43] Funk was a cattle fattener. Other landlords let farms to tenants, some of them sharecroppers.[44] By contrast with the district around Springfield, prairies in McLean County were not rapidly taken into private hands and many sections stood vacant in 1850. After that, the Swamp Land Grant, the Illinois Central Railroad Grant, and purchases by cattle kings swallowed the remainder of the public domain.

Plats and Notes of the Federal Land Survey

In addition to guides addressed to immigrants, gazetteers, travelers' journals, and letters written by pioneers, an important source of in-

41. McManis, *Initial evaluation*, 1964, 64, 66.
42. Watterson, *McLean County*, 1950, 48.
43. Ibid., 63.
44. Ibid., 46–47.

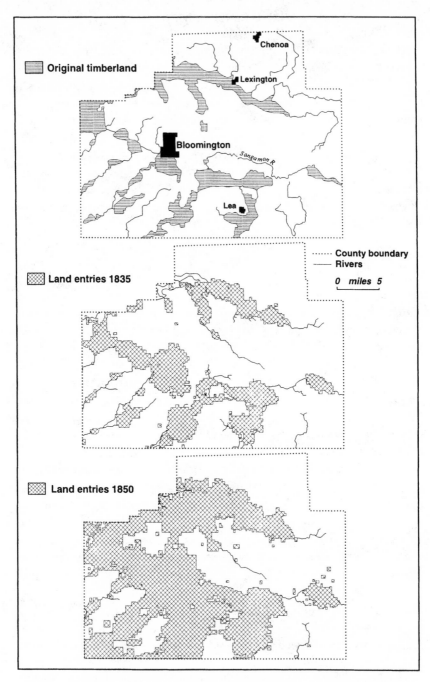

Figure 4.3 Land entries in McLean County, Illinois, in 1835 and 1850

In 1856, Peter Folsom, county surveyor, mapped timbered areas in the county. Lands entered in the tract books up to 1835 and up to 1850 show that lands close to timber were occupied earlier than open prairies. *Source:* Arthur Weldon Watterson, *Economy and land use patterns in McLean County, Illinois,* Department of Geography Research Paper 17 (University of Chicago, Chicago 1950) 31, 50, 53.

formation on land in the newly formed public domain was recorded in the plats and notes of the federal land survey. Many historians, geographers, and ecologists have valued these records for their detailed descriptions and comprehensiveness but have also criticized their numerous inaccuracies and deficiencies. It remains to ask whether or by what means observations made by federal surveyors were used by home-seekers or land companies in deciding where to locate. How readily were the maps accessible to prospective purchasers, how widely disseminated among members of the public were the surveyors' notes, and how deeply did these records influence individuals' opinions and choices?

The Land Ordinance of 1785, whose provisions were framed by a congressional committee chaired by Thomas Jefferson, set up the federal land survey. Jefferson set three aims for the survey. The first aim was to bring lands northwest of the Ohio River and south of Tennessee under direct control of the federal government until such times as new states were organized in the West.[45] A public domain was created by obtaining cessions of land from the states of Massachusetts, Connecticut, Virginia, and Georgia, each of which relinquished claims over Western territories. Additional land was brought into the public domain by obtaining cessions from native American tribes, and further cessions were obtained from foreign governments by treaty or conquest: the Louisiana Purchase from France in 1803, the annexation of Florida from Spain in 1819.[46] A second aim was to sell public land in order to reduce federal government debt. The propriety of selling land to pioneer settlers was fiercely debated. Alexander Hamilton of New York argued that the financial security of the nation must take priority over the accommodation of individuals. On behalf of pioneers, Jefferson contended: "By selling land you will disgust them and cause an avulsion of them from the common union. They will settle the lands in spite of everybody."[47] Jefferson yielded to the demand that revenue must

45. William D. Pattison, *Beginnings of the American rectangular land survey system, 1784–1800,* Department of Geography Research Paper 50 (University of Chicago, Chicago 1957) 3–36.

46. Benjamin Horace Hibbard, *A history of the public land policies* (Peter Smith, New York 1939) 7–30.

47. Ibid., 2–4; a milder view of Hamilton's attitude toward pioneers is taken in Roy M. Robbins, *Our landed heritage: The public domain, 1776–1936* (Princeton University Press, Princeton 1942) 13–15.

be raised from land sales and agreed to setting a minimum price of
one dollar an acre. He hoped this would not prevent poor people
from settling on uncultivated land. He remained committed "to
provide by every possible means that as few as possible shall be
without a little portion of land. The small landholders are the most
precious part of the state."[48] A third aim of the Land Ordinance of
1785 was to ensure that administrative boundaries and property
boundaries were clearly and efficiently demarcated and that entries
to land were accurately registered. To achieve this objective a rectangular grid was laid out by chain and compass survey, based on
north-south principal meridians and east-west coordinates. Survey
townships were marked off at intervals of six miles by six miles
along axes of the grid and these were subdivided into 36 sections,
each measuring one mile by one mile, containing 640 acres. The
lines of sections were surveyed and corners marked by blazing trees
or erecting marker stones.[49] When large areas had been surveyed,
land sales were held and entries to properties were registered in
tract books at local land offices. At a later date copies of tract books
were transferred to Washington, to state capitols, and county court
houses.

After eleven years' trial and error and changes in political
regime, a new land ordinance was introduced in 1796. The basic
principle of a rectangular survey system was reaffirmed and the de-
pendency of government on land sales as a means of paying off public debts was extended. The minimum price was raised to two dollars an acre and the minimum entry of a 640-acre section was retained. These conditions favored speculators and discouraged pioneers with little or no capital. Measures to treat settlers as trespassers
and criminals provoked resistance in the west. The 1796 act was
followed by detailed instructions to deputy surveyors on how the
field survey should be carried out. The act required surveyors to describe the land traversed by lines of survey, to record mines, salt

48. Hibbard, *History of public land policies,* 1939, 143, cites Jefferson, *Writings* (Thomas
Jefferson Memorial Association, Washington, D.C. 1904) vol. 19: 18.

49. Pattison, *American rectangular land survey system,* 1957, 37–67, credits Hugh
Williamson, a former professor of mathematics, with a key role in designing the method of
rectangular survey; Norman J. W. Thrower, *Original survey and land subdivision: A compara -
tive study of the form and effect of contrasting cadastral surveys,* AAG Monograph (Rand
McNally, Chicago 1966), describes different methods of surveying and numbering townships
and sections in Ohio.

licks, and mill sites, and to assess "the quality of the lands."[50] The instructions, amplified and amended in 1815 by the surveyor general, Edward Tiffin, provided fuller directions on what deputies were to "be careful to note" in their field books. Among other features, the instructions specified:

all rivers, creeks, springs and smaller streams of water, with their width, and the course they run in crossing the lines of survey, and whether navigable, rapid or mountainous; the kinds of timber and undergrowth with which the land may be covered, all swamps, ponds, stone quarries, coal beds, peat or turf grounds, uncommon natural or artificial productions, such as mounds, precipices, caves, etc., all rapids, cascades or falls of water; mineral, ores, fossils, etc.; the quality of the soil and the true situation of all mines, salt licks, salt springs and mill seats, which may come to your knowledge are particularly to be regarded and noticed in your note books.[51]

Detailed attention was to be paid to lines of watercourses and outlines of lakes, ponds, and swamps. The assessment of land quality was specifically related to soil characteristics, with a view to agricultural use. The surveyor general intended surveyors to compile an inventory of land and resources. The secretary of the treasury also required copies of descriptions of 36-square-mile townships to be recorded in a form that could be preserved for future inspection. Accordingly he instructed the surveyor general: "Plats of each township are to be neatly and accurately protracted according to law, on durable paper, by a scale of two inches to a mile, which are to be recorded in books to be kept in your office. Copies of the plats and field notes are to be made out on good paper, of uniform size, and with such margins as will admit of being bound in durable books, to remain in the Treasury."[52] Clearly, the secretary of the treasury and the surveyor general agreed that a full description of lands to be offered for sale ought to be provided, but neither the act nor instructions explicitly stated how, where, or when the public should be given access to the records. Letters from the surveyor general to the secretary of the treasury in 1801 and 1803 imply that the surveyors' notes were intended to furnish interested persons with means of

50. Lowell O. Stewart, *Public land surveys: History, instructions, methods* (Collegiate Press, Ames, Iowa 1935) 22.
    51. Ibid., 146.
    52. Ibid., 37.

identifying section corners and gaining an idea of "the quality of the land, & c."[53]

Joseph Schafer, William Pattison, and Hildegard Binder Johnson state that copies of plats and notes deposited in local land offices were made available to prospective purchasers or anyone who was interested when surveyed lands were offered for sale.[54] The descriptions appear to have been written expressly for the purpose of informing inquirers whether land was suitable for cultivation, whether timber or minerals were present, where springs rose or swamps or prairies covered the surface. The records are known to have been consulted by writers of guides, gazetteers, county histories, and county atlases, and the surveys themselves were directly transcribed in printed maps of states.[55]

Field surveyors were often consulted directly. Schafer remarks: "It is not to be wondered at that their advice was sought after by the great mass of land hunters or that some should have been tempted to use their special knowledge of the public lands for the purpose of advancing the interests of themselves and their speculating friends."[56] In the course of their work, surveyors gained knowledge about potential mill sites and places where towns or railways might be built. Such opinions were not recorded in the notes but might be valuable for developers or engineers. In 1837, during a congressional debate, a representative from Mississippi alleged that surveyors "note every valuable lot and sell the information thus acquired to speculators."[57] The charge was not substantiated and, within the strict terms of the public land acts, very little outright fraud was proved.[58] Fewer than 5% of all surveyors' plats and notes were

53. Pattison, *American rectangular land survey system,* 1957, 217.

54. Schafer, *Four Wisconsin Counties,* 1927, 53; Pattison, *American rectangular land survey system,* 1957, 228; Hildegard Binder Johnson, *Order upon the land: The U.S. rectangular land survey and the upper Mississippi country* (Oxford University Press, New York 1976) 78, 129.

55. Pattison, *American rectangular land survey system,* 1957, 218, cites Map of Ohio by Rufus E. Putnam, Surveyor General of the United States, in Thaddeus M. Harris, *The journal of a tour into the territory northwest of the Alleghany Mountains made in the spring of 1803* (Boston 1805) following 271; B. Hough and A. Bourne, *Map of the state of Ohio from actual survey* (Philadelphia 1815), scale 1 inch to about 5 miles; Johnson, *Order upon the land,* 1976, 16, 146–50, discusses the use of plats and notes in the preparation of county atlases and county histories.

56. Schafer, *Four Wisconsin Counties,* 1927, 54.

57. Ibid., 55.

58. Stewart, *Public land surveys,* 1935, 53–54, 63–66.

found to fall short of legal requirements.[59] On the other hand, some surveyors left the federal service to take employment with canal builders, railroad constructors, and mining companies.[60]

In the 1850s railroad companies began to receive extensive grants of public land and carried out their own surveys with the avowed purpose of attracting settlers. The Burlington and Missouri River Railroad, for example, was granted a large area in southern Iowa and Nebraska. Bernard Henn, an experienced land agent, gave detailed instructions to all his surveyors, directing them to "make your examinations as thorough & your reports as full as if on each 40 you were writing to your ladylove & describing the Paradise where you hoped to pass with her a blissful middleage."[61] The Burlington sold most of its land in 40-acre tracts, and regular sales of public land were suspended while the company made a selection of lieu lands. Lieu lands were equivalent in area and situation to lands forming part of the grant that had already been sold to other parties.[62] By 1850 the disposal of public land was subject to many acts and regulations, and had become an arena of conflict for many claimants.

For most purchasers, a surveyor's classification of the quality of agricultural soils was of less importance than a correct entry in the land office tract book and an accurate marking of lines of townships, sections, and subdivisions on the ground. Land classification, however, was crucially important to holders of military bounty warrants. In awarding warrants for bounty land in Illinois and Michigan to veterans of the War of 1812, the federal government directed that only land "fit for cultivation, not otherwise appropriated" was to be selected. The question whether large tracts of Illinois prairie were eligible became a legal issue. Taking a cautious view, registers and receivers at a land office in Edwardsville, Illinois, decided to throw "into the class of 'unfit for cultivation' much land described as first rate soil but which is prairie and distant from timber." A similar decision was taken by a land office at Springfield which stated: "Although this land is represented as unfit for cultivation, it is from the

59. Johnson, *Order upon the land*, 1976, 221.

60. Schafer, *Four Wisconsin Counties*, 1927, 54; Stewart, *Public land surveys*, 1935, 87.

61. R. C. Overton, *Burlington west: A colonization history of the Burlington Railroad* (Harvard University Press, Cambridge 1941) 143, cited in Johnson, *Order upon the land*, 1976, 70.

62. Johnson, *Order upon the land*, 1976, 145.

circumstance of there being no timber. The quality of much of this land, indeed the larger proportion, is excellent, and equal to any in the district, but its remoteness from timber and water will prevent any immediate sale."[63] Other prairie tracts in Illinois were deemed fit for cultivation. Edmund Dana concluded: "About one half of the Bounty Lands may be allowed for prairie. Of those, a large part is the finest quality of soil, being either a black vegetable mould, or a dark, sandy loam, from 15 inches to three feet deep; generally bedded on yellow clay, mixed with sand—the surface conveniently waving for cultivation, and adapted to rend the prospects more charming."[64] By the time the land survey reached Iowa and Minnesota in the 1850s, upland prairie was frequently described as first-rate, and, where timber was found, mostly second- or third-rate quality.[65] Other problems arose in selecting lands for allocation to states under the Swamp Land Act of 1850. The act required that swamp and overflowed lands be classed as "unfit for cultivation."[66] Later adjustments were made to compensate states for lands that had been sold before the act came into force. Regulations and instructions drafted by the Treasury and surveyor general closely reflected early American prejudices against open grasslands. This attitude softened as more and more settlers entered small prairies at the edges of timbered land. Extensive prairies remote from timber were mostly unoccupied in 1850. Abhorrence of wetlands remained hard and deep until late in the nineteenth century.

Swamp Land Grants

All stages in the work of the federal land survey were hindered by the presence of wetlands. Wetlands cost much time and effort to survey; they were difficult to identify, delineate, and classify in a

63. McManis, *Initial evaluation*, 1964, 37, 38, refers to U.S. Congress, *General acts of Congress respecting the sale and disposition of the public lands with instructions issued from time to time by the Secretary of the Treasury and Commissioner of the General Land Office and official opinions of the Attorney General on questions arising under the land laws* (Gales and Seaton, Washington, D.C. 1838) vol. 1, 214–16; *American State Papers. Public Lands*, vol. 5: 550, Statement by William P. McKee and James Mason, Edwardsville, 13 November 1828; vol. 5: 557, Statement by John Todd and Pascal P. Enos, Springfield, 11 October 1828.

64. Edmund Dana, *A description of the Bounty Lands in the state of Illinois* (Looker, Reynolds, Cincinnati 1819) 6–7, cited in McManis, *Initial evaluation*, 1964, 55.

65. Johnson, *Order upon the land*, 1976, 79–80.

66. United States, *Statutes at large* 9 (Act of 28 September 1850) 519.

consistent and uniform manner; and they were difficult to dispose of at the statutory minimum price.

Hardships routinely endured by survey teams in the field were multiplied when crossing wetlands. Provisions and equipment for an expedition lasting weeks or months were carried in backpacks or saddlebags. Heavy loads on firm ground became formidable burdens in watery mires. Winter frosts and blizzards were more intense over bogs and marshes than in woodlands. In summer, insect attacks were occasionally unbearable. A survey party in swamps in Michigan in 1821 refused to go on with its work because of "the suffering from mosquito bites, both men and horses being weak from loss of blood and want of sleep." In order to keep out of the water while they slept they were "frequently obliged to lay down poles and pile on them hemlock boughs" and they had to build campfires on scaffolds of logs.[67] A later survey expedition in swamps in Wisconsin in 1847 was described by Harry A. Wiltse:

During four consecutive weeks there was not a dry garment in the party, day or night.

Consider a situation like the above, connected with the dreadful swamps through which we waded, and the great extent of windfalls over which we clumb and clambered; the deep and rapid creeks and rivers that we crossed, all at the highest stage of water; that we were constantly surrounded and as constantly excoriated by swarms or rather clouds of mosquitoes, and still more troublesome insects; and consider further that we were all the while confined to a line, and consequently had no choice of ground . . . and you can form some idea of our suffering condition.

Our principal suffering, however, grew out of exhaustion of our provisions, coarse as they were.

Worn out by fatigue and hunger and nearly destitute of clothes, they had had to make a forced march of three days to the lake in search of provisions, of which during that three days, they had not eaten a mouthful.[68] Among other hardships suffered by surveyors in wetlands were bouts of ague and nameless debilitating fevers, attacks by Indians, losses of horses, blankets, and equipment, and depredations of provisions by wild animals.

67. Stewart, *Public land surveys*, 1935, cites reminiscences of Harvey Parke.
68. Ibid., 84.

Difficulties encountered by surveyors in following straight township and section lines across wetlands were compounded by uncertainties in identifying and delineating exact boundaries of waterlogged soils and shorelines of lakes and ponds. The extent of
wetlands varied from season to season and year to year. In Ohio,
swamplands covered by water for more than half the summer were
described as "slashes." Other areas in the state, recorded in surveyors' field notes as "scalded lands," were places where trees had
been killed by water standing on the surface for long periods.[69]
Where flooding was deep or extensive, surveyors were instructed to
meander or skirt around the edge of the water. A circular from the
General Land Office in 1831 directed that:

Whenever the continuation of a surveyed line is interrupted by an *impassable
swamp* or *from any other cause,* the distance of the line actually run between
the starting and finishing posts, is to be truly represented by the platting and
also by figures.

All lakes and ponds of *sufficient magnitude* to justify such expense, are to
be *meandered* and platted agreeably to courses and distances, which are also
to be exhibited by figures.

Swamps are to be represented in the ordinary method by slightly shaded
black lines and dots, and the outlines of the same should be distinctly exhibited.[70]

These instructions were difficult to carry out in the field.
Seasonal changes in the edges of wetlands are revealed on many
township plats where prairie or burnt land is indicated on one side
of a section line and marsh or swamp on the other. Here and there,
a pond is divided by a line of survey depicting open water on one
side and marsh on the other. Prairies that were dry or burnt over in
fall and winter were flooded or marshy in spring and early summer.
Positions of beaver dams, Indian lodges, and canoes clearly mark
edges of ponds and watercourses, but even the most detailed surveys
cannot be relied upon to represent the full extent of waterlogging as
it appeared at the height of a wet season in a wet year.

A general view, shared by surveyors, speculators, and prospec-

---

69. Whitney, *From coastal wilderness to fruited plain,* 1994, 272, cites H. Howe,
Historical collections of Ohio (Bradley and Anthony, Cincinnati 1848) 98; and L. J. Ives Jr.,
The natural vegetation of Lorain County, Ohio, M.A. thesis, Oberlin College 1947, 18–19.
70. Stewart, *Public land surveys,* 1935, 150.

tive settlers, was that much wetland in the Midwest was "lost or 3rd rate land" and would never be used for farming.[71] From 1826 onward, repeated attempts were made to introduce national legislation to measure the extent and map the location of swamp and overflowed lands that might be handed over to states.[72] Ohio, Illinois, Michigan, and Missouri tabled resolutions in Congress but no act was passed until 1849 when Louisiana obtained from the federal government a grant of Mississippi River bottomlands. In September 1850, other states secured grants of swamplands classed as unfit for cultivation, providing that "the proceeds of said lands shall be applied, exclusively, so far as necessary, to the purpose of reclaiming said lands by means of levees and drains."[73] In favor of the measure, it was argued that states would be interested in promoting and organizing local drainage enterprises, that malarial swamps would be made healthy, drained land would be brought into cultivation, and the value of adjoining areas of public domain would be enhanced to the benefit of the U.S. Treasury. At first, swamplands were to be selected on the basis of federal surveyors' descriptions, but this requirement was relaxed in the face of objections raised by states, largely conceded by the General Land Office. After 1854, states were offered an option of making their own selection of lands whose swampy character was to be certified by reliable citizens who lived in the locality. From their personal and exact knowledge, witnesses had to testify that such areas were "submerged and rendered useless for arable purposes in their natural condition."[74] Increasingly, state selection agents presented claims for "low" and "wet" lands and for lands occasionally flooded, including potentially some of the most productive farming lands. Disputes between states and federal government "turned on the ambiguity of the word swamp."[75] The procedure for selection was open to genuine differences of interpretation and also to flagrant abuse.

Many frauds were alleged and federal officials conducted inves-

71. Whitney, *From coastal wilderness to fruited plain,* 1994, 272, cites A. A. Lindsey, The Indiana of 1816, in A. A. Lindsey (ed.), *Natural features of Indiana* (Indiana Acad. Sci., Indianapolis 1966) xv.

72. Hibbard, *History of public land policies,* 1939, 269.

73. United States, *Statutes at large* 9 (Act of 28 September 1850) 519.

74. Roscoe L. Lokken, *Iowa: public land disposal* (State Hist. Soc. of Iowa, Iowa City 1942) 195.

75. Ibid., 204.

tigations into numerous reports of excessive claims from every state in the Midwest. Frequently, federal investigators found that over three-quarters of the land claimed was not and never had been swampy or liable to regular flooding. Of 22,000 acres in Champaign County, Illinois, listed as swamp by a state agent, the federal land office found less than half that extent was eligible. In neighboring Piatt County, 166 tracts were claimed for the state, of which the federal investigator rejected all but six.[76] In Iowa, similar discrepancies were discovered. In Dickinson County, a federal commissioner reported in 1865 that of fifteen tracts which had been selected by the state, "*nine* could not, in any sense, be claimed as Swamp land— some of it is high, dry and arable; and in successful cultivation, among the best lands in the county."[77] A comparison of 400 tracts claimed by the state as swamp and overflowed lands with descriptions recorded in land survey notes showed that 330 had been described by the surveyors as "rolling prairie," "good agricultural land," or otherwise fit for cultivation.[78] The magnitude of the differences between states' claims and federal assessments arose partly from a long time lag between the original land survey and the implementation of the Swamp Land Act. The act of 1850 set no time limit for the receipt of claims. States continued to submit new and additional claims up to the last quarter of the nineteenth century, and examination of disputed claims dragged on into the first quarter of the twentieth century.

By the time the process of selection was completed most of the public domain had been disposed of. The total extent of swamp and overflowed land ultimately patented under the act was nearly 64,000,000 acres. In addition, a very small amount of indemnity land (754,385 acres) was awarded to successful claimants after all swamplands had been taken up (table 4.3). Some claims were settled by payments of cash in lieu of land or warrants. These indemnity payments amounted to just over $2 million. Florida received the largest grant, over 20 million acres or nearly 32% of the national total. The three states occupying lower Mississippi River floodplains, Arkansas, Louisiana, and Mississippi, together obtained

---

76. Hibbard, *History of public land policies,* 1939, 278.
77. Lokken, *Iowa,* 1942, 196.
78. Ibid., 206.

Table 4.3 Swamplands and Indemnity Lands granted to states

	Swampland (acres)	Indemnity Land (acres)
Alabama	418,634	20,920
Arkansas	7,686,335	–
California	2,159,304	–
Florida	20,202,328	94,783
Illinois	1,457,399	2,309
Indiana	1,254,271	4,880
Iowa	874,094	321,977
Louisiana	9,384,626	32,631
Michigan	5,655,816	24,039
Minnesota	4,663,007	–
Mississippi	3,286,306	56,782
Missouri	3,346,936	81,017
Ohio	26,252	–
Oregon	264,069	–
Wisconsin	3,251,684	105,048
Totals	63,931,061	744,386

Source: General Land Office, *Report of the Commissioner*, 30 June 1922, 34–39.

20,357,266 acres or nearly 32% of the national total. In the Midwest, seven states, Illinois, Indiana, Iowa, Ohio, Michigan, Minnesota, and Wisconsin, together received 17,182,522 acres, plus 458,253 acres of indemnity land, about 27% of the national total. The prairie states of Ohio, Indiana, Illinois, and Iowa obtained 3,612,015 acres or less than 6% of the national total, whereas the northern lakes states of Michigan, Wisconsin, and Minnesota were granted 13,570,507 acres, more than 21% of the national total. Missouri, which had extensive areas of bottomlands, received more than the prairie states. Apart from California, which received just over 2,000,000 acres, and Florida, the largest beneficiary, almost all patented swamplands were awarded to states in the basin of the Mississippi River (fig. 4.4).

How did the states dispose of the lands allocated to them? Most states opened their own land offices and some sold lands to speculators in advance of selection. Some of these advance purchasers or their attorneys, bearing deeds from their states, presented claims to the General Land Office as agents of the states and demanded that their titles be confirmed as dues to the states rather than as private

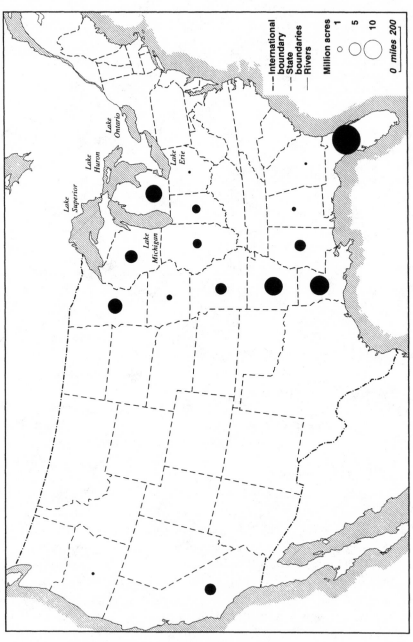

Figure 4.4 Swamp land grants in the United States

Circles are proportional to areas of swampland in millions of acres granted to states. *Sources:* B. H. Hibbard, *A history of the public land policies* (Peter Smith, New York 1939) 277; U.S. General Land Office, *Report of the Commissioner* (30 June 1922)

speculations.[79] Three states, Illinois, Iowa, and Missouri, assigned their swampland claims to counties and those counties were recognized by the federal government as agents for the states in administering the grants. The size and value of a grant depended on the vigor with which a claim was pursued. An insight into the process is revealed in a newspaper report from Iowa in 1861: "W. P. Hepburn, the District Attorney for this judicial district, has recently visited Washington City for the purpose of prosecuting the *Swamp land claims* of several counties against the government, and . . . we learn has been very successful."[80] States were pressed by speculators to release for sale swamp and overflowed lands immediately after they were authorized by the General Land Office. Large amounts were snapped up as soon as dates for sale were announced. States set a minimum price of one dollar for an acre of land periodically flooded, and very little, even of the most deeply waterlogged land, was offered for less than 50 cents an acre. Some tracts were withheld from sale or settlement by wealthy counties anticipating prices rising, but private interests clamored for early sales. In 1868, the House of Representatives Committee on Public Lands was informed that half the swamp land grant was in the hands of speculators.[81] The act failed in its aim to provide poor people with an opportunity to buy cheap land that might be turned into productive farms by hard work and careful management.

Some states drew upon proceeds of swampland sales to finance road building, to build bridges and state institutions, to establish schools and colleges, and to pay for other necessary improvements; and some used sales revenue to reduce their current debts. Some states gave part of their swampland to railroad corporations while others engaged in costly lawsuits with railroads that had received, as grants from the federal government, lands subsequently claimed by states as swamps. In principle, state claims took priority over those of railroads, but where a railroad could prove that the lands in question were not swamps, the General Land Office had the power to revoke a swamp land grant. Some states bargained with immigrant companies to put settlers on swamplands at prices ranging from 25

79. Hibbard, *History of public land policies,* 1939, 281.

80. *Fort Dodge Republican,* 2 April 1861, cited in Hibbard, *History of public land policies,* 1939, 282.

81. *House Reports of Committees,* 40th Cong., 2d Sess. 25 (1868), cited in ibid., 281.

to 75 cents an acre. Such deals led to disputes about whether private companies were being given favored treatment at the expense of the public at large. Apart from Louisiana, which gave high priority to constructing levees and digging drains, no other state honored its obligation to protect floodplains, reclaim land for agriculture, or drain malarial swamps. Funds were used for every purpose but those stipulated in the act and large sums were dissipated in fighting legal actions, paying off state debts, and carrying out a variety of public works unconnected with the improvement of wetlands. The public domain lost an area vastly larger than initially intended, an area greater than the combined territories of New York State and all the New England states put together. In return for their munificent gift, the nation gained little satisfaction. Congressmen, newspaper correspondents, and land office officials complained frequently about the meager results achieved and expressed grave disappointment at the shortcomings of the act.

Westward Migration

While prejudice against wetlands persisted, people were pouring into the Midwest, acquiring most dry lands and all other lands that were timbered. The population of the Midwest increased tenfold between 1820 and 1860 (table 4.4). In 1820, nearly 600,000 were resident in Ohio and just over 200,000 were spread over the rest of the Old Northwest. In 1860, Ohio had over two million inhabitants, and nearly five and a half million were counted in six other states. By that date, one out of every four Americans lived in the Midwest. Most of the newcomers were born on the eastern seaboard, and during the first quarter of the nineteenth century, New England welcomed a relief in pressure of population on scarce land and poor employment prospects. From the 1820s onward the Northeast began to fear that its population was rapidly draining away. Northerners were alarmed that Westerners were growing more and more powerful and their political objectives were diverging to the point of straining the bonds that held the union together. The westward march of population was signaled by the admission to statehood of Ohio in 1803, Indiana in 1816, Illinois in 1818, Missouri in 1821, Michigan in 1837, Iowa in 1846, Wisconsin in 1848, and Minnesota in 1858. Votes from "men of the western waters" carried Andrew Jackson to victory in the presidential election of 1828 and Abraham Lincoln to an electoral triumph in 1860. In the

Table 4.4 Population of the United States, 1820–60

	1820	1830	1840	1850	1860
Ohio	581,434	937,903	1,519,467	1,980,329	2,339,511
Indiana	147,178	343,031	685,866	988,416	1,350,428
Illinois	55,211	157,445	476,183	851,470	1,711,951
Iowa	–	–	43,112	192,214	674,913
Michigan	8,896	31,639	212,267	397,654	749,113
Wisconsin	–	–	30,945	305,391	775,881
Minnesota	–	–	–	6,077	172,023
Midwest total	792,719	1,470,018	2,967,840	4,721,551	7,773,820
% change		+85	+102	+59	+65
Rest of U.S.	8,845,734	11,396,002	14,101,613	18,470,325	23,669,501
U.S. total	9,638,453	12,866,020	17,069,453	23,191,876	31,443,321
% change		+33	+33	+36	+36

Source: U.S. Bureau of Census, Seventeenth Census of Population, 1950, vol. 1: 1–8, 1–9.

Senate, Thomas Hart Benton from Missouri, a "veritable champion of the West" fought tenaciously from 1822 to 1841 to release land from the public domain so as to encourage permanent settlement. To many Western politicians and newspaper writers, a resolution of the land question in favor of establishing the viability of family farms was essential for securing social, economic, and political stability in the new states.

The drift of population westward was perceived as unsettling in the East but it was also disturbing for the West. People moved from Pennsylvania to Ohio or from Virginia and Kentucky to Illinois, but many did not stay long. They moved again from Michigan to Wisconsin to Minnesota and from Indiana to Iowa to Kansas. Observers asked where and when would they stop their wagons, unpack their tools, build homes, and start cultivating the soil. The terms on which the public domain was offered for sale did not invite or encourage purchasers to settle down. Young men and women most eager to settle, build, and cultivate were least able to afford a whole square mile of land; few pioneers had thousands of dollars to take to a land office. Those with money to invest in large holdings were least willing to occupy them; they wanted to sell what they acquired for the highest profit as quickly as possible. States gained little or nothing from sales of federal land but had to find money to create an infrastructure of roads, schools, and police in order to retain settlers. The federal government, for its part, had to find money to compensate native Americans for land cessions, pay France for the Louisiana Purchase and Spain for Florida, defray expenses of the land survey, and finance the running of the General Land Office. Some of the highest expenditure was incurred on wetlands that brought very little revenue. These basic contradictions and conflicts severely strained relations between western states and defenders of the federal union. On the ground, they led to long bitter struggles between pioneer settlers, absentee landowners, and government officials.

During the first six decades of the nineteenth century, sales of public land did not follow the same trajectory as increases in population (fig. 4.5). A very small amount of land was sold in the first decade when Ohio received a huge influx of population. In the second decade, when population numbers in the Midwest more than doubled, public land sales boomed briefly in 1818 and 1819, fol-

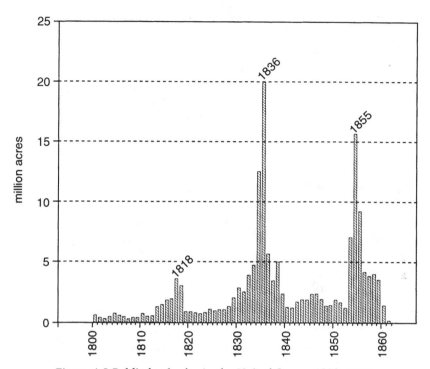

Figure 4.5 Public land sales in the United States, 1800–1862

Yearly total acreages of public land sales for the whole United States. *Source:* B. H. Hibbard, *A history of the public land policies* (Peter Smith, New York 1939) 100, 103, 106.

lowed by a collapse in the market in the great panic of 1819. From 1820 to 1830, Midwest population increased by 85% while land sales remained depressed. In the 1830s, population doubled and the area of land sold reached an all-time peak in 1836, then fell suddenly in the panic of 1837. In the 1840s, population growth slowed down while land sales failed to recover. In the 1850s, population growth picked up a little and land sales rose steeply, climbing to a high point in 1855. Another panic in 1857 was followed by a decline in the area of land sold, but no comparable slowing down occurred in the rate of population increase. Throughout the first sixty years of the nineteenth century, the sale of public land fluctuated sharply while population rose continuously: sometimes there was a surplus of land on the market, sometimes a shortage. A lack of correspondence between population growth and the uptake of public land may be explained partly by trends in the national economy, partly by changes in the terms on which land was offered for sale, partly by provision of communications and other public services.

Speculators and Settlers

The General Land Office was justifiably concerned about the accuracy of its field surveys, about the true demarcation of corners of townships and sections, and about keeping full, correct records of entries in tract books. Officers were also concerned about noting land quality, describing terrain and drainage conditions, rating agricultural values of soils, and classifying lands as fit or unfit for cultivation. Selections of military bounty lands were required by law to be fit for cultivation and selections of swamplands were required to be unfit for cultivation. But for a majority of prospective purchasers the quality of land mattered little; they bid the minimum asking price or just sufficient to obtain an entry. What mattered on site at the time of sale was the statutory minimum area of a lot, the prevailing minimum price, prior claims by occupying settlers, and any buildings or improvements carried out by present or former occupiers.

Western politicians dramatized public land sales as contests between evil speculators and heroic pioneers. Speculators were portrayed as old, mean, tightfisted Yankees, patricians, absentees, manipulators, bankers, moneylenders, real-estate salesmen, lawyers, swindlers, profiteers, parasites. Settlers were presented as young, strong, brave, ambitious, poor, upwardly mobile, thrifty, hardworking wealth creators, sons of liberty, enemies of privilege, fighters for independence. The moral tone of opposing stereotypes is exemplified by two comments from Iowa. The first condemns speculators:

Men come to this country to make money by speculating, not by steadily pursuing a course of tilling the fertile soil, of which they become the temporary proprietors, and which soon passes into the hands of others, who are equally disposed to sell out at an advance . . . The present mode of speculations is a species of gambling, leading men to rely upon uncertain events for the completion of their grasping and eager wishes for wealth. It puts a stop to the pursuit of every object worth the attainment of good and virtuous citizens. It is the moral upas which taints, with the poison of its influence, every aspiration of the mind after purity of thought and integrity of conduct. [82]

A second, contrasting comment praises settlers:

It was the work of the pioneers which gave value to the land—the value that comes from neighbors, schools, churches, mills and stores. And the pioneers were ready to defend what their enterprise and industry had earned. Beyond the reach of civil government, they made their own laws and established justice according to their needs. This was democracy—democracy as tough and uncompromising as the prairie sod, with roots deep in the soil, so deep that no storm could tear them loose. The widespread acquisition of homes by poor but ambitious settlers not only broadened the base of democracy in the United States but thereby stabilized it. [83]

In reality, speculators were not wholly evil. They prospected valuable sites, attracted public interest in the West, promoted sales, arranged loans and credit facilities, raised expectations of gain. They were boosters. On the other side, settlers were not entirely heroic. Some were lawless and violent. They seized by force or with menaces land that others hoped to win legitimately by wealth and intelligence. The distinction between settlers and speculators was not as clear-cut as propagandists claimed. Many pioneers were speculators as well as settlers. The profits gained from their first transactions, financed by borrowing or by deferring payments, were reinvested in land until they accumulated sufficient to buy a farm. For many speculators, farming was low on their order of priorities. Many hoped they might acquire a gold mine, or failing that, a lead or copper mine, or a site for a state capital, or failing that, a county seat, a bridgehead, ferry landing, important road junction, railroad depot, or factory site. A stand of merchantable timber or cattle range seemed preferable to an area of land that had to be cleared and plowed before it could be brought into cultivation. Lowest consideration would be given to lands described as unfit for cultivation, least of all, swamps which might well have carried a government health warning.

When land prices were rising, speculation reached fever pitch;

82. *Dubuque Visitor,* 9 November 1836, cited in Hibbard, *History of public land policies,* 1939, 216.

83. Lokken, *Iowa,* 1942, 269.

when the market was depressed, settlers demanded an easing of terms of entry to public lands and relief from outstanding debt repayments. Looking back from the depths of the crash in 1857, an Iowa newspaper editor surveyed the boom years from 1854 to 1857:

It is amazing how completely our citizens were filled with the desire of sudden riches. Credit was easily had—eastern currency flooded the country—imaginary towns sprung up everywhere—lands were fictitiously high—usury was unscrupulously asked and willingly promised—farms were neglected—debts were left to run on unasked about, goods and groceries being bought on credit, lands alone selling for ready money at exorbitant rates. In short, every one was a professed speculator, and the good results which always flow from a proper division of labor, skill and capital were entirely dried up. Then followed the wrecks of fortunes and the crash of business. It was natural—it could not have been otherwise. [84]

When buying and selling were brisk, large estates rapidly expanded while small properties became prohibitively expensive. During periods of depression, small proprietors struggled to keep up with mounting interest charges. Given the advantages of accumulated capital over youthful energy and ambition, it is a wonder that so many small farms were established, let alone survived.

Small farmers gained footholds by occupying vacant wild land, building log cabins, and setting to work clearing and breaking fields. Unwritten custom on the frontier sanctioned claims made by individuals and families who established farmsteads, cultivated the soil, and remained in occupation for a year or longer. From 1785 to the punitive act of 1807 that imposed penalties for squatting, federal land ordinances did not recognize occupiers' customary claims, but the government was powerless to evict them, however strongly some congressmen might denounce them as illegal squatters, trespassers, and intruders on the public domain. In the West, neighbors banded together to defend their homes and formed claim associations to ensure that their bids at land sales were successful and claim jumpers and other outsiders were deterred from participating. Political leaders in Western territories encouraged pioneers to occupy public lands because they needed resident populations to exceed 50,000 in

84. *Dubuque Daily Republic,* 14 August 1857, cited in Hibbard, *History of public land policies,* 1939, 223.

order to qualify for statehood. Thereafter, states wanted settlers to stay to broaden the economic base, carry out internal improvements, and increase tax revenues.

Preemption, Graduation, and Homesteads

Throughout the first half of the nineteenth century, westerners struggled relentlessly to change federal land laws to give pioneers a better chance of settling down and starting family farms. Legislative reforms advanced in four directions: first, to secure a legal right of preemption for genuine settlers on the public domain; secondly, to obtain a graduation, that is reduction, of prices for lands that remained unsold after ten years or more; thirdly, to offer for sale parcels of land smaller than full square-mile sections; fourthly, to grant to seasoned farm makers and their families who worked the land for five years continuously, a 160-acre homestead free of charge. All these measures were introduced again and again, but were fiercely contested and repeatedly rejected either in the Senate or the House of Representatives.

Urging the right of preemption for the western settler, Thomas Hart Benton asked the Senate to acknowledge that a person who built a cabin and created a farm added value to the land and was entitled to enjoy without let or hindrance the fruits of his own efforts and enterprise. He declared: "It is nothing but a right of first purchase. It is no donation . . . it is no gratuitous distribution of the land."[85] In 1830, an act was passed permitting preemption of a quarter-section, 160 acres, of surveyed land by a settler who had occupied and cultivated it during the preceding year. The property had to be paid for at the time of a regular land sale and the act was to run for one year. It was reenacted in 1832 and again in 1834 but rejected in 1837. The crash of 1837 strengthened arguments in favor of farmers. Banks closed, manufacturers laid off workers, overseas trade shrank, but most farmers managed to remain solvent. They demanded a permanent guarantee of the right of preemption. Benton called it the "Log Cabin Bill"; its official title was the Permanent Prospective Preemption Law. It was passed in 1841 after long debates in both houses of Congress. It conferred a right of preemption on a head of a family, either a man over the age of 21 years

85. Thomas Hart Benton, *Thirty years' view* (New York 1858), vol. 1: 102, cited in Hibbard, *History of public land policies,* 1939, 154.

or a widow, to enable him, her, or them to settle on a piece of land
160 acres in extent and buy that land without competitive bids at
the minimum government price.[86] Three problems remained: did
the act permit preemption by immigrants who had not yet become
U.S. citizens; did the act cover lands not yet platted by federal surveyors; how long might a settler be allowed to hold a preemption
claim before having to pay for it? Of these problems the question of
payment was the most pressing and contentious.

A proposal to graduate the price of land that failed to attract a
buyer at the prevailing minimum rate was first debated as early as
1790. Following the panic of 1819, fresh attempts were made to
implement the principle of graduation. In 1820, a bill presented to
the Senate was heavily defeated but Senator Benton returned with
new measures in 1822, 1824, and 1826, all of which were strongly
opposed by spokesmen from the East. Arguments for graduating
land prices were threefold. First, imposition of a fixed minimum
price prevented the free play of market forces. It tended to inflate
land prices by creating an artificial scarcity: land worth less than the
set minimum was rendered unsaleable, so competition for more
valuable land was keener. Secondly, the poor man was denied an
opportunity of buying poor land cheaply. He was excluded from
the bottom end of the market and thereby prevented from climbing
a farming ladder by adding further parcels to a small core of low-
priced land. Thirdly, the federal government originally fixed the
minimum price in order to pay off its debts. In 1835, the national
debt was discharged and new Western states were practically bankrupt. The states invoked the principle of distribution: public lands
lay within their territories; it was wrong to sell them and take away
the proceeds to add to already full coffers of a remote treasury in
Washington. The money needed to be spent locally on building
roads, canals, railroads, schools, and other institutions. The states
wanted the remaining public land assigned to them for these purposes but Congress would not agree. In 1841, partial cessions of
federal lands, the 500,000-acre grants, were handed over to the new
states. In 1854, after a long struggle, a Graduation Act was passed. It
provided for a progressive reduction in prices proportional to the

---

86. United States, *Statutes at large* (Act of 4 September 1841) 457; Hibbard, *History of
public land policies,* 1939, 158.

length of time the land had been offered for sale without a bidder. After ten years, unsold land would be marked down to one dollar an acre; after fifteen years the price would be reduced to 75 cents; after twenty years, 50 cents; after twenty-five years, 25 cents; and after thirty years, 12 cents. Immediately after the act passed, the amount of public land sold in Illinois and Iowa soared, since no limit was imposed on the quantity an individual was allowed to purchase. Graduation fueled speculation rather than settlement.

A third way of trying to assist small farmers was to reduce the minimum size of entry. An act of 1804 lowered the minimum amount that could be purchased to 160 acres; in 1820, it was further cut to 80 acres and, in 1832, to 40 acres. Hildegard Johnson discussed how forties were put together to make larger holdings and how different configurations of four adjoining forties qualified for entry as quarter-sections under the provisions of the preemption acts.[87] By offering smaller parcels for sale, the land office enabled pioneers with little capital to start smallholdings. In practice, all purchasers were encouraged to be more discriminating in picking and choosing the most valuable parcels and avoiding ponds, marshes, and swamps. Settlers were given no special competitive advantages over speculators.

The ultimate inducement to settlement was to offer a holding free of charge to a person who put up a dwelling and brought land into cultivation. In 1862, the Homestead Act granted a homestead not exceeding 160 acres to an individual or family who resided on it and made improvements continuously for five years. The act provided that such a holding was not to be forfeited for debts incurred before the grant nor was the property to be used for the benefit of any person other than the homesteader or the homesteader's family. An occupier who wished to sell out before the five-year term expired first had to pay the land office the minimum entry price in order to acquire freehold possession. A determined attempt was made to close all foreseeable loopholes that might be exploited by speculators. It was the first measure intentionally drafted to promote the formation of family farms and prevent those farms from falling into the hands of speculators. For the Midwest it came too late to affect patterns of ownership. Almost all public lands in the prairie states of

87. Johnson, *Order upon the land,* 1976, 66–72, 109–10, 136–42.

Ohio, Indiana, Illinois, and Iowa had been disposed of before 1862. In northern Michigan, Wisconsin, and Minnesota some boreal forests were still available for homesteading.

The overall disposition of settlement and landownership in the Midwest around 1860 is that family farms occupied formerly wooded areas and many small prairies in Ohio, Indiana, Illinois, and Iowa. Large prairies, more than a mile or two from timber and running water, were almost all in the hands of large landowners and had few settlers. Wet prairies in northern Indiana, northeast Illinois, and Iowa belonged to the largest estates and were mostly uninhabited. In the northern lakes states of Michigan, Wisconsin, and Minnesota, timbered prairies were beginning to be settled and cultivated but most northern marshes, swamps, and bogs were still unoccupied. Attitudes toward open grasslands had softened appreciably since 1800, but Americans continued to fear and hate wetlands.

5

Landowners, Cattlemen, Railroads, and Tenants on Wet Prairies

[In the middle decades of the nineteenth century, high expectations for economic growth in the Midwest arose from superficial appraisals of potential productivity of different soils and linear projections of high rates of population increase. Many believed that, in the long run, holdings of even the poorest soils would reward patient investors. Small investors did not enjoy that choice. They selected lots that appeared likely to yield the highest returns in the shortest time. Large investors could exercise a range of options; some may not have been clear how they would use their property until it became ripe for development or disposal. Investment decisions were taken in a political and cultural milieu that abhorred speculators and loan sharks. Family farms, on the other hand, were viewed in a warmly sentimental light, admired for promoting arts of cultivation and livestock husbandry, praised for providing healthy outdoor work, encouraging virtues of self reliance, and offering just rewards for effort and enterprise. In reality, most Midwesterners were not inclined to follow the plow or engage in any wealth-creating activity other than prospecting for gold and precious metals or cruising for lumber.

Of the hundreds of thousands of newcomers who migrated to the Midwest in the first half of the nineteenth century only a small minority, if the testimony of journalists and novelists is to be relied upon, wanted to be sodbusters or family farmers.[1] Many dreamed of founding banks, establishing estate agencies, opening general stores,

---

1. Intimate portraits sketched by Sherwood Anderson, *Winesburg, Ohio* (B. W. Huebsch, New York 1919); and Sinclair Lewis, *Main Street* (Harcourt Brace, New York 1920), dis - close the secret yearnings and ambitions of ordinary Midwesterners half a century or more after settlement. From the beginning, land deals were important steps toward the realization of people's dreams. The appalling hardships of trying to wrest livings from the soil are re - counted in Ole Rolvaag, *Giants in the earth* (Harper, New York 1927); Hamlin Garland, *A son of the middle border* (Macmillan, New York 1917).

159

keeping saloons, hotels, or barber shops, becoming steamboat captains, or editing newspapers. Some hoped to practice as lawyers, doctors, architects, professors, or schoolteachers. Some wanted the adventure of outdoor life and imagined they might strike riches as miners, lumbermen, surveyors, or ranchers. An initial aversion to settling on a small patch in the midst of great open spaces was reinforced by the fact that most land was soon taken up as private property. No frontier custom permitted latecomers to squat on land belonging to others. Inevitably, some claims overlapped and quarrels and misunderstandings flared over who owned what and who held prior title, but entries in land office tract books settled most disputes. A few notorious conflicts "reached heights of excitement and open warfare" that bore little relation to the value of the property.[2]

In the early nineteenth century almost all who had money to invest bought land in the West. An English observer remarked: "Speculation in real estate . . . has been the ruling idea and occupation of the Western mind. Clerks, laborers, farmers, storekeepers merely followed their callings for a living, while they were speculating for their fortunes . . . The people of the West became dealers in land rather than its cultivators."[3] Financial panics in 1819, 1837, and 1857 only briefly checked a rising tide of hope. "Every one was imbued with a reckless spirit of speculation," wrote a wealthy New York businessman. "The mania, for such it undoubtedly was, did not confine itself to one particular class, but extended to all. Even the reverend clergy doffed their sacerdotals and entered into competition with mammon's votaries, for acquisition of this world's goods, and tested their sagacity against the shrewdness and more practiced skill of the professed sharper."[4] The inclusion of clerks, laborers, and clergymen among frontier capitalists may have been exaggerated. On the other hand, many frontiersmen with little or no training and without previous experience may have styled themselves re-

2. Paul W. Gates, *Landlords and tenants on the prairie frontier: Studies in American land policy* (Cornell University Press, Ithaca 1973) 43.

3. D. W. Mitchell, *Ten years in the United States: Being an Englishman's views of men and things in the North and South* (London 1862) 325–28, cited in Gates, *Landlords and tenants,* 1973, 50.

4. Levi Beardsley, *Reminiscences: Personal and other incidents* (New York 1852) 252, cited in Gates, *Landlords and tenants,* 1973, 52.

altors, bankers, and merchants. A detailed analysis of those engaged in land speculation in Iowa lists in-state investors, in descending order of importance, as realtors, bankers, lawyers, farm-stock raisers, merchants, manufacturers, county and town officials, physicians, and attorneys. Bankers combined with specialized professionals, lawyers, surveyors, and land office registers were the principal speculators.[5]

Large Landowners Acquire Wet Prairies

Many small purchasers borrowed heavily to acquire carefully selected lots that were expected to rise rapidly in value. They wanted quick returns to pay their debts and were content with short-term gains. Analyses of sources of mortgage credit in wet prairie counties in Illinois, Indiana, and Wisconsin from 1865 to 1880 indicate that a majority of loans were raised locally, were for relatively small amounts, bore high interest charges, and were repayable in three to five years.[6] Large investors acquired increasing shares in the market after 1870. They sought large quantities of low-priced land. They hoped to gather a few windfalls but their main aim was to retain possession of their property until prices appreciated. They looked for long-term gains. Neither small nor large purchasers were prepared to spend much on improving their estates as long as they saw greater opportunities for profit from reselling and reinvesting. In the early years of settlement the short-term strategy of small investors yielded higher returns than the long-term strategy of large investors. Public opinion and government policy favored small investors; the opening of fresh tracts of public domain further west prevented large landowners from gaining anything approaching a monopoly position. Many succeeded in making modest profits, a few made fortunes, and others lost money, either being forced to sell sooner than intended or failing to clear their debts after a long term.

In popular imagination large speculators were rich Easterners,

5. Robert P. Swierenga, *Pioneers and profits: Land speculation on the Iowa frontier* (Iowa State University Press, Ames 1968).

6. Robert F. Severson, The sources of mortgage credit for Champaign County, 1865–1880, *Agric. Hist.* 36 (1962) 150–55; Jay Ladin, Mortgage credit in Tippecanoe County, Indiana, 1865–1880, *Agric. Hist.* 41 (1967) 37–43; Sean Hartnett, The land market on the Wisconsin frontier: An examination of landownership processes in Turtle and La Prairie townships, 1839–1890, *Agric. Hist.* 65 (1991) 38–77.

either possessors of great personal wealth or executives of banks, insurance companies, and investment trusts. Paul Wallace Gates has shown that very large sums of money were advanced by Eastern banks and finance houses to purchase land in the West.[7] Stereotypical New York bankers rarely appeared in person at public land sales. They were represented by local land agents or arranged loans for purchases by local clients. Some of these intermediaries came from the East. Henry W. Ellsworth, son and partner of Henry L. Ellsworth, Connecticut-born, Yale-educated federal commissioner of patents, conducted a very successful land agency in Lafayette, Indiana. Henry L. Ellsworth, the father, attracted hundreds of Easterners, mostly from New England, to buy land in the prairies of northwest Indiana and north central Illinois in the 1830s. The business maintained close connections with the "Yale crowd." Members of the Ellsworth family intermarried with daughters of Yale alumni, took advice from Yale law school graduates, and bequeathed thousands of acres to their alma mater.[8] Not only Yale but Harvard alumni and other highly educated individuals, including doctors, lawyers, writers, and a prosperous publisher from Philadelphia, acquired extensive tracts of prairies in northern Indiana and Illinois.[9] A most successful relationship between a New York speculator, William A. Woodward, and a land agent in Wisconsin, Henry C. Putnam, flourished in the 1850s. Putnam dealt in university lands, school lands, and swamp lands, acted as agent for a land-grant railroad, was elected county surveyor, was later appointed deputy assessor to the U.S. land office in Eau Claire, and founded a bank. On behalf of Ezra Cornell, Woodward and Putnam helped to select a million acres granted to New York State under the Agricultural College Act. In this way, Cornell University came to derive part of its endowment income out of profits arising from land speculation in the Chippewa valley and elsewhere in the lakes states.[10] Members of elite groups, notably alumni of old universities and families of well-established New York bankers, figured prominently among purchasers of extensive tracts of prairies. Some of the

7. Gates, *Landlords and tenants,* 1973, 55–57, 150.

8. Ibid., 53, 115–26, 172, 248.

9. Ibid., 149–51.

10. Ibid., 55; Paul Wallace Gates, *The Wisconsin pine lands of Cornell University* (Cornell University Press, Ithaca 1944).

largest speculative ventures were organized by combinations of New York and New England capitalists with resident agents in the Midwest.

Edward Higbee focused on the role of folk from German and Swiss Rhineland origins, coming from southeastern Pennsylvania, in the initial settlement of the corn belt and the introduction of corn-growing. Higbee stated: "The crop and animal husbandry practices of these farmers from early Pennsylvania set the style for the modern Corn Belt."[11] Joseph Spencer and Ronald Horvath doubted whether corn was as dominant among crops or hogs were as dominant among livestock in early Pennsylvania as on new farms beyond the Appalachians.[12] John Hudson claimed that Southerners were largely responsible for bringing corn-growing and hog-fattening together and for importing distinctive social customs. The corn belt originated in the first quarter of the nineteenth century in five "islands": the Pennyroyal plateau just north of the border between Tennessee and Kentucky; the Nashville basin in Tennessee; the bluegrass country of Kentucky; the Scioto valley in Ohio; the Miami valley further west in Ohio. These islands were settled by people who traced their families back to Virginia, North Carolina, or Maryland. From Kentucky they carried Southern farming practices to Ohio and into Indiana.[13]

In the second half of the nineteenth century the center of the corn belt shifted westward and northward into wet prairies in northern Illinois and later to Iowa and adjoining parts of Wisconsin

11. Edward Higbee, *American agriculture: Geography, resources, conservation* (John Wiley, New York 1958) 233; a similar view was proposed by J. T. Lemon, *The best poor man's country: A geographical study of early southeastern Pennsylvania* (Johns Hopkins Press, Baltimore 1972) xiii.

12. J. E. Spencer and Ronald J. Horvath, How does an agricultural region originate? *Annals Assoc. Amer. Geog.* 53 (1963) 76.

13. John Hudson, *Making the corn belt: A geographical history of Middle-western agriculture* (Indiana University Press, Bloomington 1994) 3–10; a Yankee-determinist view of Midwestern history is also criticized by Allan G. Bogue, *From prairie to corn belt: Farming on the Illinois and Iowa prairies in the nineteenth century* (University of Chicago Press, Chicago 1963) 20; Richard Lyle Power, *Planting corn belt culture: The impress of the upland Southerner and Yankee in the Old Northwest* (Indiana State Hist. Soc., Indianapolis 1953), has argued for a blending of Southern and Northern cultures; further discussion of the cultural origins of corn belt settlement is provided in Robert D. Mitchell and Milton B. Newton, *The Appalachian frontier: Views from the East and the Southwest,* Historical Geography Research Series 21 (Cheltenham 1988).

and Minnesota. Hudson does not follow Southern influences in these areas but Southerners were engaged in land speculation in wet prairies alongside Northerners. Paul Wallace Gates has estimated the size of Southern investments in the Midwest before the Civil War.[14] A substantial amount of capital from Virginia and Kentucky was invested in the Virginia Military Tract in Ohio before 1820, but the largest investments of Southern capital were made in Indiana, Illinois, and Michigan between 1834 and 1837 and in Illinois, Wisconsin, and Iowa between 1854 and 1857. Merchants and planters from Virginia, South Carolina, and Louisiana invested heavily in wet prairies in Indiana in 1835. Between 1833 and 1837 many Southerners from Virginia, Kentucky, Mississippi, and Alabama acquired large tracts of Illinois prairies. In the 1850s, investors from every part of the South, not only shipowners, traders, bankers, railroad promoters, and landed families, but politicians, wanted to get their money out of the Southern states. Some were trying to hide money gained dishonestly; others considered Northern states safer places for investment than the South; a few were political adherents of the abolitionist cause. In the 1850s Southerners purchased large areas of prairie in Illinois, Iowa, Wisconsin, and Kansas, and some in Minnesota and Nebraska. During the Civil War repeated attempts were made to confiscate lands owned by Southerners, especially those who supported the Confederacy. By the end of the war, very little land held by Southerners had been seized and most of that had been forfeited for nonpayment of taxes.

In addition to Northerners and Southerners, a few foreign investors acquired interests in prairies, especially in Illinois. Communal land settlement schemes such as those organized by English radicals Morris Birkbeck, George and Richard Flower, Elias Pym Fordham, and Robert Owen bought prairie lands in southern Illinois in the early decades of the nineteenth century. In the 1830s, some land companies sold shares in Britain and other industrializing European countries. In addition, one or two individual investors from overseas made large purchases in prairie lands. The most notorious was an Irishman, William Scully, who became the largest absentee landlord in Illinois. Between 1850 and 1852 he acquired 38,000 acres of open prairies in Logan and Grundy counties. Scully

14. Gates, *Landlords and tenants,* 1973, 72–107.

intended to keep these lands permanently and lease them as farms while offering tenants prospects of compensation for improvements they carried out. In practice, very few tenants qualified for stipulated benefits, and hostility was inflamed by evictions of settlers in arrears with rents and by raising of rents for those who could afford it. Between 1875 and 1886, Scully embarked on a second phase of expansion, adding a further 4,000 acres to his possessions in Illinois, mostly in the Grand Prairie in Logan, Sangamon, and Livingston counties. In 1878, a protest meeting by tenants was reported sympathetically in the Chicago *Times:* "All over central Illinois a strong feeling is growing up against such immense estates, especially when operated by persons outside the State." Popular agitation brought no immediate redress for aggrieved tenants. Not until 1887 were limitations imposed on the acquisition of land by aliens and exemptions from payment of state taxes removed. Scully responded by revising the terms of leases to pass the burden of taxes to his tenants in the form of rent increases and, in 1900, he took out American citizenship so that his heirs would be free to inherit and add to the estate.[15]

Beneath the xenophobia directed at foreign landowners lay much deeper feelings of hostility toward all absentees whether they were from the North, South, or overseas. All were perceived to be out of touch, slow to listen to local opinions, unapproachable except through agents. Absentee owners of large tracts of unimproved prairies were especially vilified as miserably tightfisted and cautious about putting money into buildings, roads, fences, and other fixed assets.

Wet Prairies as High-Risk Speculations

Speculation in land was profitable only when prices were rising. Small investors depended on short-term gains; they needed to resell to get rich quickly. Large investors could afford to take longer views. They reckoned that in the long term the supply of public land would run out while demand for farms and sites for town development would continue to rise. As long as they were prepared to wait, they could not fail to profit; the safety of their investment seemed assured. In the meantime, portions of estates might be man-

15. Ibid., 266–97, quotation from 284.

aged as ranch land, improved and cultivated by hired laborers, or let to tenants, yielding a regular rental income. The compelling logic of the law of supply and demand appealed to educated Easterners because it implied that however ill-favored the land and however neglectful its management, it would gain in value as its scarcity increased.

In the Midwest, the value of large wet prairies did not follow the general trend. Prejudice against wetlands remained strong. The menace of insects and the incidence of fires, storms, and floods sustained a mood of constant anxiety. In cold accounting terms, costs of improvements were proportionately higher than for woodlands or small prairies. Excess charges for transporting timber, constructing roads and railroads and, above all, draining had to be added to normal costs incurred in clearing land, building, and fencing. Prospective settlers had to make allowances for initial outlays in assessing the value of wet prairies and ask themselves whether they could afford to carry these charges as overheads in estimating production costs. Speculators discounted at least part of the excess costs by putting low bids on unimproved land. In considering whether to spend money on improvements, landowners had to take account of prevailing prejudices. Would improved accessibility and improved soil drainage be sufficient to allay fears among intending settlers? This was a big risk for pioneer improvers. It would take time to demonstrate benefits of costly works, but until an intrepid investor ventured to take an initiative, prejudice could not be dispelled. Even the wealthiest and boldest hesitated. They had to grapple with another problem. Their predictions of long-term price rises were correct, but bouts of inflation and widespread panics such as those of 1819, 1837, and 1857 could destroy credit overnight. Land prices fell, loans secured against land could not be repaid, and banks collapsed. Among the hardest hit were owners of land in the course of improvement, where money had been borrowed to carry out works but productivity had not yet been raised. The predicament for owners of wet prairies was that no large profits could be expected while land remained unimproved, but gains from improvement depended on favorable timing and the vagaries of market sentiment, neither of which could be guaranteed in advance.

Small farmers with large debts suffered worst from financial crises, but large landowners also found their reserves drained away at an alarming rate. Once banks began to call in loans the largest landowners had nowhere to turn. Paul Wallace Gates reported that following the panic of 1837, abstracts of conveyances "show a tremendous volume of mortgage foreclosures of large estates."[16] In 1837, unproductive lands that became tax delinquent were sold to pay amounts outstanding on them. In prairie counties, much land was in the hands of a few large landowners, some of whom voted against raising taxes, thereby inhibiting the building of public roads and schools. Lack of public services deterred settlement.[17] Henry L. Ellsworth's vast, ill-drained estate in the Wabash valley in Indiana outlasted the 1837 crisis, acquired more wet prairies in north central Illinois in the 1840s and 1850s, and collapsed after the 1857 crisis. A near monopoly of prairie country in Benton County, Indiana, enabled the Ellsworths to defy county officials to collect taxes eight years in arrears. When the crash came, 110,328 acres in Indiana and Illinois were sold. The purchasers were mostly local men who had accumulated wealth from cattle trading, railroad building, banking, and speculating in town development. "They were part of the community, had grown up with it, and were now showing their confidence in it by investing their surplus at home." They "had the capital and the driving force to make the prairies productive."[18] In the 1860s and 1870s, absentee landowners continued to allow large areas of wet prairies to lie dormant and another crisis in 1873 forced some to disgorge their holdings.[19]

Many large proprietors who bought land as a simple speculation, expecting to profit from rising prices, neither making an effort to improve their own property nor assisting in the development of the prairie region, ended by losing their capital. In depression years in the 1840s, equity was gradually eroded by interest charges, taxes, agents' fees, legal expenses, and costs of fire and flood damage. Paul Wallace Gates has echoed Joseph Schafer's conclusion that "land speculation was on the whole an unprofitable business."[20] Over a

16. Ibid., 61.

17. Ibid., 130, 232.

18. Ibid., 135, other references on the rise and decline of the Ellsworth estate on 124–26, 130, 139.

19. Ibid., 161–62.

20. Ibid., 62, cites Joseph Schafer, *Wisconsin Domesday Book: Town studies* (Wisconsin State Hist. Soc., Madison 1924), vol. 1: 10.

long span of time from 1820 to 1890, the value of unimproved wet prairies failed to increase at rates prevailing for timbered lands or improved lands.

Cattlemen

Speculators hoped that without spending additional money on improvements, increasing scarcity of land and increasing demand for it would raise the value of their properties. In the prairies that hope was not fulfilled. Grass was the sole natural resource to be obtained free of charge. Timber had to be imported at great cost, water had to be drawn from wells that were expensive to dig, soil could not be utilized until tough prairie sod had been laboriously broken. In 1834, on open prairies in Illinois, it was noted that "acres by the million were wasting. Grass upon which cattle would become much more fat than upon the best clover pastures in Pennsylvania; on which good horses could be raised at a cost of ten dollars a year."[21] Grass could be grazed in the summer or cut for hay as winter feed. Paul Wallace Gates and Allan and Margaret Beattie Bogue sketch geographical characteristics of tallgrass prairies as areas for ranching but make few references to problems associated with water-logging.[22] In spring, stock could not be turned loose until floods subsided and the ground was firm enough to be trodden. In summer, swarms of insects had to be destroyed by fire and smoke. Gates draws upon reminiscences written by George Ade, whose father was a contemporary of cattle kings in Indiana and himself came to own 2,400 acres in Newton County, for a description of his part of the prairie range, "so shaggy and waterlogged that it was commonly regarded as a hopeless proposition." In detail, Ade recalled:

Every low spot on the prairie was a slough rank with reeds and cattails, and breeding ferocious gallinippers by the millions. Also a large kind of horse-fly, called the "green head," which was so warlike and blood-hungry that when it

21. Power, *Planting corn belt culture,* 1953, 156, cites John M. Peck, *Gazetteer of Illinois in three parts* (R. Goudy, Jacksonville 1834), 41.
22. Bogue, *Prairie to corn belt,* 1963, 86–99, 197; Margaret Beattie Bogue, *Patterns from the sod: Land use and tenure in the Grand Prairie, 1850–1900* (Illinois State Hist. Doc., Springfield 1959), contains a chapter on the cattlemen's domain, 48–84. Gates, *Landlords and tenants,* 1973, contains chapters on Hoosier cattle kings and cattle kings in the prairies, 170–237.

attacked a horse, in swarm formation, it would either kill him or weaken him so much that he had no value as a work animal. Oxen were used in breaking the raw prairie, and even these tough and thick-skinned animals suffered torture when attacked by armies of green heads.[23]

These wetlands were among the last tracts of open range in the Midwest to be occupied by cattlemen.

Cattle were first grazed on prairies in central Kentucky as early as 1784. They were driven there from the south branch of the Potomac for summer pasture. Within a few years, stock were being grazed on extensive prairies in the Scioto valley in Ohio during the summer. In winter they were stall-fed on locally grown corn and wheat straw.[24] By 1840, improved breeds were raised and fed on the Kentucky bluegrass, in the Scioto and Miami valleys in Ohio, and along the Wabash in Indiana and Illinois. They were kept for up to five years then driven 400 or more miles across wide rivers and the Allegheny Mountains to Baltimore and Philadelphia. In 1820, native prairie grasses in Illinois were grazed and stock cattle produced, but most of the uplands remained "an unsettled wilderness because of marshiness."[25] In the Grand Prairie of Indiana before 1830, "the sloughs filled with a rank growth of cattails, the swarms of greenhead horseflies, and the distance from markets discouraged corn farmers from entering."[26] Much of the Grand Prairie was open range, supplying thin cattle to feeding regions. Stock enjoyed the coarse, strong prairie grass in the spring. Flies were killed by burning the grass in June and the aftermath was grazed again until winter.[27]

Upgrading stock by introducing shorthorn, Devon, and Hereford bulls began in Kentucky at the end of the eighteenth century. In the Kentucky bluegrass and Scioto valley, impressive weight gains in fatstock were achieved in the 1830s. In the 1840s, numbers of cattle increased but prices fell and production spread to new areas, notably the Miami, Wabash, and Sangamon valleys. On the

23. George Ade, Prairie kings of yesterday, *Saturday Evening Post* 204 (4 July 1931) 14, 76, cited in Gates, *Landlords and tenants,* 1973, 174.

24. Paul C. Henlein, *Cattle kingdom in the Ohio valley, 1783–1860* (University of Kentucky Press, Lexington 1959) 4–7.

25. Ibid., 16, also 9–12.

26. Ibid., 18.

27. Ibid., 19.

Grand Prairie, open range grazing remained profitable in the 1830s and 1840s and stock were driven up to 800 miles from Illinois to the eastern seaboard. In the 1840s, a new group of cattlemen, including well-to-do fatteners of German origin such as Isaac Funk, began to acquire large areas of prairie in McLean and adjoining counties in Illinois and Indiana. For winter feed, hay production was essential.[28] After 1850, grazing and grass fodder gradually gave way to corn feeding as the range was fenced in and the sod broken (fig. 5.1). Some sloughs were dried out by exposure to sun and air following cutting of wild hay.[29] By 1878, in Hamilton County, Iowa, it was remarked that prairie hay "will soon be a thing of the past."[30] When ranchers moved to short grass prairies on the Great Plains, open range grazing and scrub cattle disappeared from tall-grass prairies in the Midwest.

A changeover to corn feeding was accompanied by an enlargement of herds and improvement in quality of stock. In the mid-nineteenth century, most Midwestern cattlemen were mixed farmers, not open-range herders. Their operations were capital-intensive and also more labor-intensive than ranches. Many of the new farmers were of British or German origin.[31] Paul Wallace Gates has examined how large estates in Indiana and Illinois transformed their enterprises

---

28. Terry G. Jordan, *North American cattle-ranching frontiers: Origins, diffusion, and differentiation* (University of New Mexico Press, Albuquerque 1993) 200–203.
29. Ibid., 62–63.
30. Bogue, *From prairie to corn belt,* 1963, 197.
31. Jordan, *Cattle ranching frontiers,* 1993, 267–74.

Figure 5.1 *(see facing page),* Key:

1 Thomas Armstrong holdings (1895), 1,506 acres

2 Hiram H. and Harold Catlett purchases (1849–98), 4,000 acres

3 Joseph S. and Thomas F. Christman purchases (1879–1927), 3,200 acres

4 Francis A. and Thomas F. Collison purchases (1855–1917), 2,990 acres

5 William Fithian purchases (1832–69), 4,500 acres

6 Willy Fowler purchases (1863–91), 4,000 acres

7 John W. Goodwine purchases (1848–91), 5,440 acres

8 Thomas Hoopes purchases (1853–92), 11,250 acres

9 Ira G. Jones purchases (1862–1908), 1,499 acres

10 Abraham Mann purchases (1835–75), 5,720 acres

11 David and W. A. Rankin purchases (1867–84), 4,870 acres

12 James S. Sconce purchases (1860–85), 2,430 acres

13 John Sidell holdings (1875), 4,170 acres

14 Sullivant purchases (1853–67), 4,600 acres

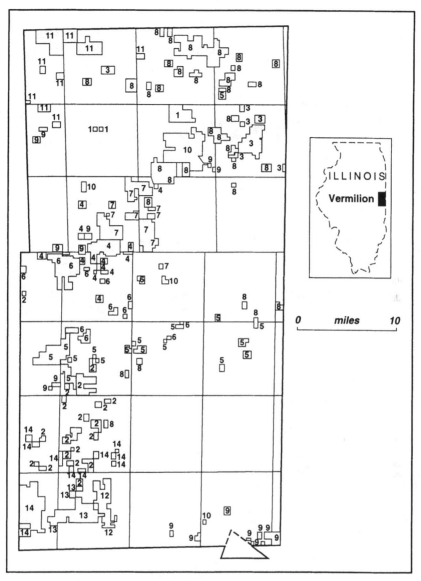

Figure 5.1 Cattle ranches in Vermilion County, Illinois

Source: Margaret Beattie Bogue, *Patterns from the sod: Land use and tenure in the Grand Prairie, 1850–1900* (Illinois State Hist. Soc., Springfield 1959) 50–51. *(See key on facing page.)*

from grazing to feeding between 1850 and 1890. Successful individuals added to their landholdings and their herds (table 5.1). In
the 1850s and 1860s, they established tenant farms and began moving stock by railroad. When they reached open grasslands in Ohio,
railroads charged high rates for shipping cattle and provided unsuitable boxcars, with the result that animals suffered more injuries and
loss of weight than when driven on the hoof. By 1860, a grid of
competing lines crossed all prairies east of the Mississippi, tariffs
and journey times were reduced substantially, and rolling stock was
designed to transport cattle cheaply and efficiently. Grasslands west
of the Mississippi in Nebraska, Kansas, and Texas became breeding
grounds for feeder stock and an emergent corn belt specialized in
fattening.[32] A new struggle with railroad companies arose in the
1870s when techniques of ice-cooling and the invention of refriger-
ated railroad cars enabled dressed meat to be transported rapidly to
Eastern cities. Under pressure from complaining Eastern butchers,
railroad executives initially set high rates on this traffic, but competition and political action brought reductions.[33] It also brought producers on the prairie into close contact with consumers.

In the 1870s and 1880s, cattle kings in Indiana and Illinois introduced purebred Shorthorns and Herefords and built grand man-
sions in spacious grounds.[34] In the heyday of fatstock production,
leading proprietors flaunted their wealth, spending lavishly on luxu-
rious living and pedigree stock. They amassed hoards of costly furniture, art treasures, and moveable goods, but they were reluctant to
sink money into the land itself. They fought proposals to increase
taxes, disputed demands by tax collectors to pay appropriate dues,
and were often in arrears. They also resisted attempts to compel
them to fence in their stock. In Illinois, a herd law was delayed until
1867 and proved ineffective against the most powerful graziers. In
many localities, cattlemen opposed the organization of drainage districts on the grounds that they "would increase their obligations

---

32. Henlein, *Cattle kingdom,* 1959, 124–25; Gates, *Landlords and tenants,* 1973, 198–99;
Paul W. Gates, *The farmer's age: Agriculture 1815–1860,* Econ. Hist. of the United States III
(Harper Row, New York 1960) 212–14.

33. Bogue, *From prairie to corn belt,* 1963, 282.

34. Gates, *Landlords and tenants,* 1973, 170–237; Gates, Hoosier cattle kings, *Indiana
Magazine of History* 44 (March 1948) 1–24; Gates, Cattle kings in the prairies, *Miss. Vall.
Hist. Rev.* 35 (December 1948) 379–412.

Table 5.1 Number of cattle other than working oxen
or milch cows, 1850–90

	1850	1860	1870	1880	1890
Ohio	749,067	895,077	758,221	1,084,917	968,554
Indiana	389,891	588,144	618,360	864,846	932,621
Illinois	541,209	970,799	1,055,499	1,515,063	1,975,233
Iowa	69,025	293,322	614,366	1,755,343	3,397,132
Michigan	119,471	238,615	260,171	466,660	549,160
Wisconsin	76,293	225,207	331,302	622,005	855,327
Minnesota	740	51,345	145,736	347,161	779,671
Midwest total	1,945,696	3,262,509	3,783,655	6,655,995	9,457,698
% change		+68	+16	+76	+42
Rest of U.S.	7,747,373	11,516,864	9,782,350	15,832,555	31,679,144
% change		+49	−15	+62	+100
U.S. total	9,693,069	14,779,373	13,566,005	22,488,550	41,136,842

Source: Department of Interior, Census Office, Compendium of the tenth census, 1880 (Washington, D.C. 1883), pt. 1, 679; U.S. Census Office, Twelfth census, 1900, vol. 5, Agriculture (Washington, D.C. 1902) 705.

without bringing in commensurate returns."[35] One of the most serious conflicts with small farmers followed the importation of longhorn cattle from Texas in 1866, 1867, and 1869. Ticks carried by these animals infected local herds that had no immunity to Texas or Spanish fever. The plague killed thousands of cattle, claims for compensation led to ruinous lawsuits, and the constitutional validity of measures taken by states to ban or control movements of livestock across state boundaries were contested all the way to the U.S. Supreme Court. In 1889, the federal government instituted quarantine regulations that effectively solved the problem.[36] Cattlemen acquired a reputation for bad neighborly behavior toward small farmers and a lack of respect for the law. In order to retain their estates after the panic of 1873, they had to bring more and more land into cultivation. Their interests drew them closer to the farming community and drove them to consider means of improving their land.

Railroads Extend into Wet Prairies

After experiencing the panic of 1837, surviving large landowners examined their assets and liabilities, taking account of a sharp drop in the value of unimproved wetlands. They were in no mood to put more funds of their own into improvements, but they were eager to encourage railroad companies to project lines across the prairies. Speaking in the Senate in 1850 in support of the Illinois Central Railroad bill, Henry Clay argued that "this road will pass directly through the Grand Prairie lengthwise, and there is nobody who knows anything of that Grand Prairie who does not know that the land is utterly worthless for any present purpose—not because it is not fertile, but for want of wood and water and from the fact that it is inaccessible. . . . And now, by constructing this road through the prairie, through the center of Illinois, you bring millions of acres of land immediately into the market which will otherwise remain for years and years entirely unsalable."[37] Railroads promised to provide means of transporting heavy building materials, fencing timber, fuel, and farming equipment and also carry grain and livestock from

35. Gates, *Landlords and tenants,* 1973, 232–34, quotation on 234 from *National Live-Stock Journal* 10 (May 1879) 195.

36. Bogue, *Patterns from the sod,* 1959, 59–68.

37. Carlton J. Corliss, *Main line of mid-America: The story of the Illinois Central* (Creative Age Press, New York 1950) 18.

farms. Interiors of prairies were remote from the Great Lakes and navigable rivers. Constructing canals across interfluves was expensive and costs of transporting bulky goods by road were prohibitively high. Both state and federal governments wanted to overcome the dispiriting parsimony and inertia that gripped private landowners in order to set in motion processes of development. As inducements to opening up thinly settled territories, railroad companies might be offered sites joining with existing rail routes or waterways and the federal government might be persuaded to make grants of land from the public domain.

Precedents for subsidizing railroad construction were federal acts granting rights of way and donating public lands for building roads in Ohio in 1802 and 1823 and for laying a wagon road across Indiana from Lake Michigan to the Ohio River in 1827. Land grants had also been made to support the cutting of a canal linking Lake Erie to the Wabash River in 1824 and to inaugurate the Illinois and Michigan Canal in 1833. The Midwestern states of Ohio, Indiana, Illinois, Michigan, and Wisconsin succeeded in securing all federal grants awarded for canal construction, totaling 4,598,668 acres.[38] In 1837, Illinois' own state legislature passed an Internal Improvement Act authorizing the building of roads, bridges, canals, and also railroads. The state issued bonds to cover $10,250,000 allocated to these projects. This was too heavy a burden to be borne by a state whose population was less than 400,000. Following the financial crash of 1837, state revenues were insufficient to pay interest charges on the debt, and work on the schemes was suspended.[39]

Ten years later, a fresh proposal was launched to build an Illinois Central Railroad running from Chicago across the Grand Prairie to Cairo, thence to Mobile, Alabama, on the Gulf coast. Another line from Dubuque to Galena, begun in 1837, was to be connected with the main line at Centralia. The promoters of this scheme applied to the federal government for assistance. The request was contested on the grounds that, unlike roads and canals, which were public rights of way vested in the state, a railroad was privately owned and could not be given money from the public

38. Benjamin Horace Hibbard, *A history of public land policies* (Peter Smith, New York 1939) 228–42.

39. Paul Wallace Gates, *The Illinois Central Railroad and its colonization work* (Harvard University Press, Cambridge 1934) 22.

purse. It was also contended that the State of Illinois should not enjoy preferential treatment. The objection to subsidizing private interests was partly met by awarding a land grant along a strip six miles wide on each side of the proposed line of railroad, even-numbered sections being assigned to the company, odd-numbered sections in the public domain to be sold for at least double the government's minimum price, that is not less than $2.50 per acre. The railroad company likewise agreed not to sell its land for less than double the minimum price. The provision was to ensure that the Treasury would eventually receive as much money from the sale of every alternate section as it would have obtained from a pair of adjoining sections. The question of Illinois being given undue preference was resolved by a majority of senators and representatives from other states voting in favor of the measure. Mississippi and Alabama shared an interest in the scheme; New Englanders and New Yorkers hoped that a financial recovery of Illinois might enable them to recoup some of the money lost in 1837; Eastern manufacturers hoped to win contracts for supplying rails and locomotives and, in general, benefit from an expansion of their market areas. In September 1850, Congress voted for a bill under whose terms over 2,589,498 acres were to be granted to the Illinois Central Railroad, the first company to receive a federal land grant.[40]

A period of conflict and resistance to railroad land grants ensued, only two bills gaining approval in the next three years. Resistance then gave way to acceptance: in 1856, Congress conceded grants amounting to 14,599,000 acres and in 1857, another 5,118,000 acres were given away.[41] In May 1856, land grants were awarded to construct four railroads from east to west across Iowa: the Burlington and Missouri, the Mississippi and Missouri, the Iowa Central Air Line, and the Dubuque and Pacific. Each company had to complete its work in ten years; alternate sections to a width of six miles on each side of the lines were to be selected; 230,724 acres of public land reserved for the Burlington company were to be transferred as successive lengths of line were opened; and the State of Iowa was made responsible for regulating safety and tariff structures.[42] At this time many similar schemes won support in Congress

40. Ibid., 26–43.
41. Hibbard, *History of public land policies,* 1939, 246.
42. Richard C. Overton, *Burlington west: A colonization history of the Burlington Railroad*

with little or no opposition. Toward the end of the 1860s attitudes hardened. Homesteaders and home-seekers called for a halt to further land grants, and state legislatures in Indiana, Illinois, and Wisconsin passed resolutions urging Congress to uphold the rights of settlers against the demands of corporations.[43] Between 1837 and 1870 opinions swung back and forth: at first, Western landowners hoped that railroads would make wet prairies attractive for settlement while Eastern financiers and manufacturers looked forward to expanding their businesses; then the South feared that they might be injured by government-subsidized Western competitors, and landless home-seekers complained that opportunities to settle on the public domain were denied because all the land was given to big corporations.

From the 1830s, families and associates of New York bankers, Boston merchants, and New England manufacturers engaged both in large-scale speculation in wet prairies and also in railroad promotion. New Yorkers Jonathon Sturges, Hiram and Morris Ketchum, and Franklin Haven, who held large tracts of land in Illinois, were among the earliest subscribers to the Illinois Central Railroad. Morris Ketchum was also a member of a firm manufacturing locomotives. Abbot Lawrence, a Massachusetts textile manufacturer who owned lands on the Fox River in Wisconsin, backed several railroad projects. In 1835, a group of New England capitalists organized the Boston and Western Land Company. They purchased extensive holdings in Illinois, Wisconsin, and Missouri.[44] Through agreements with John Murray Forbes, a Boston merchant who had prospered from trade with China, the Boston and Western Land Company acquired interests in the Michigan Central and the Chicago, Burlington and Quincy railroads.[45]

Railroad directors as well as Northern landowners recruited young executives from Ivy League universities or from among their own kinsmen. In 1858, John Murray Forbes, director and leading stockholder in the Burlington and Missouri Railroad, invited 23-

(Harvard University Press, Cambridge 1941) 73, 83–85, 111.

43. Hibbard, *History of public land policies,* 1939, 252–54.

44. Gates, *Illinois Central Railroad,* 1934, 37–40, 49–51; Arthur M. Johnson and Barry E. Supple, *Boston capitalists and western railroads: A study in nineteenth century railroad investment process* (Harvard University Press, Cambridge 1967) 67–68.

45. Overton, *Burlington west,* 1941, 26, 42–44.

year-old Charles Russell Lowell, who had a brilliant record at
Harvard and was making a successful start in the iron industry, to
become assistant treasurer at the railroad headquarters in Burlington. A year later, Lowell was put in charge of land sales and
Forbes's 18-year-old cousin, Charles Elliott Perkins, was brought in
to assist in the land department. Lowell returned to iron and steel
making during the Civil War and in 1881 Perkins succeeded Forbes
as president of the company. Easterners, however well-educated or
loyal to their families, suffered serious handicaps in dealing with
Midwesterners. David Neal, a New England merchant engaged in
trade with China in association with Forbes, was another leading
stockholder in the Illinois Central Railroad. For the first five years of
the company's operations, Neal was in charge of the selection,
management, and sale of lands in Illinois. This was a task for which
he was poorly qualified, being unfamiliar with land problems in the
Midwest and inexperienced in running a real-estate business. He
had to be removed from office as vice president before the handling
of land sales could be put on a proper footing. Geographical remoteness, social distance, and lack of local knowledge had to be
overcome in order to satisfy potential settlers who sought informed
advice about climate, soils, and farming conditions, and also required financial assistance. Formulas approved in Boston board-
rooms were no substitute for practical experience learned in the
field, on prairies in Illinois.

Railroad Land Sales and Freight Rates

What effect did railroads have on settlement in wetlands? A year or
two after grants had been allotted to railroads, there was a scramble
for land, and prices more than doubled. Was this a direct result of
benefits brought by railroads in providing access to places hitherto
not reached by roads or waterways, or was it attributable to perceived scarcities of land, artificially created by handing vast areas to
schools, land grant colleges, and counties for building roads and
draining swamplands, as well as to railroad companies? What effects
did railroads have on economic development in wet prairies: did
they fulfill expectations of their promoters, landowners, and local
politicians in assisting agriculture and generating town growth?[46]

46. A more fundamental question about effects of railroads on economic growth has been
studied at the national level by Robert Fogel and Albert Fishlow, but special features of wet

Land values rose before other sectors of the economy showed signs of growth, and will be discussed first.

Railroad companies had little room for maneuver in selling land. First, not all lands designated in a federal grant, such as alternate sections up to six miles on each side of a line, were available when an act was passed. Some had already been purchased, some were preempted by settlers awaiting confirmation of titles, some were subject to other grants, some were claimed by states under the Swamp Land Act. Overton has reported protracted disputes between the Burlington Railroad, county authorities, the State of Iowa, and the federal government over rival claims to swamplands.[47] Federal land offices closed while railroads selected lands they had been granted and searched for compensatory indemnity lands within a band up to 15 miles on each side of the line. Secondly, railroad companies had to carry out their own surveys, value land, compile accurate land registers, and enter into negotiations with prospective purchasers. Money to build lines ultimately came from sales of land. Some companies raised mortgages, some issued bonds secured against land, but all these borrowings had to be repaid out of the proceeds of sales. Successful management of land departments was crucial for railroads. The Illinois Central, Gates remarks, "in the first decade of its existence, was primarily a land company and secondarily a railroad company."[48] In order to meet heavy outgoings the company had to insist that purchasers pay high prices for land, but to make such deals attractive it offered long-term credit at low rates of interest. An Iowa newspaper editor described this as "a sort of hot-bed forcing system"; it resulted in many farms reverting to the railroad or being taken over by speculators.[49]

The Burlington learned from mistakes made by the Illinois Central but faced new perils in Iowa. The company was enmeshed in a web of hostility, fraud, and corruption in the swampland imbroglio; many speculators attempted to establish false preemption

prairies have not been examined. Robert William Fogel, *Railroads and American economic growth: Essays in econometric history* (Johns Hopkins Press, Baltimore 1964); Albert Fishlow, *American railroads and the transformation of the ante-bellum economy* (Harvard University Press, Cambridge 1965).

47. Overton, *Burlington west,* 1941, 122–38.

48. Gates, *Illinois Central Railroad,* 1934, 149.

49. Overton, *Burlington west,* 1941, 109, cites *Burlington Weekly Hawk-Eye* (23 November 1858).

rights; and construction of the line, having reached Ottumwa on the
Des Moines River, stopped abruptly in 1862 when men and materials were diverted to the war effort. To pay outstanding interest
charges, directors had to take out a new mortgage in 1863 and obtain a new act in 1864, granting additional indemnity land and extending time limits allowed for completing the work. Genuine farmers had to be reassured "that we are a *railroad* company & not a
land company—that settlers are more important to us than a high
price for our land."[50] Instructions were sent to surveyors to keep on
good terms with established settlers and "avoid giving unnecessary
trouble and offense" to those who were willing to stay.[51] On the
other hand, squatters and fraudulent claimants had to be firmly
discouraged. In 1865, when construction work was about to be resumed, it was considered essential to notify local residents

that these lands are given as a fund to construct this road—that sold at present
prices they would produce nothing for such purpose & this can only be by a
mortgage on time & then when the lands have been made more valuable by
the road's being built, selling them at their then value to repay the money. The
community should understand that we are struggling to build the road & need
all the credit we can get from all sources in deeding these lands & that therefore we cannot waste or sell them at present prices & must realize all the
value the road will give them.[52]

Applicants for railroad lands were strongly advised "that *no improvements* should be made upon the land, so as to avoid all
chance of hard feelings on the score of misunderstanding and difference of opinion" regarding claims for fixed improvements. In order
to build the line to Missouri the company had to drive hard bargains with buyers. Like the Illinois Central, it was willing to advance
credit for up to ten years at low or moderate rates of interest but it
was forced to sell much of its land for less than it had budgeted.[53]

Although railroads repeatedly proclaimed that they were doing
all in their power to help genuine settlers and exclude speculators,

50. Ibid., 150, cites a letter from Charles Lowell to Henry Thielsen, engineer in charge of
survey work in Iowa, 8 December 1859.

51. Ibid., 149, cites a letter from Lowell to John Ames, field surveyor and examiner, 14
December 1859.

52. Ibid., 207, letter from president James Joy to Charles Perkins, head of land depart -
ment, 10 July 1865.

53. Ibid., 186–217.

neither Overton nor Gates produce substantive evidence of companies refusing to sell land to affluent speculators. Clare Cooper has noted that settlers delayed moving to counties in west, northwest, and north central Iowa for some years after railroads arrived, by which time speculators had snatched large tracts. As in north central Illinois, these included areas of wet prairie requiring additional capital outlays before cultivation could begin.[54] In the postbellum years, railroads expected to sell the poorest land, over five miles from a line, at not less than three dollars an acre and better lands for at least five to ten dollars an acre.[55] Farm-seekers complained that they were eligible for free 160-acre homesteads from government land offices and that wetland would cost at least an additional five dollars an acre to bring into cultivation. In negotiations, painful compromises were made by both parties, but most agreements resulted in substantial increases in the price of wet prairies. In general, lands reached by railroads were expected to fetch two to four times as much as lands of similar quality over five miles from a line.[56] Allan and Margaret Bogue's analysis of profits made by land speculators indicates a sharp increase in the rate of returns from prairie land in Illinois from 1855 to 1870, starting just before the outbreak of war and the onset of inflation. The study concludes that the rise in price was influenced by railroad construction.[57]

How did owners, occupiers, and purchasers themselves assess benefits that railroads brought to wet prairies in Illinois, Iowa, Wisconsin, and Minnesota? Were they primarily interested in costs of transporting goods to and from farms? Would they have asked questions such as those posed by Robert Fogel: "Did the interregional distribution of agricultural products depend on the existence of the long-haul railroad?" A supplementary question was, how dependent were farmers on railroads for "intra-regional or short-haul distribution of agricultural products," carrying many different items of freight over thousands of lines to local markets?[58] It was generally agreed that some form of cheap transportation was essential for

54. Clare C. Cooper, The role of railroads in the settlement of Iowa: A study in historical geography, M.A. thesis, University of Nebraska 1958, 136–38.

55. Overton, *Burlington west,* 1941, 239.

56. Ibid., 236–56; Gates, *Illinois Central Railroad,* 1934, 107–9, 153–57.

57. Allan G. Bogue and Margaret B. Bogue, "Profits" and the frontier land speculator, *Journ. Econ. Hist.* 17 (1957) 13.

58. Fogel, *Railroads and American economic growth,* 1964, 15–16.

prairie farmers to compete effectively in national and local markets. Fogel claimed that railroads offered no more than very small reductions in charges over long distances compared with those offered by waterways, simply because waterway distances between two points were much greater than those traversed by railroads.[59] In the mid-nineteenth century farmers and landowners would have been intrigued by Fogel's analysis of the technical and economic feasibility of constructing canals across previously remote prairies. By drawing profiles of their courses and referring to Army Corps of Engineers' water consumption formulas he deduced that the terrain over which proposed canals would have been built and their water supplies drawn were more favorable than those used by canals already operating and competing successfully with railroads in Ohio and New York.[60] Furthermore, costs of wagon transport to local markets might have been lowered by extending the road network and improving road surfaces.

Fogel acknowledged evidence for a "boom psychology" urging the building of railroads across sparsely settled tracts, but rashly inferred that "the same favorable mental disposition could have been created by the construction of a new canal or the introduction of any new mode of transportation that unexpectedly and drastically reduced the cost of transportation in a given area."[61] That was not the case. Wherever choices were presented, railroads were preferred and in most places railroad schemes alone were considered. Fogel conceded that by the time the prairies were settled, "the railroad had achieved clear technological superiority over canals."[62] Technological superiority probably had more to do with speed than with freight rates. For weeks or months each year, ice stopped all movements on rivers, lakes, and canals. Waterways were lengthy and circuitous. The Great Lakes offered few straight-line connections between major ports. The Mississippi, Ohio, and other navigable rivers meandered over wide floodplains and were frequently choked with logs and other floating debris. Canals were contoured or divided into sections by flights of locks. Steamboats and steam tugs shortened journey times, but no ship could equal the speed of a

59. Ibid., 110.
60. Ibid., 92–107.
61. Ibid., 11.
62. Ibid., 80.

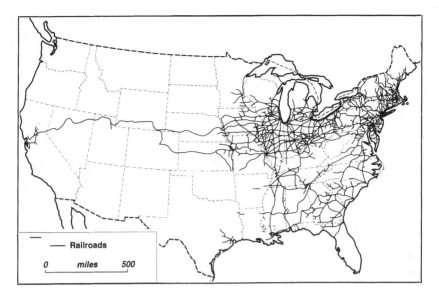

Figure 5.2 Railroads in the United States, 1870

Source: Charles O. Paullin and John K. Wright, *Atlas of the historical geography of the United States* (Carnegie Institute and American Geographical Society, Washington, D.C. 1932), plate 140A.

locomotive (fig. 5.2). On the prairies, railroads had optimum conditions for fast running: tracks laid on almost level grades followed straight lines from point to point. The most expensive engineering works were bridges spanning wide rivers. From the outset railroads captured all passenger traffic and carriage of perishable commodities, including fruit, vegetables, livestock, and meat. Railroads accelerated the conveyance of mails and were accompanied by the opening of telegraph lines. An era of rapid movement of people and nearly instantaneous long-distance communications abolished the isolation of prairie settlers. Railroad companies advertised land sales and promised special services to their customers. They invited settlers to travel by rail and visit farm lots, town sites, and surrounding country. They held banquets and festivities to celebrate openings of new sections of track. For the first time, wet prairies were open to personal inspection by interested observers. After initially unsuccessful attempts, land departments began to employ experienced local agents and published informative guides for immigrants. Railroads succeeded in arousing a feeling that these lands were desirable and valuable. The sound of a locomotive whistle reminded prairie settlers that they were not far from great cities and lines of communication that stretched around the globe.

Breaking Prairies and Fencing

As land prices rose, owners of large estates were driven to find ways of increasing their incomes. For agricultural land, the aim was to use the land more intensively than for rearing and feeding cattle. This objective was achieved mostly by converting pasture to arable for growing corn and other crops. The prairie sod had first to be broken and fields fenced. Prairie breaking was a spring and summer occupation. Done by hand, it was grim, arduous work, described as "deviling." A large team of oxen might be harnessed to a massive iron plowshare weighing from 60 to 125 pounds. Teams required skillful direction to draw straight furrows at an even depth.[63] In 1837, John Deere, a blacksmith at Grand Detour, Illinois, constructed a steel breaking plow using saw steel to make a one-piece share and moldboard. The implement was advertised in local newspapers as cutting a cleaner furrow than others: "The whole face of the mold board and share is ground smooth, so that it scours perfectly bright in any soil, and will not choke in the foulest ground. It will do more work in a day and do it much better and with less labor, to both team and holder, than the ordinary ploughs that do not scour."[64] By 1846, Deere and his partner, sawmill owner Leonard Andrus, were producing about 1,000 plows a year. In 1847 Deere moved to larger premises in Moline, Illinois, and ten years later he was selling over 10,000 plows annually. They were used on thousands of prairie farms.[65]

No early or easy solution was found to problems of fencing out or fencing in livestock on treeless grasslands. Up to 1860, split-rail fences, similar to those made in Virginia, predominated in the Midwest. Large quantities of timber were required for their con-

63. David E. Schob, Sodbusting on the upper Midwestern frontier, 1820–1860, *Agric. Hist.* 47 (1973) 47–50.

64. Wayne D. Rasmussen (ed.), *Readings in the history of American agriculture* (University of Illinois Press, Urbana 1960) 78–79, quotes *Rock River Register*, 10 March 1843.

65. Leo Rogin, *The introduction of farm machinery in its relation to the productivity of labor in the agriculture of the United States during the nineteenth century*, University of California Publications in Economics 9 (University of California Press, Berkeley 1931) 32–35.

struction. The coming of railroads and exploitation of northern pineries enabled lighter posts and boards to be imported. These provided cheaper but less durable fences.[66] Alternative methods of fencing had drawbacks. Osage orange hedges grown from seeds raised in Texas were introduced to Illinois in the 1840s. By 1850, hedges were established around Jacksonville, Illinois; by 1860, they had spread to most prairie areas; by 1870, about three-quarters of fences on the wet prairies of Kankakee County were osage orange.[67] Hedges took about four years to grow into stock-proof enclosures. After that, they required frequent cutting to prevent them growing out of control; field crops were injured by their extending shade, smothering snowdrifts, spreading roots, and proliferating pests. By 1880, hedges had outlived their usefulness and were being grubbed up everywhere. They were superseded by barbed wire. Early experiments with plain wire and wire netting failed. In the 1840s, plain wire stretched between posts was tried on Illinois prairies but did not keep stock out. Ungalvanized wire and wire netting rusted after a year. Single strands of wire expanded and went slack in summer and contracted and snapped in winter.[68] By twisting two or three strands of galvanized wire around a wire prong or barb, the defects of early experiments were remedied. The inventor, Joseph Glidden, was a large farmer on Illinois prairie at De Kalb. He began selling barbed wire in 1874, then formed a partnership with I. L. Ellwood to patent and manufacture his wire. In 1876, Washburn and Moen, manufacturers of steel wire in Worcester, Massachusetts, designed a machine that would produce barbed wire automatically. They acquired Glidden's patents for a lump sum plus royalties and installed the new machines in a greatly enlarged factory at De Kalb. Production soared and by 1900 it was being sold throughout the world. Prescott Webb claims: "Barbed wire made the hundred-and-sixty-acre homestead both possible and profitable on the Prairie Plains."[69]

Little by little, wet prairies appeared ready to receive farming settlers, but each step toward their full improvement cost more than

66. Clarence H. Danhof, The fencing problem in the eighteen-fifties, *Agric. Hist.* 18 (1944) 168–86.

67. Walter Prescott Webb, *The Great Plains* (Ginn and Co., Boston 1931) 290–91.

68. Earl W. Hayter, *The troubled farmer, 1850–1900: Rural adjustment to industrialism* (Northern Illinois University Press, De Kalb 1968) 101.

69. Webb, *Great Plains,* 1931, 295–318, quotation from 318.

the last. The coming of the railroad raised the price of unimproved land from the government minimum $1.25 to at least $3, rising to over $50 an acre. Sodbusting cost from $2 to $4 an acre. Barbed wire fencing cost between $2 and $5 an acre, depending on the size and shape of enclosures. Artificial draining might add another $5 to $20 an acre or more.[70] Investments in fixed assets were over and above farmers' basic expenditure on moveable stock: machinery, implements, livestock, and expendable materials. These were large sums, beyond the means of most pioneers and not easily afforded by most landowners.

Large-scale Farming

Speculators poured money into buying more and more land and saved little for reclamation and improvement. In 1860, farms over 1,000 acres in size were concentrated in Ohio, Indiana, and Illinois. Their numbers increased most rapidly in Illinois during the 1860s and in Ohio, Indiana, Illinois, and Iowa during the 1870s. In 1870 they were most characteristic of the wet prairies (fig. 5.3). During the last twenty years of the nineteenth century the number of very large farms decreased a little in Ohio and Indiana, decreased considerably in Illinois, and increased considerably in Wisconsin and Minnesota. In 1900, thousand-acre farms were more widely and thinly spread but were most characteristic of areas west of the Mississippi River (table 5.2). In their heyday in Illinois, many large farms contained extensive areas of pasture with some arable land. A few very large farms, put together by wealthy landowners, were brought into cultivation by their owners and farmed on a large scale with the help of large numbers of hired hands. Some were planned as showplaces to publicize new techniques.

An early model farm was created on prairies in Indiana by Henry L. Ellsworth, federal commissioner of patents. He acquired about 18,000 acres in Benton and Tippecanoe counties in Indiana, and Vermilion and Iroquois counties in Illinois. He began farming in Tippecanoe County in 1836, putting his son Edward in charge of managing the business. Many laborers were employed on works of reclamation and improvement. Large sums of money were invested in labor-saving machines and equipment. Ellsworth was among the

---

70. Bogue, *From prairie to corn belt,* 1963, 286.

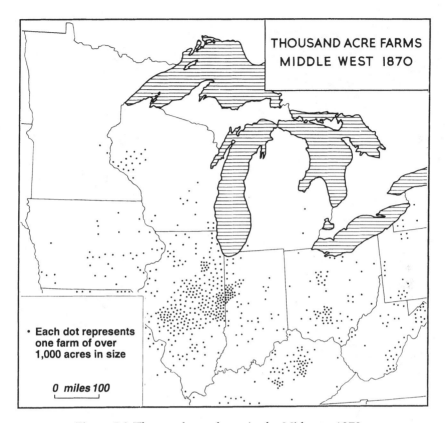

Figure 5.3 Thousand-acre farms in the Midwest, 1870

Source: U.S. Census Office, *Ninth census: Productions of agriculture, 1870* (Washington, D.C. 1874).

first to use a steel breaking plow and pioneered the use of machines for seeding, cultivating, and harvesting. In 1838, he published a booklet drawing attention to benefits to be gained by improving prairies.[71] The guide described at length the making of large farms and cultivation of specialized cash crops such as flax, sugar beet, tobacco, and hemp, while neglecting staple crops of corn and small grains grown by average settlers. It was addressed to people with substantial capital, arguing that prairies could be cultivated most ef-

71. Henry W. Ellsworth, *Valley of the upper Wabash, Indiana, with hints on its agricultural advantages: Plan of a dwelling, estimates of cultivation, and notices of labor saving machines,* 1838; Gates, *Landlords and tenants,* 1973, 118–21.

Table 5.2 Number of farms 1,000 acres and over, 1860–1900

	1860	1870	1880	1890	1900
Ohio	112	69	252	163	164
Indiana	74	76	275	241	224
Illinois	194	302	649	383	282
Iowa	10	38	364	428	340
Michigan	3	5	84	87	136
Wisconsin	11	32	109	109	365
Minnesota	–	2	145	282	365
Midwest total	404	524	1878	1693	1876
% change		+30	+258	–10	+11
U.S. total	5,364	3,720	28,578	31,546	42,276
% change		–31	+668	+10	+34

Source: U.S. Census Office, *Tenth census, 1880, Compendium,* 1883, pt.1: 653; *Twelfth census: Agriculture,* 1902, pt. 1, summary.

ficiently and profitably at a large scale using machinery. The panic in 1837 forced the Ellsworths to curtail costly experiments in farming in hand, but the estate survived. In the 1840s, as prosperity returned, more wet prairie was acquired. Between 1847 and 1852, another 73,500 acres in western Indiana were bought for 65 cents to $1.10 an acre and in the 1850s Henry Ellsworth greeted the arrival of a steam plow with undiminished optimism. He was reported to be "as delighted as a child with a new toy" and predicted widespread advances in prairie farming.[72] Unfortunately, he had insufficient funds to carry out ambitious projects he had planned. The estate was badly hit by the panic in 1857 and was broken up after Henry Ellsworth's death in 1858.

A succession of publicity-seeking ventures was launched by Michael Sullivant. In the 1830s he ran a successful 5,000-acre farm near Columbus, Ohio, raising 2,300 acres of corn, feeding high-quality cattle. He helped to organize the Ohio Company for Importing English Cattle and paid record prices for a pedigree Durham bull and two Durham cows. In the 1850s he rented this profitable business on a share basis and, with his brother, set out to buy unimproved land on the Grand Prairie in Illinois. From federal land offices, from the newly incorporated Illinois Central Railroad, and

72. Gates, *Landlords and tenants,* 1973, 119.

from speculators willing to take a quick return for cash, they acquired 80,000 acres in Champaign, Ford, Livingston, and Vermilion counties. They marked out a compact block containing 22,000 acres in Champaign County for development as a large farm. Michael Sullivant's flair for putting on a show was exhibited in his marshaling a convoy of nine heavy wagons that were to serve as temporary accommodation for the pioneers. Accompanied by 35 men, 30 horses, and cartloads of implements, the convoy left Columbus in February 1855, arriving at the site of the planned farm, appropriately named Broadlands, in time for spring plowing and sowing. Additional laborers were recruited on the spot as well as fresh teams of horses and mules to speed the task of breaking the prairie, seeding corn, and setting up fences. From 1855 to 1866 the farm operated on a giant scale. In the early 1860s between 100 and 200 laborers were employed during the summer; 200 horses and mules and a large herd of oxen were kept for plowing and cultivating; 5,000 cattle and 4,000 worn-down government horses were pastured. Vast fields were laid out; 1,800 acres were sown to corn, 340 acres to other grains; cultivation was performed by the latest machines. From the outset, the project carried huge debts: $225,000 was borrowed for land purchases, some on generous credit terms from the Illinois Central Railroad, some bearing high interest charges; further loans totaling over $250,000 were raised to pay for breaking, fencing, and ditching as well as for stock and materials. During the Civil War, business flourished; after paying all charges, it yielded a profit of $80,000. Then prices of produce fell and in 1866, Broadlands had to be sold to pay outstanding interest. Undaunted by this blow, Michael Sullivant immediately embarked on an even bolder scheme. He moved to a 40,000-acre tract, more distant from railroads, in a largely undeveloped area in Ford and Livingston counties. Again, he began reclaiming land on a prodigious scale, laying out fields thousands of acres in extent cultivated with the largest array of machinery ever deployed on a single farm. In the course of ten years, Burr Oaks became the largest and most celebrated farm in America. In 1871, a reporter from *Harper's Weekly* noted that 16,000 acres were growing grain; corn, the leading crop, yielding a harvest of 600,000 bushels. He added:

The machinery in use at Burr Oaks would handsomely stock two or three agricultural implement stores: 150 steel plows, of different styles; 75 breaking

plows; 142 cultivators, of several descriptions; 45 cornplanters; 25 gang har-
rows, etc. The ditching-plow, a huge affair of eighteen feet in length, with a
share of eleven feet by two feet ten inches, is worked by sixty-eight oxen and
eight men. These finish from three and a half miles of excellent ditch each day
of work.[73]

A labor force of between 200 and 400 single men, mostly
Swedes and Germans, engaged from April to January, were housed
in rough barracks. All the buildings looked starkly utilitarian and
meanly furnished in contrast to the imposing power of the machines. As with the first venture, the panic of 1873 precipitated a
crisis of falling agricultural prices, a mounting burden of debt, and a
decline in the value of land. Other unwelcome troubles beset the estate in the 1870s, including quarrels with tax assessors in both
Livingston and Ford counties. In Livingston County, a dispute arose
over Sullivant's failure to pay his share of a tax levied to award
bounties to Civil War volunteers. In Ford County, small farmers ob-
jected that Sullivant gained an unfair advantage in that valuations
placed on his horses were about 30–50% lower than theirs.[74] A series of poor harvests combined with management difficulties finally
forced the farm into liquidation in 1879. It was broken up into
small parcels.

The fundamental weakness of large-scale farming operations
was their dependence on increasingly heavy borrowing. They failed
to grow to a point where loans could be repaid. Whenever they
made profits, they needed to invest in further improvements.
Whenever they made losses, the value of land on which loans were
secured fell and they sank into negative equity. The problem was
partly caused by short-term planning. Lenders were reluctant to extend credit over a sufficiently long period to earn sustained returns
on their investments. Borrowers were constantly reviewing their capital requirements, attempting to reschedule loans, struggling to meet
rising interest charges. Intensification of land use was directly related
to capital investment. It was cheaper to break the sod with a steel
plow than by ax, cheaper to fence with barbed wire than with laboriously planted and cut hedges, cheaper to sow and reap by machine than by hand, yet all these machines were costly investments.

73. Ibid., 249–55, quotation on 252–53 from *Harper's Weekly* 15 (23 September 1871)
897–901.
74. Bogue, *Patterns from the sod,* 1959, 243–44.

Wet prairies suffered from strong and persistent prejudices and they remained distant and difficult to reach from large cities. They were doubly handicapped, rating high in need for investment and low in public esteem.

Wheat Growing

In attempting to obtain large returns in the shortest time from the smallest outlay on preparing the soil and sowing seed, some farmers decided to grow wheat. In the 1830s and 1840s, on prairies in Illinois cultivation was notoriously slovenly. An apocryphal account tells of a farmer riding through a cornfield on horseback broadcasting seed from a basket on uncultivated ground. From the beginning, serious doubts were expressed about the suitability of prairie soils for wheat. Were they too rich? Was drainage too poor? Were winters too severe for winter wheat? Farmers themselves most frequently complained about ravages caused by pests and diseases. In Illinois, rust, blight, Hessian fly, and to a less extent, chinch bugs, afflicted winter wheat between 1847 and 1852. In Iowa and Minnesota in later years grasshoppers destroyed spring-sown crops.[75] Crops were frequently damaged by hard winters or by waterlogged soils in spring and early summer. On the other hand, higher than average yields were obtained from wet prairies in dry seasons when other areas produced less than average. Exceptionally good harvests tended to occur in years of general scarcity. Added to uncertainties of soil, climate, and diseases, violent fluctuations in market prices heightened financial risks. Prices were exceptionally high between 1853 and 1856 when supplies from the Russian steppes were cut off during the Crimean War. They sank to unprofitably low levels from 1857 to 1862 and rose again from 1863 to 1865 during the Civil War, to slide down again in the postwar years. A gradual recovery in the 1870s was accompanied by an increase in output, reaching a peak in 1880.[76] In the face of keen competition from new lands growing spring wheat in Kansas, Nebraska, the Dakotas, and Canada, production in Iowa and Illinois declined rapidly. From the late 1840s until its final demise as a cash crop in the 1880s, wheat growing had been recognized as a big gamble. In 1847, Henry L. Ellsworth, commenting on the performance of his model farm at

75. Bogue, *Prairie to corn belt,* 1963, 123–29.
76. Bogue, *Patterns from the sod,* 1959, 123–29.

Lafayette, Indiana, wrote: "The profits of wheat appear well in ex-
pectation on paper, but this prospect is blasted by a severe winter,
appearance of insects, a want of harvesting, bad weather in harvesting, in threshing . . . and lastly, a fluctuation of the market itself."[77]
Looking back at a boom-and-bust cycle from the perspective of
1859, the editors of *Illinois Farmer* noted: "The high prices obtained for wheat and the abundance of wheat [in the years 1854 to
1856] stimulated the purchase of large tracts of land, on credit, and
hiring large amounts of help, and all this mostly on credit . . . We
know that the wheat crop has failed for two years to a great extent—
the farmers are unable to pay their debts constantly accumulating."[78] In the prairies of Indiana and Illinois, few if any farmers relied solely on wheat as a cash crop and few emerged as leading
wheat producers even in boom years in the mid-1850s and mid-
1860s. Many lost money when prices failed to cover production
costs, and a few, including Michael Sullivant, attributed their down-
fall to harvest failure and market weakness.[79]

Hired Labor

A deeper problem for large-scale farming was labor supply. Between
1860 and 1880, population increase in the Midwest slowed down
(table 5.3). In the Civil War decade, the rate of increase was still
ahead of the national average, but during the 1870s the rate of increase in the Midwest fell behind that for the nation as a whole.
Growth in rural areas failed to keep pace with urban growth, and a
reserve army of young people seeking homes and work on farms
trekked westward to the Great Plains and across the Rockies to
California.

In the prairies, increasing numbers recorded in the population
censuses as working in agriculture were not farm owners, but laborers. In 1860, Iowa, a state scarcely open to settlement, whose density of population was less than two persons per square mile, with
less than one-third of its land in farms, most of whose public domain was already disposed of, reported 40,827 farm laborers. In
this frontier state, 23% of those engaged in agriculture were hired

77. Commissioner of Patents, *Annual Report, 1847,* 530, cited in Gates, *Landlords and tenants,* 1973, 120.

78. Bogue, *Patterns from the sod,* 1959, 96.

79. Gates, *Landlords and tenants,* 1973, 227.

Table 5.3 Population of the United States, 1860–80

	1860	1870	1880
Ohio	2,339,511	2,665,260	3,198,062
Indiana	1,350,428	1,680,637	1,978,301
Illinois	1,711,951	2,539,891	3,077,871
Iowa	674,913	1,194,020	1,624,615
Michigan	749,113	1,184,059	1,636,937
Wisconsin	775,881	1,054,670	1,315,497
Minnesota	172,023	439,706	780,773
Midwest total	7,773,820	10,758,243	13,612,056
% change		+38	+27
Rest of U.S.	23,669,501	27,800,128	36,577,153
U.S. total	31,443,321	38,558,371	50,189,209
% change		+23	+30

Source: U.S. Bureau of Census, Seventeenth Census of Population, 1950 (Washington, 1952), vol. 1: 1–8, 1–9.

hands. In Ohio, Indiana, Illinois, and Wisconsin, farm laborers accounted for between 20% and 28% of those enumerated in agricultural occupations.[80] A growing number of people in prairies were immigrants from northern Europe: Irish, Germans, Swedes, and Norwegians. Some were employed in railway construction and other building work; some sought work on farms, hoping that they might learn local farming methods and save enough to equip themselves to set up as farmers on their own account. By working for people born in the United States or earlier immigrants from their own countries, they offered themselves a period of acculturation and practical training.[81] In reality, not all laborers lodged with family farmers or worked the whole year round. Some were hired by large estates during the summer and fall, housed in bare barracks affording virtually no privacy and few amenities. Others joined teams of sodbusters, fence erectors, well diggers, barn raisers, or gangs of migratory haymakers and harvesters. These casual workers received no formal instruction in the arts of husbandry and gained little experience of farm management. Financially they fared badly. Many were paid on a piecework basis, hired by the day, dismissed at the

80. Ibid., 304.
81. Bogue, Prairie to corn belt, 1963, 238.

end of a job. Those who enlisted on monthly terms were charged for food, accommodation, and breakages.[82] Employers combined to keep down wages and rates for piecework but would not allow employees to join trade unions.[83] Few laborers were able to save sufficient out of their earnings to buy a farm. Those who aspired to climb a farming ladder took the first step by acquiring tenant holdings, spending all their savings on stock and tools. In 1860, a 30-year-old laborer in the Midwest might hope to become a tenant at the age of 36.9 years, a part owner at 40.2 years, and an owner-occupier at 42.8 years.[84] Accumulated wealth enabled workers to move slowly up the ladder. Poverty denied them that opportunity.

Employers had their own catalogues of grievances against laborers. They regarded hired hands as irresponsible and idle, careless with tools, cruel to animals, neglectful of their tasks, frequently drunk, brawling, insubordinate, demanding high wages, threatening to quit. Isaac Funk, a leading landowner in McLean County, Illinois, declared in 1861 that "no one could afford to hire men to grow and market grain at prices then prevailing."[85] Neighbors noted a depressing aspect of the country owned by Sullivants and Alexanders in Ford and Champaign counties, Illinois: "without roads, but few dwellings, cultivated by hirelings who have no interest in the work."[86] Many large employers decided to dispense with unreliable hirelings, substituting machines for hands. Early machines required much labor to operate and service and not until the 1870s were real savings in labor achieved. On some estates, tenants were paid to help in cultivating, haymaking, and harvesting. Some small farm owners worked part-time on neighboring farms and some workers from towns came in summer to spend a few days or weeks in the fields. After brief spells of labor shortages in the 1850s and 1860s, the labor market was flooded with job-seekers during most of the 1870s and 1880s. In the late 1870s, some unemployed laborers in southern Wisconsin, central Iowa, and southern Minnesota resorted

82. Bogue, *Patterns from the sod,* 1959, 81, 95; Bogue, *Prairie to corn belt,* 1963, 182–84, 247–48.

83. Gates, *Landlords and tenants,* 1973, 324.

84. Jeremy Atack, The agricultural ladder revisited: A new look at an old question with some data for 1860, *Agric. Hist.* 63 (1989) 10–11.

85. *New York Tribune,* 30 July 1861, cited in Gates, *Landlords and tenants,* 1973, 316.

86. J. S. Lothrop, *Champaign County directory, 1870–1871* (Chicago 1871) 419, cited in Gates, *Landlords and tenants,* 1973, 227.

to breaking machines and setting fires.[87] At the end of the nineteenth century, competition from manufacturing and other employment absorbed the surplus. Once again, farmers had difficulty in recruiting suitable workers at rates they could afford.[88]

Tenant Farmers

In the 1860s and 1870s, railroad companies and state and county governments were exerting pressure to raise prairie land prices: railroads in order to pay for extending their lines, local governments in order to build roads and schools. Efforts by corporate bodies and state agencies to increase land prices were generally welcomed by private landowners. Individuals were able to borrow more money on the security of more valuable land and might expect public authorities to lower, or at least not raise, taxes. All landowners were tax resisters, opposing introductions of new imposts, striving to abolish or reduce rates of old taxes. Some large landowners were accused of tax avoidance and evasion. Some were in arrears with payments and some were forced to sell land in order to pay outstanding liabilities.[89] In the third quarter of the nineteenth century, landowners were tightly squeezed for cash: interest charges and taxes continued to move upward; prices for store cattle and wheat moved up and down but more sharply downward than upward. Ranchers from Texas to Kansas were driving out of business cattlemen from the Grand Prairie of Illinois, and wheat growers in the Dakotas and Canadian prairies were underselling competitors in the Midwest. Large estates had to reorganize their activities as a matter of urgency. Model farms and experimental farms were too costly to maintain and had to be broken up. New machinery was expensive but labor was far less efficient. Large farms worked with hired labor had to cut their workforces or convert to tenant farming. Cattle ranches had to sell or be divided into tenant holdings. The most frequent adjustment was to create tenant farms.

From the early years of the federal land system many critics feared that a right to enter and purchase unlimited quantities of pub-

---

87. Peter H. Argersinger and Jo Ann Argersinger, The machine breakers: Farmworkers and social change in the rural Midwest of the 1870s, *Agric. Hist.* 58 (1984) 400–407.

88. Bogue, *Prairie to corn belt,* 1963, 182–87; Bogue, *Patterns from the sod,* 1959, 139; Gates, *Landlords and tenants,* 1973, 165, 313–14.

89. Bogue, *Patterns from the sod,* 1959, 223–52.

lic land for investment purposes would encourage large-scale speculation, cause partial monopolization, and lead to tenant farming and
sharecropping. Some large purchasers set out with a deliberate intention of setting themselves up as landlords of tenant-farmed estates, among them Henry L. Ellsworth, Matthew T. Scott, Daniel
Webster, and the notorious Irishman William Scully. Most of these
new estates were situated on prairies in Indiana, Illinois, Iowa,
Michigan, and Wisconsin, the largest of all occupying the Grand
Prairie of Illinois and adjoining parts of Indiana. From the outset,
Ellsworth planned to improve a large part of his property by inviting
tenants to break prairies, erect fences, put up buildings, and pay
rents amounting to one-third of the crops they produced. He looked
for "smart, enterprising young men from the New England states to
take the farm on shares."[90] He was prepared to make allowances
for fencing or ditching, stipulating that the capital must be repaid
before profits were shared. In 1858, Ellsworth's own model farm in
Tippecanoe County, Indiana, was subdivided into large but more
manageable tenant holdings.[91] Matthew Scott acquired nearly
10,000 acres in Livingston and McLean counties, Illinois, between
1853 and 1876. Like Ellsworth, he attempted to farm in hand,
growing wheat, and sustained heavy losses. He learned the advantages of substituting tenant labor for wage laborers, taking rents in
the form of shares of profits. During the second half of the nineteenth century, the number of tenants on Scott's estates in Illinois
and Iowa multiplied tenfold and the family retained possession of
the land.[92]

A transition from large-scale grazing to growing corn as feed for
cattle was frequently accompanied by a breakup of ranches into
several large farms rented to tenants. A subdivision of large estates
coincided with a general reduction in average size of farms (table
5.4). In wet prairie areas in Ohio, Indiana, Illinois, and Michigan,
farms became smaller and more frequently occupied by tenants than
in Wisconsin or Minnesota. In Madison County, Ohio, tenant farms
were established on former grasslands as early as the 1840s.[93] In
Indiana and Illinois, Paul Wallace Gates has traced the conversion

90. Letter from Henry L. Ellsworth to his son, Henry W. Ellsworth, 1 January 1837,
quoted in Gates, *Landlords and tenants*, 1973, 241.
91. Gates, *Landlords and tenants*, 1973, 5–6, 124, 131.
92. Bogue, *Patterns from the sod*, 1959, 99.
93. Hudson, *Making the corn belt*, 1994, 65.

Table 5.4 Average size of farms (acres), 1850–80

	1850	1860	1870	1880
Ohio	125	114	111	99
Indiana	136	124	112	105
Illinois	158	146	128	124
Iowa	185	165	134	134
Michigan	129	113	101	90
Wisconsin	148	114	114	114
Minnesota	184	149	139	145
Midwest average	152	132	120	116
U.S. average	203	199	153	134

Source: U.S. Census Office, *Tenth Census, 1880: Compendium*, 1883, pt. 1: 657.

of ranches into tenant farms back to the 1850s. When former cattle kings Isaac Funk and Jacob Strawn died in 1865, their estates had ceased to be ranches and were occupied by tenant holdings.[94] New landlords were not only freed from responsibility for day-to-day management decisions, but they were rewarded with shares in increasing production that tenants succeeded in winning. Between them, Gates and Margaret Beattie Bogue followed the declining fortunes of over twenty cattlemen on the Grand Prairie of Illinois and Indiana, all of whose estates between about 1860 and 1880 were split among tenants.[95] Cattlemen who could afford to build spacious farm buildings and put in fences and ditches were more likely to stay in business as landlords than those who relied on tenants to carry out improvements.

Tenants were also recruited from farmers who had fallen behind with their mortgage payments. Lenders who foreclosed on mortgages were sometimes willing to take on defaulting borrowers as tenants. Some farmers owed money for improvements they had made and claimed compensation from their creditors for the value added. If these former debtors were allowed to occupy the land as tenants, they continued to enjoy the use of improvements, and questions of compensation could be taken into account in fixing rents and terms of leases. Many early settlers who bought more land from

94. Gates, *Landlords and tenants,* 1973, 206.
95. Gates, Hoosier cattle kings, 1948, 1–24; Gates, Cattle kings in the prairies, 1948, 379–412; Bogue, The cattlemen's domain, in *Patterns from the sod,* 1959, 48–84.

government land offices than they could afford or had the physical strength to farm, eventually leased or mortgaged one or two unimproved quarter sections. Some of these newly created holdings were taken by friends or kinsfolk. On retirement, many small farmers passed farms to their children in return for shares in the produce. During the lifetime of their parents, such heirs were nominally tenant farmers.[96] Tenancy for some was a step down the farming ladder, a step between ownership and dispossession; for others, including heirs to family farms, it was a step toward full ownership.

The burdens of tenancy were onerous and it was very difficult for sharecroppers to climb the farming ladder to freehold ownership. In 1852, urging Congress to reform federal land laws, Richard Yates declared: "There are thousands of tenants in the Western country with large families, who are unable to make a dollar over and above the amount required for the support of themselves and families, after paying the owners of the soil one third of the proceeds of their annual labor."[97] In the middle years of the nineteenth century, the numbers of tenants may be deduced from manuscript census returns, but no totals for states, counties, or townships were compiled or published.[98] Contemporary commentators noted that tenancy was increasing, and increasing most rapidly on wet prairies. The subdivision of large estates and mounting debts among large and small farmers were concentrated in these districts. By 1880, when statistics were first published in the census, Illinois had 31% of its farms operated by tenants, followed by Indiana and Iowa, each having 24%.[99] In prairie counties, the incidence was much higher than elsewhere, rising above 40% in some parts of northern Illinois and Indiana. In Benton County, Indiana, where Henry Ellsworth founded his estate, now subdivided, 45% of all farms were rented. In Ford County, Illinois, the tenancy rate was 45% and in neighboring Champaign County, 40%. This was where Michael Sullivant once farmed Broadlands and Burr Oaks on a grand scale. Logan County, Illinois, the heart of Scullyland, as William Scully's vast estate was often called, reported 50% of its farms occupied by

96. Bogue, *Prairie to corn belt*, 1963, 60.
97. *Congressional Globe*, 32d Cong., 1st Sess., 1851–52, 473, cited in Bogue, *Patterns from the sod*, 1959, 158.
98. Bogue, *Prairie to corn belt*, 1963, 63–65.
99. Gates, *Landlords and tenants*, 1973, 166.

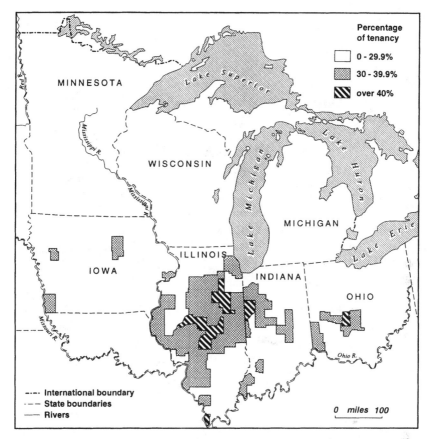

Figure 5.4 Farm tenancy in the Midwest, 1880

Source: U.S. Census Office, *Tenth census: Productions of agriculture, 1880* (Washington, D.C. 1883), vol. 3: 28–101.

tenants. Logan County was surpassed only by Madison County, in a prairie area in Ohio, where 52% of farms were tenanted. In Illinois, over 20% of all farms were reported to be in the hands of tenants in all but four of 102 counties. In stark contrast, no county in Michigan, Wisconsin, or Minnesota had more than 20% and most had under 10% of farms occupied by tenants (fig. 5.4). From 1880 through to the 1930s, tenancy continued to increase. Debt forced some farmers to step down the farming ladder, while young families and immigrants swelled the numbers struggling to get a foot on the bottom rung. An increase in tenancy at the end of the nineteenth century may be attributed to the formation of new small farms out of larger holdings and also to an influx of settlers with insufficient funds to buy farms.

Conflicts of interest between landlords and tenants inflamed widely held objections to tenancy as an institution that deprived Americans of their promised freedom and independence. Landlords felt cheated and injured; tenants felt insecure and exploited. Some landlords described bad tenants as idle, untrustworthy, dishonorable scoundrels who deliberately stole their property. In 1866, a contributor to the *Prairie Farmer* complained:

There is however a growing—and I think well founded—aversion to the letting of land. Men prefer to let their land lie idle, rather than run the risk of letting to unworthy, lazy, stealing tenants. I know there are worthy tenants, but they are few. Men who have for ten or twenty years rented farms, moved from pillar to post, sometimes having a team of their own, and sometimes not having any, I say such men are generally a hard set.[100]

Some tenants complained that landlords let homes unfit for beasts to live in, offered no security of tenure, and took extortionate shares in harvests won by farmers' efforts and skills. Neither tenants nor owners took much care of prairie soils during the second half of the nineteenth century and observers began to comment on signs of accelerated erosion and depletion of plant nutrients. In 1883, a local historian, reviewing prevailing attitudes toward wet prairies in Indiana, noted: "Jasper County is still too new, its soil too little exhausted, to encourage or feel the necessity of a regular system of agriculture."[101] In other words, farmers were prepared to mine the soil until crop yields began to suffer. Also in 1883, an agent for an Eastern landlord writing from Boone County, Iowa, remarked that an unbroken tract of prairie was too valuable "to be butchered by Iowa tenants."[102] Some great estates degenerated into rural slums through both landlords and tenants attempting to extract maximum returns from minimum outlays. Some tenants constituted a floating population, moving from one estate to another, exhausting soils,

---

100. *Prairie Farmer* (31 March 1866) 207, quoted in Bogue, *Patterns from the sod,* 1959, 177.

101. *Counties of Warren, Benton, Jasper, and Newton, Indiana: Historical and biographical* (Chicago 1883), cited in Gates, *Landlords and tenants,* 1973, 138.

102. Bogue, *Prairie to corn belt,* 1963, 62, cites a letter from Stephen A. Foley to John and Ira Davenport, 6 December 1883.

neglecting repairs of buildings and equipment. A few conscientious landlords attracted a superior class of tenants by providing first-rate housing, insisting on a proper rotation of crops, including clover or alfalfa one year in four, clean weeding, and applications of manures, lime, and phosphate at regular intervals.[103] At best, tenant farming enabled a few landowners to survive and provided some hardwork-ing, young immigrant families with an opportunity of making a start in farming. On the other hand, realization of the full potential of wet prairies as productive farmland awaited the beginning of systematic draining. A very small minority of enlightened landlords took that initiative on behalf of their tenants.

Speculation and Disappointment

From about 1830 to about 1880, large tracts of wet prairies in northwestern Ohio, northern Indiana, east central and northern Illinois, much of Iowa, and southern Minnesota were acquired by large landowners, many of whom were absentees unfamiliar with the physical characteristics of the region and inexperienced in local farming practices. Paul Wallace Gates estimates that in the United States between 1835 and 1837, some 38 million acres of public land were sold, of which 29 million acres were bought as specula-tive investments. Sales reached a higher peak between 1854 and 1858, when over 130 million acres were disposed of. About half went to private purchasers, mostly speculators, many of whom resold the land without adding to its value by development. Another half went to states and other bodies to pay for public improvements. In 1870, in the three prairie states of Indiana, Illinois, and Iowa alone, speculators held not less than 20,500,000 acres.[104] Specu-lators treated deeds, conveyances, scrip, warrants, and solid acres of prairie as currency. Clive Bush has written: "Land was seen *as money* and became a primary agent in economic development. It paid off soldiers and debts. It was sold to investment companies to raise revenue and it financed local and state government. It gener-ated working capital in land grants to canal and railroad developers who then resold to other developers."[105] Unlike cash, land gained

103. Gates, *Landlords and tenants,* 1973, 181, 190, 236.
104. Ibid., 56, 63, 155.
105. Clive Bush, 'Gilded backgrounds': Reflections on the perception of space and land-scape in America, in Mick Gidley and Robert Lawson-Peebles (eds.), *Views of American*

added value through development and wet prairies required exceptionally large inputs of capital to develop them. High risks that attended these additional investments in fixed improvements deterred
many speculators. Speculation based on an arithmetic of rising population and diminishing public domain failed to provide safeguards
against sudden financial panics, nor could it overcome deep-seated
prejudices against wetlands. While favored timberlands rose in
value, wet prairies remained stubbornly low-priced. They stayed at
the bottom of the market until draining was carried out. Extensive
use of grasslands by cattlemen and successful cultivation of wheat in
exceptionally dry years were no more than temporary palliatives. As
long as resident settlers were few in number and widely dispersed,
provision of roads, schools, churches, and general stores came
slowly and farmers were burdened with high taxes and endured
shortages of labor. Railroads brought substantial gains in accessibility but they encountered the same problems as other absentee
landowners in attempting to sell unattractive, undrained wetlands.

---

landscapes (Cambridge University Press, Cambridge 1989) 21; Richard N. L. Andrews (ed.),
Land in America: Commodity and natural resource (D. C. Heath, Lexington, Mass. 1979) 30–
38.

6

Draining and Agricultural Change on Wet Prairies

In the early nineteenth century it was widely believed that extensive tracts of wet prairie would remain largely uninhabited, but by mid-century attitudes were changing. Faith was growing that draining could destroy breeding grounds of greenhead flies and mosquitoes, eliminate miasmas associated with the ague, and remove stagnant waters thought to transmit dysentery, typhoid, and cholera. Where wet spots had been successfully drained, insect populations were observed to have been greatly reduced and health of local inhabitants and their domestic animals appeared to have improved. Speculators welcomed rising confidence although they continued to doubt whether they themselves could afford to pay for digging ditches and laying tiles. Newcomers from upper New York State and other areas that had been drained were responsive to the mood of optimism and were willing to make an effort to accomplish the task. Books and farmers' journals helped to disseminate information and practical advice.

Draining was the last in a succession of technological advances that contributed to the transformation of wet prairies, but no cheap, labor-saving method of doing the work was to be found. In 1870, poor drainage remained the most formidable obstacle to further development of upland prairies. A variety of techniques for removing excess water from soils had proved effective in different localities but all were costly in money, materials, and employment of labor. It was evident that all heavy, upland soils might benefit from artificial draining: some could not be cultivated without it and most gained some improvement from it. From a farmer's point of view, all prairies were wet prairies. The wettest spots were first avoided, then used seasonally as open-range grazing land, then ditched and drained for permanent cultivation. As the value of newly reclaimed farmland increased, other, partially waterlogged, heavy soils were provided with drains until the edges of the prairies were reached. By 1920, draining extended beyond the limits of Clyde and Webster soils to cover much of the flat, upland surface. Soils composed of clay, silt, or other fine-grained material might be cultivated with varying degrees of success without draining, but draining ensured regular crops and high yields.

Open Ditches

On an upland prairie, a drainage engineer was not faced with problems of arresting and controlling floodwaters invading the area from upstream. Prairies needed no protection from their streams. Embanking was unnecessary and water could be evacuated from ditches without pumping. The first task was to provide a closely spaced network of ditches to increase the speed at which surplus water was discharged into existing streams. Open-ditch draining alone was often sufficient to ensure great increases in crops, and on many prairies yields were multiplied.

To assist the reclaiming and improving of millions of acres of potentially productive wetlands, the federal government granted to states swamp and overflowed lands within the public domain. States were empowered to sell swamplands granted to them and apply the proceeds to the work of draining and improvement. In practice, the funds were devoted to many other public works, including the construction of roads and the building of bridges, schools, and courthouses, and scarcely any was spent as Congress intended on the provision of ditches and drains. A belief expressed by an Indiana commentator in 1853, that "speculators will buy the land, and will have it drained with the money they pay for it," had little basis in fact.[1] In most localities, costs of providing drainage were greater than states were likely to recover from sales of land. The State of Illinois, following the collapse of an ambitious program of state-funded public works in 1837, was in no position to raise further loans to carry out a state-funded program of draining swampland after 1850. The state devolved responsibility for administering the swamp land grant to county authorities, which had difficulty in borrowing large sums of capital. Most counties passed responsibility for draining to private investors. Counties offered swamplands for sale

1. Harvey L. Carter, Rural Indiana in transition, 1850–1860, *Agric. Hist.* 20 (1946) 110.

at less than the government minimum price, some for 50 cents, some for as little as 10 cents an acre, in the hope that purchasers might undertake improvements. Most of the land was acquired by speculators who added to their large holdings. In the 1850s, the State of Iowa did not have a sufficiently large tax base to fund drainage schemes. Like Illinois, Iowa devolved to counties responsibility for handling the swamp land grant. Vexatious disputes with the federal land office and others over claims to swampland served to discourage improvement. In Minnesota, a state law of 1883 conferred powers upon county commissioners to construct drainage works, and four years later, individual landowners were authorized to set up drainage districts. Throughout the Midwest, higher authorities passed responsibility for draining to lower authorities, and ultimately to interested individuals. The Swamp Land Act accomplished little or no improvement.

Before 1860, individual farmers, without the help of expensive equipment, completed a small amount of open-ditch draining to remedy wet conditions in scattered fields and small depressions. Digging ditches by hand was itself an expensive operation. It was also laborious, dirty work, affording only temporary relief from waterlogging because ditches soon crumbled and became filled with sediment or choked with weeds. In 1818, the earliest ditches were dug on Morris Birkbeck's and George Flower's English settlements in southern Illinois. Morris Birkbeck's son recalled ditching being his father's "favourite operation, and he would never rest until he had completely secured any spot which appeared to be ever wet."[2] Newly arrived immigrants carried out the work. George Flower praised the skills of a team of three brothers, Joseph, Thomas, and Kelsey Crackles, "able-bodied farm labourers from Lincolnshire, who came with a full experience in the cultivation of flat, wet land; and brought with them the light fly-tool for digging ditches and drains, by which a practiced hand can do double the work that can be done by a heavy steel spade."[3] In 1831–32, William Sewall, who

2. Gladys Scott Thomson, *A pioneer family: The Birkbecks in Illinois, 1818–1827* (London 1953) 64, cites a letter dated 29 December 1819.

3. George Flower, *History of the English settlement in Edwards County*, Collections of the Chicago Historical Society (Fergus, Chicago 1882), vol. 1: 130–31, cited in David Schob, *Hired hands and plowboys: Farm labor in the Midwest, 1815–60* (University of Illinois Press, Urbana 1975) 112.

farmed near Jacksonville in central Illinois, had part of his land ditched by a professional ditcher, assisted by two of his own farmhands. Those most adept with digging tools were English, Scottish, German, and, above all, Irish ditchers. Itinerant gangs of Irishmen were regularly engaged in digging canals, and individuals would undertake contract work for farmers to supplement their income in order to buy plots of land. In the 1840s, these men dug ditches for farmers in and around the Black Swamp in Ohio and along the Wabash in Indiana and Illinois; in the 1850s, they were reported at work all over wet prairies in southern Michigan, southeast Wisconsin, and central Iowa, as well as northern Illinois.[4] In addition to cutting ditches, farmers had to build bridges and culverts to provide access to their fields, but the greatest difficulty was to obtain permission to open outlet channels across the land of others. It was not sufficient for a farmer to drain the surplus water from a field if, by doing so, neighboring fields were rendered liable to flood. Indeed, piecemeal ditching could not provide a permanent solution.

Drainage Laws and Public Benefits

Large-scale operations were needed to bring under control the drainage of entire catchment areas. Ditch laws or drainage laws authorized the organization of drainage undertakings which required groups of farmers to participate. Such laws were passed by state legislatures in Ohio and Michigan as early as 1847, and during the next 25 years all other Midwestern states approved similar measures (table 6.1). Ben Palmer has examined the legal history of corporate drainage enterprises with special reference to Minnesota.[5] The earliest drainage law in Minnesota, passed in 1858, simply enabled individuals who associated together for the purpose of digging a drainage ditch to enjoy the status and rights of a corporate body. Later acts from 1883 to 1905 defined powers of drainage enterprises to finance projects, acquire land, where necessary by compulsory purchase, construct and maintain works, and allocate charges among beneficiaries by way of levies and taxes. In 1901, a state drainage commission was set up, both to carry out a topographical

---

4. Schob, *Hired hands and plowboys*, 1975, 113–21.
5. Ben Palmer, *Swamp land drainage with special reference to Minnesota*, University of Minnesota Studies in Social Sciences 5 (Minneapolis 1915).

Table 6.1 Date of first state law authorizing public drainage enterprises
and date of first organized drainage enterprises in the Midwest

State	First law	First enterprises
Ohio	1847	1850s
Michigan	1847	1890s
Indiana	1852	1850s
Minnesota	1858	1880s
Wisconsin	1862	1870s
Illinois	1865	1870s
Iowa	1872	1870s

Source: U.S. Bureau of Census, *Fourteenth Census of the United States, 1920: Agriculture,* vol. 7, *Irrigation and Drainage* (Washington, D.C. 1922) 354.

survey of watersheds as a basis for planning a statewide drainage system and to lay down rules and regulations governing the construction of ditches.

In order to qualify as bodies exercising powers of eminent domain and taxation, drainage enterprises had to show that the proposed improvement was clearly to the benefit of the public at large, not merely to the advantage of a section or private interest. In 1888, the Michigan Supreme Court defined public interest specifically as the protection of public health.

Drain laws which take from the citizen his private property against his will, can be upheld solely on the ground that such drains are necessary for the public health. They proceed on the basis that low, wet, marshy lands generate malaria, causing sickness and danger to the health and life of the people; that when they are of such character as to injure the health of the community . . . the Legislature has the right to protect the citizens . . . and to preserve the public health.[6]

Later judgments found that the building of public highways also contributed to the welfare and convenience of the public at large, and in 1900 the Minnesota Supreme Court decided: "The fact that large tracts of otherwise waste land may thus be reclaimed and made suitable for agricultural purposes is deemed and held to constitute a public benefit."[7] The ruling affirmed that a general public interest was served by reducing areas of wasteland and extending

6. Ibid., 34.
7. Ibid., 52.

productive agriculture, an advancement of private interests being regarded as incidental to this wider purpose. In reality, as Palmer recognized, most drainage enterprises were initiated by landowners
and farmers for their own private gain and "for no public purpose
whatsoever so far as the legal definition of that term is concerned."
He considered that "the courts have seized upon the possible benefits to the public health or welfare as a mere pretext for upholding
legislation which practically everyone agrees to be wise."[8] State legislatures and the judiciary claimed to reflect the views of many, if
not most, people in the Midwest in the early twentieth century.
Landowners who opposed drainage schemes were treated as reac-
tionaries and obstructionists; they were not to be allowed to stand in
the way of agricultural and economic progress. The idea that pre-
serving wetlands for ecological purposes might be in the public
interest was not entertained.[9] A little protection was afforded to
meandered lakes, whose sinuous outlines were recorded on plats of
the original land survey. These lakes were defined as public waters,
reserved for enjoyment by everyone, and excluded from areas subject to artificial drainage. In practice, it was difficult to draw boundaries between wetlands and open water or drain wetlands without
lowering the level of water in lakes. In the light of recent interest in
undoing the work of drainage enterprises, Mary McCorvie and
Christopher Lant have critically reappraised the aims and organization of drainage districts in the Midwest from 1850 to 1930.[10]

Drainage laws provided for a lengthy process of public consulta-
tion and expert scrutiny. A petition required support from a majority of landowners or holders of more than 60% of the area to be
drained. An engineer appointed by the county prepared a preliminary survey and reported on the merits of the proposal. A public
hearing reviewed the scheme, considered amendments, and heard
objections. Plans and costs had to be agreed by all parties and a
panel of viewers was appointed to assess benefits and damages and
apportion shares of costs. Arrangements were then made to carry
out the work and subsequently maintain and repair it. Judicial decisions and later drainage laws strengthened procedures for public

8. Ibid., 56–57.

9. Robert T. Moline, The modification of the wet prairie in southern Minnesota, Ph.D.
diss., Geography, University of Minnesota 1969, 188.

10. Mary R. McCorvie and Christopher L. Lant, Drainage district formation and the loss
of Midwestern wetlands, 1850–1930, *Agric. Hist.* 67 (1993) 30–36.

hearings and clarified provisions for compensating property owners for their losses and damages. Legislation alone did not call into existence large-scale drainage enterprises; it enabled organized bodies to construct works and levy contributions from those who were to benefit from improved drainage conditions. Large outlays of capital were needed to launch an enterprise and returns were uncertain or, at least, deferred.

Large Landowners Invest in Ditching

The initial cost and risk involved in experimenting with different draining techniques, in adopting new methods of cropping and management, were beyond the means of small farmers, such as those already occupying the southern margins of the prairies, notably in Indiana.[11] On the other hand, large landowners in Illinois, hitherto dependent on cattle or cash crops of small grains, might profitably diversify their economy by bringing wetlands into production. Confidence in wheat as a cash crop was shaken by a harvest failure in 1858, which fell at a time when the supply of well-drained land was nearly exhausted. In east central Illinois, a few of the wealthiest landowners took the initiative in cutting ditches across wet prairies. Margaret Bogue remarks that large landowners took the lead "because they already had invested much in the land" and now realized that in order to remain profitable, "further development depended on proper drainage, for which large amounts of capital were required, far more than the average farmer had to gamble."[12] In the 1850s, Matthew Scott, owner of about 5,000 acres of wet prairies, purchased specially built ditching plows, drawn by up to 40 head of cattle, with which he excavated 250 miles of ditch. In 1859, the Illinois Central Railroad began to promote the sale of its wetlands. It organized a competition for a mechanical ditcher, recognizing that only large landowners could afford such expensive machines. It supported the draining operations of Solomon Sturges on 17,000 acres of prairie in 1863 and an equally ambitious venture by A. H. and G. W. Danforth in 1866.[13] The Danforths em-

11. Richard Lyle Power, Wet lands and the Hoosier stereotype, *Miss. Valley Hist. Rev.* 22 (1935) 48.

12. Margaret Beattie Bogue, *Patterns from the sod: Land use and tenure in the Grand Prairie, 1850–1900* (Illinois State Hist. Soc., Springfield 1959) 136.

13. Paul Wallace Gates, The promotion of agriculture by the Illinois Central Railroad, 1855–1870, *Agric. Hist.* 5 (1931) 64; Gates, *The Illinois Central Railroad and its colonization*

ployed an expensive ditching machine capable of cutting three- to four-foot ditches. An early American revolving wheel ditch-digger, patented by R. C. Pratt of Canandaigua, New York, was demonstrated at the New York State Fair in 1854. It is illustrated in Henry French's handbook on farm drainage.[14] By 1860, ditching machines were at work in McLean County, Illinois, and during the next few years most wet prairies there were sold.[15] A few landowners dug ditches across their wet prairies and farmed them on their own account, with the aid of machinery and gangs of hired labor; but a majority converted their large holdings into smaller farming units of 80 to 160 acres. Some rented these to tenant farmers; others sold them as improved farms. Draining was generally initiated by great landowners who could raise capital and obtain equipment necessary to carry out work on a large scale. By draining and farming intensively they not only made larger profits than from cattle ranching or wheat growing but also enhanced the value of their property. As the value of land increased, it became easier to finance further improvements by raising rents or mortgaging parcels of land.

Before 1870, progress in ditching was intermittent and uneven. English observers concluded that it was practically unknown west of Chicago, and even to the east it was regarded as prohibitively expensive on all but the best and most accessible lands.[16] Most ditches were still dug by hand and only the wealthiest landed proprietors in Illinois were able to afford ditching machines or feed the large teams of horses or oxen required to draw them. Close to Chicago, ditches across the Kankakee Marsh were dug by hand in 1870, but a steam dredge greatly accelerated progress after 1884.[17] The patchiness of early draining efforts owed less to considerations of cost and labor

work (Harvard University Press, Cambridge 1934) 239, 240, 291, 292; Bogue, *Patterns from the sod,* 1959, 136.

14. Henry F. French, *Farm drainage* (A. O. Moore, New York 1859) 249.

15. Arthur Weldon Watterson, *Economy and land use patterns in McLean County, Illinois,* Department of Geography Research Paper 17 (University of Chicago, Chicago 1950) 75.

16. Harry J. Carman, English views of Middle Western agriculture, 1850–1870, *Agric. Hist.* 8 (1934) 16.

17. Alfred H. Meyer, The Kankakee "marsh" of northern Indiana and Illinois, *Papers Mich. Acad. of Science, Arts and Letters* 21 (1935) 381; G. L. Gillespie, Examination of a route for a canal from Lake Michigan to the Wabash River, Indiana, in U.S. Congress, House of Representatives, *Report of the Chief Engineers,* 44th Cong., 2d Sess., Ex. Doc. 1, pt. 2, 1744 (GPO, Washington, D.C. 1876) 454–63.

than to the inadequacy of drainage procured by open ditches. Only the narrowest strips were satisfactorily drained by ditches alone. What was needed, above all, on stiff clays and silts, on the broad open prairies, was an effective means of underdraining to provide large fields with supplementary networks of subsurface drains.

There is no evidence that plowing land into ridges and furrows, once practiced in New York and Pennsylvania, was carried out in the Midwest.[18] In a few areas of heavy soils in Illinois, Iowa, and Minnesota, closely spaced random ditches, about one foot deep, were drawn by plow. A widely publicized and more effective method of draining impervious subsoils was to lay stones, brickbats, rails, or brushwood at the bottom of a trench and backfill the excavated earth. The cost of transporting aggregate was high, however, and the drains silted up after a few years.

A new method, mole draining, introduced in 1854 to Macon and Piatt counties, Illinois, was highly praised in the *Illinois Farmer*. A mole drainer was described as "a wedge of iron . . . attached to a sharp coulter, some three or four feet long . . . fastened to a frame, so as to work above the surface of the ground. In lands inclined to be wet . . . this instrument is plunged into the ground the desired depth, and with two yoke of cattle attached to a windlass, it can be forced readily through the earth at the rate of one-half mile a day."[19] In the late 1850s, seven firms manufactured and advertised such implements for sale at prices ranging from $100 to $175 each. The machine bored a hole about four to six inches in diameter at a depth of between three and four feet. The walls of the bore would remain smooth and hard for several years unless broken by roots or burrowing animals or disturbed by frost and thaw. On stiff clays in northern Illinois, mole draining was claimed to have had varying success. In southern Illinois it was reported to be of doubtful value. In claylands in Michigan, where frost not infrequently penetrated to the subsoil, the operation was ineffective because the bores collapsed or filled with mud. At best, mole draining offered only a temporary respite from waterlogging, and a majority of landowners made no attempt to reclaim wetlands by this method. By 1863, more than half the mole-draining schemes in Cedar County, Iowa,

18. John R. Haswell, Drainage in the humid regions, in U.S. Department of Agriculture, *Soils and men: Yearbook of agriculture, 1938* (GPO, Washington, D.C. 1938) 726.

19. Bogue, *Patterns from the sod,* 1959, 122.

had failed and the practice was discontinued.[20] In the last quarter of the nineteenth century, it was superseded by the more effective remedy of tile draining.

Tile Draining

From impermeable strata were dug the very materials with which the draining of claylands was accomplished. From these clays were manufactured cylindrical tiles, a foot or more long and from two to eight inches in diameter. Tiles were more durable than either open ditches or mole drains. They rarely collapsed and hardly ever required cleaning. They were laid end to end at the bottom of carefully leveled trenches generally not less than four feet deep. The most efficient spacing of tile lines was calculated on the basis of the gradient of the outfall, the depth of the trench, and the permeability of the subsoil. On a more or less level surface, at a depth of four feet, lines no more than 40 feet apart might be required where the subsoil was clay and 60 to 100 feet apart for loams. The earliest tiles were two inches in diameter but experience showed that bores of from four to five inches performed much more effectively and, in the twentieth century, concrete tiles with diameters of six inches or more became standard. Many early schemes were underdesigned and less than optimally efficient. Tiles manufactured in the late nineteenth century were of higher quality than earlier products and tilers were much more highly skilled. Trenches had to be accurately surveyed and leveled and tiles laid with care to ensure that their ends fitted closely together and rested firmly on a smooth surface. Water entered tiles at the joints from the lower side. The upper side was blinded with topsoil containing organic matter packed tightly over the joints to seal them. Unlike ditches, which were required now as main outlet channels, tile lines did not interfere with cultivation: they occupied no space at the surface; they presented no danger to machines turning at the edges of fields, nor did they sever fields from access roads, thus saving the need for bridges and culverts. The cost of laying and replacing tiles was high by any reckoning, more expensive than digging ditches and considerably more expensive than mole draining. Such investment could be justified only on land whose potential value was exceptionally high. The progress of

20. Allan G. Bogue, *From prairie to corn belt: Farming on the Illinois and Iowa prairies in the nineteenth century* (Chicago University Press, Chicago 1963) 84.

tile draining in the Midwest was inevitably slow and unspectacular but the total achievement was gigantic.

John Johnston, who is claimed to have been the first to lay tile drains in the United States, began work on his 300-acre farm at Geneva, New York, in 1835.[21] In 1850, the first tiles were laid in Indiana but little was done there until 1870 and the peak period for tiling did not occur until the last quarter of the century.[22] In 1858, small-scale experiments with tile drains were begun in Champaign County, Illinois, and three years later firms in Chicago and Joliet were advertising tiles—but at prices which, with the addition of freight charges, few farmers could afford.[23] The bold decision of the N. B. Haefer Tile Company to build a factory in the middle of wet prairies in 1862 brought immediate rewards and soon it was one of the largest suppliers in east central Illinois. By 1869, tile factories were operating in Polk and Henry counties in Iowa.[24] In 1870, tiling began in McLean County, Illinois, and reached a maximum in the following decade.[25] The chronology is similar in the Black Swamp of Ohio, beginning about 1870, expanding during the 1880s, and reaching a height of activity during the first decade of the present century.[26] A succession of abnormally wet years gave an added impetus to draining at the end of the 1870s. Ditching and dredging were carried out with renewed vigor, and after 1879 new laws were passed, strengthening the authority and financial powers of organized drainage districts. By 1882, tile factories had become one of the most distinctive features of wet claylands east of Chicago: of 1,140 firms in the United States listed at that date, over 90% were located in Indiana, Illinois, and Ohio[27] (table 6.2). During the next twenty years, hundreds of thousands of miles of tile were laid in the

21. John Johnston, Draining, *Trans. New York State Agric. Soc.* 15 (Albany 1855) 257–59.

22. W. LeRoy Perkins, The significance of drain tile in Indiana, *Econ. Geog.* 7 (1931) 382; over 31,000 miles of tile drains were laid in Indiana before 1882; H. H. Wooten and L. A. Jones, The history of our drainage enterprises, in USDA, *Water: Yearbook of Agriculture, 1955* (Washington, D.C. 1955) 478.

23. Margaret Beattie Bogue, The Swamp Land Act and wet land utilization in Illinois, 1850–1890, *Agric. Hist.* 25 (1951) 178–79.

24. Bogue, *Prairie to corn belt,* 1963, 84.

25. Watterson, *McLean County, Illinois,* 1950, 77.

26. Martin R. Kaatz, The Black Swamp: A study in historical geography, *Annals Assoc. Amer. Geog.* 45 (1955) 23, 30.

27. S. H. McCrory, Historical notes of land drainage in the United States, *Proc. Amer. Soc. Civil Engineers* 53 (1927) 1629.

Table 6.2 Tile factories in operation in the United States, 1 January 1882

State	Number of factories
Indiana	486
Illinois	320
Ohio	230
Michigan	63
Iowa	18
Wisconsin	13
New York	8
Pennsylvania	2
Total U.S.	1,140

Source: S. H. McCrory, Historical notes of land drainage in the United States, *Proc. Amer. Soc. Civil Engineers* 53 (1927) 1629.

Midwest. In Illinois alone, some 117,763 miles were laid between 1880 and 1895, increasing the total length of subsurface field drains from 8,500 to 126,263 miles.[28]

To the east of Chicago great landowners continued to play a leading role in organizing drainage districts and laying tiles during the 1880s and 1890s as they had done in ditching activities in earlier decades. They possessed the means not only to organize large-scale enterprises but also to control outlet channels; the greater the enterprise, and the fuller its control over the outfall, the higher were its chances of success. In 1879, Benjamin Gifford began draining and improving 7,500 acres in Champaign County, Illinois. To supply this vast project, he built his own tile factory. By selling and mortgaging land thus improved, he was able to acquire another 30,000 acres, organize the Big Slough drainage district in 1885, then extend his activities to Ford County and, in 1887, move on to tile drain lands in Jasper County, Indiana.[29] From 1870 to the end of the 1880s, cattle kings in Indiana and Illinois began to dig ditches and lay tiles in order to convert pastures into cornfields. The process accelerated following wet seasons in the late 1870s, and tiling reached a peak in the mid-1880s. Margaret Bogue traces the conversion of a few large-scale cattle ranches in east central Illinois whose owners had made sufficient money from selling livestock to have

28. Ernest L. Bogart and Charles M. Thompson, *The industrial state, 1870–1893*, vol. 4 of C. W. Alvord (ed.), *The centennial history of Illinois* (Chicago 1922) 230.
29. Bogue, Swamp Land Act in Illinois, 1951, 180.

accumulated capital for investment in draining. Unsuccessful cattlemen were forced to sell to newcomers who possessed additional
funds needed for draining. In 1870, wealthy rancher Lemuel Milk
began an elaborate drainage program on his wetlands in Iroquois
and southern Kankakee counties, Illinois, and Newton County,
Indiana; in 1875, another cattleman, John Sidell, began tiling his
6,000-acre tract in Vermilion County; by 1880, the Foos family
were investing heavily in draining and fencing their lands in
Champaign County, where draining continued for over ten years; in
the 1880s, Willy Fowler tile-drained the wetter parts of his 3,000-
acre estate in Vermilion County; also in the 1880s, James Sconce
began tiling his Fairview Farm and, in the late 1880s, the ranch-
owning Fithian family sold part of their land to tile the remainder.[30]
     A shining example of tile draining succeeding in the 1880s,
where attempts to drain by means of ditches and mole drains had
failed in the 1850s, was demonstrated on the estate of Matthew
Scott. Scott's program to lay 26 miles of tile on 3,600 acres in Ford,
Champaign, and McLean counties enabled rents to be increased to
two-fifths of all corn grown, one-third of all small grains, and four
dollars an acre for pastureland. Commenting on this improvement,
Margaret Bogue noted: "It was poor economy indeed to pay heavy
assessments for the construction of dredge ditches and then to neglect the tile which would make them effective. Tile soon more than
repaid the initial expense, for it resulted in increased yields and
larger rental shares."[31] Some landowners, including Thornton K.
Prime, who took a strong lead in draining the Kankakee Marsh in
the 1880s, were able to increase the size of their holdings at the expense of laggard neighbors.[32] Draining large wet prairies in Illinois
and Indiana led to territorial aggrandizement by great estates; it fos-
tered landed monopolies.
     West of Chicago, draining began later than in the eastern sector
and was organized differently. In north central Iowa and southern
Minnesota work began in earnest in the 1880s and 1890s and continued apace during the first and second decades of the twentieth
century. Slow progress was made in many areas in north central

30. Bogue, *Patterns from the sod,* 1959, 73–78.
31. Ibid., 107.
32. Chester McArthur Destler, Agricultural readjustment and agrarian unrest in Illinois,
1880–1896, *Agric. Hist.* 21 (1947) 110.

Iowa before 1884, but in Story County the first schemes were not started until 1895. Tile production in Iowa rose sharply between 1894 and 1907 when the peak was reached.[33] In Blue Earth County, Minnesota, scarcely any draining was done before the end of the nineteenth century. The first drainage district was organized in 1898 and work continued until 1918.[34] In Iowa and Minnesota most farmers were owner-occupiers and the average size of farms in wet prairie counties was about 150 acres in 1880. Cooperation between farmers was essential for the formation of drainage enterprises and borrowing was essential for carrying out tile drainage. Contractors played a much greater part in the West than they had east of the Mississippi. Outlet ditches were dug with giant plow ditchers or Swamp Angels, drawn by large teams of oxen or horses. Steam-powered dredgers, mounted on barges, were introduced at the end of the nineteenth century. Mechanical trench diggers were widely used for laying tiles. These machines were improved designs based on Pratt's model exhibited at the New York State Fair in 1854.[35] Blickensderfer's tile drain ditching machine, manufactured in Decatur, Illinois, dug a four-foot trench in a single pass over the ground. Drawn by one horse, it was operated by two attendants.[36] A steam-powered trencher, invented in 1883, was a prototype for modern trenching machines.[37] In Minnesota, where winter freezing and frost heaving were more severe than in Illinois, deeper trenches were dug and tiles of larger diameter were laid (figs. 6.1 to 6.4). Some contractors were part-time farmers, some were employed by public agencies, others took commissions from tile manufacturers. In wet prairies in southern Minnesota, tiles were produced in small factories employing about ten workers. They supplied farms within a radius of thirty or forty miles. Output of tiles increased rapidly after 1885, attaining a maximum in the decade 1900–1909. Many

33. W. J. Berry, The influence of natural environment in north central Iowa, *Iowa Journ. Hist. and Politics* 25 (1927) 294; Leslie Hewes and Phillip E. Frandson, Occupying the wet prairie: The role of artificial drainage in Story County, Iowa, *Annals Assoc. Amer. Geog.* 42 (1952) 35.

34. Bert Earl Burns, Artificial drainage in Blue Earth County Minnesota, Ph.D. diss., Geography, University of Nebraska 1954, 98.

35. Marion Weaver, *History of tile drainage (in America prior to 1900)* (M. M. Weaver, Waterloo, N.Y. 1964) 165.

36. Ibid., 173.

37. McCrory, Historical notes 1927, 1630.

Figure 6.1 Plow ditcher and sixteen horses near Madison,
Lac Qui Parle County, Minnesota, circa 1915

Source: Collection of Minnesota Historical Society. Photo by F. T. G. Dawson.

firms founded at the beginning of the twentieth century were still in business when Robert Moline conducted a survey of tile manufacturers in 1967.[38]

At the northern margins of this western sector, settlers included many immigrants from northern Europe seeking the cheapest available land to establish homesteads for their families. Large areas of wet prairies and marshes in Michigan, north central Iowa, and southwest Minnesota were acquired by Germans, Irish, Swedes, Norwegians, Finns, Danes, Dutch, Poles, and Bohemians.[39] They maintained close links with their birthplaces, encouraging relatives, church members, and former neighbors to join them in building

38. Moline, Wet prairie in southern Minnesota, 1969, 141–49.
39. Hildegard Binder Johnson, The location of German immigrants in the Middle West, *Annals Assoc. Amer. Geog.* 41 (1951) 23; W. R. Mead, A Finnish settlement in central Minnesota, *Acta Geographica* 13 (Helsinki 1954) 3–16; George J. Miller, Some geographic influences in the settlement of Michigan and in the distribution of its population, *Bull. Amer. Geog. Soc.* 45 (1913) 347; Ralph H. Brown, *Historical geography of the United States* (Harcourt, Brace, New York 1948) 339; Hewes and Frandson, Story County, Iowa, 1952, 32–33.

Figure 6.2 O. F. Doyle dredging outfit from St. Cloud, digging a drainage ditch near Hogdenville, Minnesota, circa 1910

Source: Collection of Minnesota Historical Society. Photo by Carl Graff.

new settlements on the prairies. They formed social structures in the likeness of homeland communities. They were more cohesive and less mobile than native Americans. Cooperation in building churches and schools and establishing drainage districts demonstrated their confidence in, or at least their hope for, lasting unity. Inheritance of family farms expressed their attachment to the land.[40]

40. John G. Rice, The role of culture and community in frontier prairie farming, *Journ. Hist. Geog.* 3 (1977) 155–75; Robert Ostergren, A community transplanted: The formative experience of a Swedish immigrant community in the upper Middle West, *Journ. Hist. Geog.* 5 (1979) 189–212; Jon Gjerde, The effect of community on migration: Three Minnesota townships, 1885–1905, *Journ. Hist. Geog.* 5 (1979) 403–22; Jette Mackintosh, Ethnic patterns in Danish immigrant agriculture: A study of Audubon and Shelby counties, Iowa, *Agric. Hist.* 64 (1990) 59–77; Mackintosh, Migration and mobility among Danish settlers in southwest Iowa, *Journ. Hist. Geog.* 17 (1991) 165–89; Brian Q. Cannon, Immigrants in American agri-culture, *Agric. Hist.* 65 (1991) 17–35. Attachment to land and family succession among communities of German and Yankee origin are discussed in Sonya Salamon, *Prairie patrimony:*

In economic activities, these newcomers had little choice but to pursue American goals. Almost all were small farmers with very little capital. To enable them to drain their farms, drainage districts were financed by banks, land companies, railroads, and a few well-established private landowners. A controlling interest in drainage enterprises soon came to be held by absentee owners, financiers, and city mortgage corporations.

In hard times, immigrants showed greater tenacity than American speculators. Rice Lake Marsh covered about 15,000 acres in Freeborn County, southeast Minnesota. Much of the wetland was granted originally to the Southern Minnesota Railroad. In 1895, the railroad sold 6,500 acres to a Minneapolis businessman, P. D. McMillan. Between 1898 and 1906, an Illinois banker, Bryant Barber, bought adjoining tracts and together with McMillan began draining Turtle Creek. Draining was carried out mostly from 1907 to 1909. After the deaths of McMillan and Barber, the enterprise was bought by George H. Payne of Omaha, Nebraska. In 1921, the Payne Investment Company mounted a campaign to attract Dutch settlers from Michigan and Iowa as well as from Holland itself to take up farms in what was now called Hollandale. A Dutch Reformed church was dedicated in 1922, a school opened the following year, and in 1924, about forty farms produced enormous quantities of sugar beet, onions, potatoes, and other vegetables. During the next few years, the area suffered serious floods and droughts, and the company's financial position deteriorated until it went into receivership in 1929. The Dutch community, united by strong cultural and religious ties, held on and individuals bought their farms from the receivers. In the 1940s, the survivors emerged as freehold owners of family farms. Vegetable growing was abandoned in favor of corn and soybean production.[41]

Perceived Effects of Tile Draining Wet Prairies

The colossal effort of tile draining attracted wide acclaim and inspired a spirit of competition. When mechanical diggers and trench-

---

Family, farming, and community in the Midwest (University of North Carolina Press, Chapel Hill 1992).

41. Janel M. Curry-Roper and Carol Veldman Rudie, Hollandale: The evolution of a Dutch farming community, *Focus* 40 (1990) 13–18.

Figure 6.3 P. G. Jacobson's ditch and tile machine at work on a farm near Madison, Minnesota, circa 1910

Source: Collection of Minnesota Historical Society. Photo by M. J. Viken.

ing machines were introduced, contractors went out looking for new lands to drain. Local tile manufacturers sought new customers for their products. County commissioners and district courts encouraged farmers to organize drainage districts. Fainthearted scrimpers and waverers were coopted willy-nilly into collective enterprises. Many caught the prevailing mood of confidence and were swept along by a new wave of enthusiasm. Neighbors who watched tiles being laid in adjoining fields hurried to join new drainage enterprises. Commentators exaggerated some benefits of draining much as earlier writers exaggerated dangers of undrained wetlands.

From 1880 to 1920 was a period of active propaganda. Farming journals and newspapers published practical notes on tile draining, and important books on draining ran to several editions.[42] A con-

---

42. Moline, Wet prairie in southern Minnesota, 1969, 151–67.

Figure 6.4 Steam trench digger and tile layer in the Red River valley, western Minnesota, circa 1918

Source: Minnesota Historical Society. N.P. League Collection.

certed effort was made to win minds of farmers to the advantages of tile draining. Publications addressed to farmers dealt exclusively with benefits to agriculture; no mention was made of public health, improved roads, or other conveniences. John Klippart's *The principles and practice of land drainage,* a third edition of which appeared in 1888, listed twelve benefits to be derived from draining:

1. Removes stagnant water from the surface
2. Removes surplus water from the undersurface
3. Lengthens the seasons
4. Deepens the soil
5. Warms the undersoil
6. Equalizes the temperature of the soil during the season of growth
7. Carries down soluble substances to the roots of plants
8. Prevents freezing out or heaving out
9. Prevents injury from drought
10. Improves the quality and quantity of crops

11. Increases the effect of manures
12. Prevents rust in wheat and rot in potatoes.[43]

Probably the greatest benefit conferred by artificial draining resulted from aeration of the soil. Organic matter decomposed more
rapidly, roots penetrated more deeply, growing seasons lengthened,
effects of frost, floods, and droughts diminished.[44] Draining, by
lowering the water table, induced roots of plants such as corn to
reach greater depths. In undrained soils, where a high water table
persisted into late spring and early summer, roots spread laterally,
little below the surface (fig. 6.5). When a drought occurred in late
summer the level of groundwater fell, but the rooting system developed by fully grown plants extended no deeper than the now desiccated topmost layers of the soil. In a deeply drained soil, seedlings
had to put down roots to the level of groundwater in early spring.
As young plants they grew a mass of strong hairy roots which sustained them through dry periods. Roots also drew nutrients from
greater depths and formed capillary channels through which air,
moisture, and organisms moved up and down through the soil.

High expectations that draining would eliminate malaria were
not fulfilled. It was claimed that on Delavan and other prairies,
where the disease had been endemic, it became almost unknown by
1880 as a result of widespread draining.[45] The whole of northern
Illinois was said to be largely rid of it by 1890.[46] Before draining,
outbreaks were intermittent; after draining, its incidence was reduced but deaths still occurred.[47] Data on mortality from malaria
for 1870, 1880, and 1890 in sixteen counties in and around the
Grand Prairie in Illinois have been examined by Roger Winsor.[48]

43. John Klippart, *The principles and practice of land drainage,* 3d ed. (R. Clarke,
Cincinnati 1888), 42.

44. Nyle C. Brady, *The nature and properties of soils,* 8th ed. (Macmillan, New York
1974) 233–34, 254.

45. John K. Rose, Delavan Prairie: An Illinois corn belt community, *Journ. Geog.* 32
(1933) 2.

46. John Duffy, *The healers: A history of American medicine* (University of Illinois Press,
Urbana 1979) 309.

47. Erwin H. Ackerknecht, *Malaria in the upper Mississippi valley, 1760–1900* (Baltimore
1945) 29.

48. Roger A. Winsor, Environmental imagery of the wet prairie of east central Illinois,
1820–1920, *Journ. Hist. Geog.* 13 (1987) 389.

Undrained

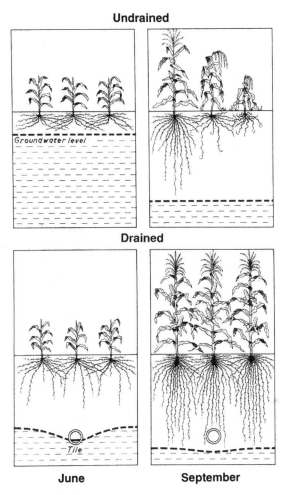

Drained

June   **September**

Figure 6.5 Effect of tile draining on corn root growth

Tile drains lower ground water in spring. In undrained soils, root growth is stunted by high levels of ground water in spring and plants wilt in late summer. On drained soils plants develop deep rooting systems that sustain them through the summer. Drawing by John Bryant.

The absolute number of deaths reported hardly decreased at all between 1870 and 1880, at the height of draining activity. From 1880 to 1890, the number of deaths declined steeply but did not disappear. Expressed as rates per 100,000 population, malarial mortality fell from 29.7 in 1870 to 28.9 in 1880 to 10.9 in 1890. While ditch-digging was in progress more people were exposed to mosquito bites than when tiles were buried and the surface was being cultivated. It was not definitely known that the disease was transmitted by anopheles mosquitoes until 1898, and the destruction of mosquitoes by spraying pyrethrum, DDT, and other insecticides did not begin until the twentieth century. Windows and doors of new houses were fitted with insect screens that protected people at night while mechanization of farm operations reduced the number of workers in fields open to attack by day.[49]

Commentators repeatedly marveled at the rapidity and magnitude of the transformation wrought by tile draining, converting malarial wetlands into America's most bountiful cornfields. Maize, the leading crop, yielded twice as much food per acre as any other cereal and could be relied upon to produce good harvests year after year.[50] By 1879, the highest proportion of land growing corn and the highest yields gathered were from former wet prairies (fig. 6.6). In the 1880s, the term "corn belt" was first used in print to describe this highly productive region.[51] The cost of drainage works and rising taxes were met by increasing returns from sales of corn and hogs. Land values soared, farm tenancy increased, rents rose, and cash grain farming expanded.[52] In the end, greater productivity was the decisive argument in favor of tiling. Margaret Bogue aptly remarked: "Drainage improvements were expensive, but experimentation soon showed that increased yields made them distinctly worthwhile."[53]

Draining and Economic Change, 1880 to 1920

After 1880, financiers increased their control over drained wet prairies and the number of independent small farmers declined. In

49. Gordon Harrison, *Mosquitoes, malaria, and man: A history of hostilities since 1880* (John Murray, London 1978) 197–98, 210–11.

50. Richard Lyle Power, *Planting corn belt culture: The impress of the upland Southerner and Yankee in the Old Northwest* (Indiana Hist. Soc., Indianapolis 1953) 151; John Fraser Hart, The Middle West, in Hart (ed.), *Regions of the United States* (Harper and Row, New York 1972) 258–82, esp. 267.

51. William Warntz, An historical consideration of the terms "corn" and "corn belt" in the United States, *Agric. Hist.* 31 (1957) 43.

52. John C. Hudson, *Making the corn belt: A geographical history of Middle-western agri-culture* (Indiana University Press, Bloomington 1994) 189.

53. Bogue, *Patterns from the sod,* 1959, 88.

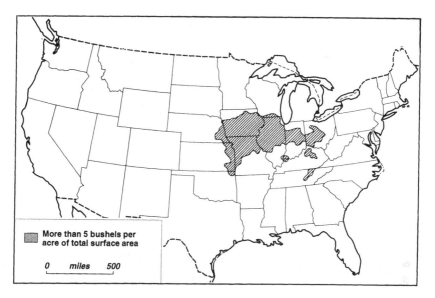

Figure 6.6 Corn production in the United States, 1879

Shaded areas represent production of more than 5 bushels per acre of total surface area. *Source:* U.S. Census Office, *Tenth Census: Productions of agriculture, 1880* (Washington, D.C. 1883).

Illinois farm mortgage debts increased by 19% between 1880 and 1887 and in the latter year no fewer than 12 of the 21 most heavily indebted counties were situated in newly drained prairie areas.[54] Most farm mortgages were granted for periods of five years or less, at 7% interest, excluding commission and fees. They were particularly onerous for small family farmers, many of whom could not afford to finance improvements on these terms. In Iowa, the proportion of owner-occupied farms fell from 76.2% in 1880 to 62.2% in 1910. By 1914, more than half the drained wet prairie was farmed by tenants[55] In all states in the Midwest, the proportion of farms occupied by tenants increased between 1880 and 1900 and continued to increase more slowly in the early years of the twentieth century (table 6.3). Owner-occupiers suffered more than tenants, but the conditions of tenancy also deteriorated. Long leases were almost

54. Destler, Agricultural readjustment in Illinois, 1947, 108, 109.
55. John D. Hicks, The western Middle West, 1900–1914, *Agric. Hist.* 20 (1946) 70.

Table 6.3 Tenant farmers as a percentage of all farmers, 1880–1920

	1880	1900	1920
Ohio	19.3	27.4	29.5
Indiana	23.7	28.6	32.0
Illinois	31.4	39.3	42.7
Iowa	23.8	34.9	41.7
Michigan	10.0	15.8	17.7
Wisconsin	9.1	13.5	14.4
Minnesota	9.1	17.3	24.7
Midwest average	18.1	25.3	29.0
% change		+7.2	+3.7
U.S. average	25.5	35.5	38.1
% change		+10.0	+2.6

Source: L.C. Gray et al., Farm ownership and tenancy, in USDA, *Agricultural Yearbook, 1923* (Washington, D.C. 1923) 507–600.

completely replaced by annual agreements, sharecrop rents rose steeply, and sharecropping itself was superseded by money payment, safeguarding the landlord's income against effects of falling prices. Small farmers paid higher taxes than city businessmen and their assessments were raised as lands were improved. Illinois farmers also complained that railroads were offering preferential rates to Western farmers and that industrial combines were overcharging them. In support of their grievances they adduced as clear evidence for oppression of small farmers the depopulation of rural areas and the engrossment of large farms. The greatest decrease in the number of small farms occurred in the most extensively drained areas, whence many joined the exodus to growing industrial districts as well as to Kansas, Nebraska, and California. In 1892, emigration from central Illinois exceeded all previous records; emigration from Iowa reached a peak in 1895. Large farms were enlarged or consolidated while small farms were amalgamated. While average sizes of farms in most parts of the United States declined after 1880, they increased throughout the wet prairies, more in the west than in the east (table 6.4).

If distress among small farmers was a faithful indicator of economic depression, its cause would not be far to seek, because the first signs of distress coincided with the beginning of an unprece-

Table 6.4 Average size of farms (acres), 1880–1920

	1880	1890	1900	1910	1920
Ohio	99	93	89	89	92
Indiana	105	103	98	99	103
Illinois	124	127	124	129	135
Iowa	134	151	152	156	157
Michigan	90	86	86	92	97
Wisconsin	114	115	117	119	117
Minnesota	145	160	170	177	169
Midwest average	116	119	119	123	124
U.S. average	134	137	147	139	149

Source: U.S. Bureau of Census, *Census of Agriculture, 1969* (Washington, D.C. 1969), vol. 2: 23.

dented slump in the price of farm products. By 1882, the world market was glutted with the immense produce of vast areas of fertile arable land beyond the Mississippi, brought into cultivation and harvested for the first time with the aid of machinery, served by over 100,000 miles of new railroads. Down to 1896, the depression of prices deepened, until wheat fetched scarcely half the price for which it had been sold in the late 1870s. Many well-established arable farmers found cash-crop farming unremunerative, but farmers on newly drained wet prairies suffered no loss. The improved lands yielded them greater quantities of grain per acre for their labor than any arable lands in North America and they could afford to reduce their prices by increasing their output (fig. 6.7). Moreover, they could abandon wheat and grow oats for feeding horses or convert their entire corn crop into more valuable animal products. Corn belt farming was sufficiently versatile and expansive to meet changing demands in the market.

Far from being a symptom of agricultural malaise, the decline of small farmers can only be regarded as a sign of prosperity. Mounting debt was a measure of confidence shown by investors in the ability of farmers to meet their obligations and, in the absence of foreclosures, it represented a rising level of capital investment.[56] Increases in rent were economically justified by the continued de-

56. Roy V. Scott, *The agrarian movement in Illinois, 1880–1896* (University of Illinois Press, Urbana 1962) 13.

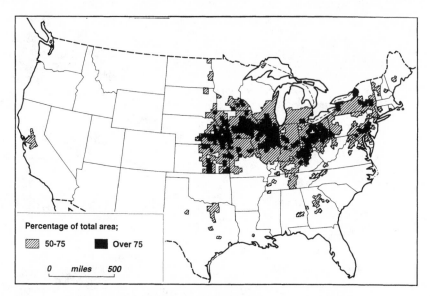

Figure 6.7 Improved land in the United States, 1889

Improved land is shown as a percentage of total surface area. *Source:* U.S. Census Office, *Eleventh Census of Population, 1890* (Washington, D.C. 1892).

mand for wet prairie land. Holdings of the dispossessed did not fall vacant; they were eagerly sought after by neighboring farmers. Successful farmers were evidently able to afford these increased impositions and bear heavy taxes, discriminatory railroad charges, and high prices for some manufactured goods. Such families were not victims of an agricultural depression.

Owners of drained wet prairies benefited not only from great increases in productivity and opportunities to change their farming enterprises but also from substantial and entirely uncovenanted gains in the value of land itself. Before 1880, unimproved wetland could be bought for as little as $2 to $3 and not more than $12 an acre, $7 being an average price. When drained, the value of such land might be multiplied fivefold, but during the 1880s the price of unimproved land advanced rapidly, ranging from $13 to $40, averaging $25 an acre. In the 1890s, drained land was selling at $60 to $75 and rising in value every year. Chester Destler observed that, "occurring as it did in the face of a declining agricultural price level, this increase in land values was striking proof of the profits with which systematic drainage rewarded investors."[57] But that is not all. Land values continued to rise for years after draining had been completed. This further advance was a response to threatened exhaustion of accessible fertile land in the West. In 1888, Lord Bryce predicted a time of trial early in the coming century, when all cultivable land in the public domain would be appropriated. In the 1890s, a prolific Kansas publicist, C. Wood Davis, repeatedly asserted that good farmland was already so scarce that farmers were assured of high prices for all the future.[58]

As nearly all remaining arable land in the West was occupied and the urban population in the East and in Europe continued to grow, it seemed inevitable that the market for foodstuffs would expand and land values rise yet higher. These expectations were fulfilled as the depression lifted at the end of the 1890s. People continued to pour into the Midwest from the East and from Europe. With the exception of largely rural Iowa, which recorded a net loss of 7,000 in the decade 1900–1910, all Midwestern states gained in numbers through the whole period from 1880 to 1920. Population increase was closely associated with rapid urban expansion on the southern shores of Lake Michigan, around St. Louis, Detroit, and Minneapolis–St. Paul (table 6.5). Between 1896 and 1914, the price of farm produce doubled and the value of improved wet prairie more than doubled. Drained farmland again doubled in value during the First World War and by 1920, it was worth about $400 per acre. By holding land for a decade or more a landowner multiplied his wealth not only from the profitable sale of his produce from good husbandry but also from the unearned increment accumulating on the value of his land. To a considerable extent, he was, as J. D. Hicks remarked, "only a successful speculator."[59]

Laying Tiles to the Edge of the Prairies

The draining of large areas of formerly waterlogged wasteland proved so successful that tiling was extended to many other tight, hard-to-drain soils, to improve their productivity rather than bring

57. Destler, Agricultural readjustment in Illinois, 1947, 109.
58. Lee Benson, The historical background of Turner's Frontier essay, *Agric. Hist.* 25 (1951) 66–77.
59. Hicks, Western Middle West, 1946, 73.

Table 6.5 Population of the United States, 1880–1920

	1880	1890	1900	1910	1920
Ohio	3,198,062	3,672,329	4,157,545	4,767,121	5,759,394
Indiana	1,978,301	2,192,404	2,516,462	2,700,876	2,930,390
Illinois	3,077,871	3,826,352	4,821,550	5,638,591	6,485,280
Iowa	1,624,615	1,912,297	2,231,853	2,224,771	2,404,021
Michigan	1,636,937	2,093,890	2,420,982	2,810,173	3,668,412
Wisconsin	1,315,497	1,693,330	2,069,042	2,333,860	2,632,067
Minnesota	780,773	1,310,283	1,751,394	2,075,708	2,387,125
Midwest total	13,612,056	16,700,885	19,968,828	22,551,100	26,266,689
% change		+23	+20	+13	+16
Rest of U.S.	36,577,153	46,278,881	58,423,340	69,677,396	79,754,848
U.S. total	50,189,209	62,979,766	76,212,168	92,228,496	106,021,537
% change		+25	+21	+21	+15

Source: U.S. Bureau of Census, Seventeenth Census of Population, 1950 (Washington, 1952), vol. 1: 1–8, 1–9.

them into cultivation for the first time. Tiling was often undertaken as a form of crop insurance, to secure regular and dependable yields from land of indifferent quality, to prevent frost heaving in the soil, to eliminate frost pockets, to make wet spots firm enough to carry machinery, and to remove obstructing patches of waterlogged land at the corners of fields. Much of this work was carried out before the First World War, but in 1918 it was considered economically desirable to provide even average soils in Iowa with closely spaced tile lines.[60] In Indiana, tile production slackened at the beginning of the present century; the number of tile factories decreased from a maximum of 143 in 1904 to 24 in 1931.[61] To the west of Chicago, tiling continued for another twenty years. In Blue Earth County, Minnesota, it declined after 1918; in Story County, Iowa, it came to a standstill in 1925.[62]

By 1920, all but a few, very small patches of wet prairie soils lay within the boundaries of organized drainage enterprises. They occupied a continuous block of territory north of the Bloomington moraine, extending west from Cleveland, Ohio, to Springfield, Illinois, covering all areas of Webster soils in north central Iowa and southern Minnesota. The Red River valley, the Wabash lowlands, and many small tracts of wetlands also lay in drainage districts. In 1930, the acreage of farmland provided with drains was greater than the acreage classed as having been too wet to raise a normal crop without draining (fig. 6.8). It extended beyond the margins of waterlogged soils to the edge of the prairies. It included much impervious clayland that had been improved by piecemeal tile draining by individuals without the construction of open ditches or organization of drainage enterprises.

Draining in the prairies produced a remarkable uniformity of landscape and land use. It obliterated what was formerly an important regional division. It led to the reclamation of almost all agriculturally unproductive land, to the elimination of all but a small area of land devoted to the production of wild hay, and to a considerable reduction in the area of pasture. A few undrained spots on upland plains remained unimproved grassland in 1920 and, at the southern

60. W. J. Sclick, *The theory of underdrainage*, in *Iowa Engin. Expt. Sta. Bull.* 50 (1918), cited in USDA, *Soils and men*, 1938, 730.

61. Perkins, Drain tiles in Indiana, 1931, 380.

62. Burns, Blue Earth County, 1954, 98; Hewes and Frandson, Artificial drainage in Story County, 1952, 35.

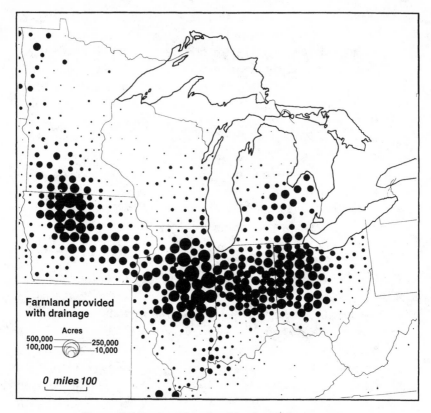

Figure 6.8 Drained farmland in the Midwest, 1930

The area of each circle represents the area of farmland provided with drainage in each county. *Source:* U.S. Bureau of Census, *Fifteenth Census of the United States, 1930*, vol. 4, *Drainage of agricultural land* (Washington, D.C. 1932).

margin of the Illinois prairies, on poorly drained acidic soils, a native grass, Redtop, was gathered both for hay and seed in 1930.[63] The stubborn poverty of these unreclaimed patches presented a striking contrast to the highly cultivated aspect of all the surrounding country.

The well-drained condition of almost all prairie surfaces was largely artificial. The land that had been very poorly drained was

63. Robert G. Buzzard, Redtop production of south-east Illinois, *Annals Assoc. Amer. Geog.* 20 (1930) 24; Stanley D. Dodge, Sequent occupance on an Illinois prairie, *Bull. Geog. Soc. Philadelphia* 29 (1931) 209.

made to produce heavy crops, and land once considered fairly well drained was also tiled and its drainage raised to the high standard prevailing on most prairies. Where there was hope of improvement tiles were laid, even to the edge of the prairies. The most important consequence of this activity was that the prairies were made uniform. The orderly appearance of large fields of regular shape, efficiently worked by machinery, may be traced to the period of tile draining. It is a paradox that where draining was most successfully accomplished, a casual observer can find least trace of it. The present landscape hides its past; tile lines are hidden from view, main drains are unobtrusive, and farming is distinguishable from that practiced on naturally well drained, but rapidly eroded, slopes only by its superiority.

Farming on Drained Prairies

Draining on wet prairies has been accompanied invariably by increases in the area of tillage. In Story County, Iowa, the proportion of land in crops increased from 40% before draining to 70% in 1947.[64] A map of Ohio representing the amount of land in crops as a proportion of the acreage of farmland shows clearly that the Black Swamp became the most completely cropped area in the state. By the middle of the twentieth century, the whole level, drift-covered land which had benefited most from draining had more than 60% of its area in crops by contrast with the driftless hill land in the southeast which had less than 40%.[65] A similar comparison, made in 1913, between the driftless hill country and flat glaciated areas in Wisconsin indicated that about 44% of the driftless land but over 63% of the flat glaciated lands were in crops.[66] In the most intensively drained parts of Illinois as much as 90% of farmland was in crops.[67] There can be little doubt that when this land was success-

64. Hewes and Frandson, Artificial drainage in Story County, 1952, 42.

65. Guy Harold Smith, The relative relief of Ohio, *Geog. Rev.* 25 (1935) 284; J. Sitterley and J. Falconer, Better land utilization for Ohio, *Ohio State Univ. Agric. Expt. Sta. Bull.* 108, Department of Rural Economy, mimeo (Columbus 1938) 39, cited in Kaatz, The Black Swamp, 1955, 33.

66. Ray Hughes Whitbeck, Economic aspects of the glaciation of Wisconsin, *Annals Assoc. Amer. Geog.* 3 (1913) 71.

67. Wallace E. McIntyre, Land utilization of three typical upland prairie townships, *Econ. Geog.* 25 (1949) 269.

fully reclaimed it was converted almost entirely from unimproved
grazing land to first-class arable land.

On most newly drained land an arable rotation was adopted in
which the emphasis was overwhelmingly on the production of corn.
In the Black Swamp of Ohio a monoculture of corn was followed
with little variation for over half a century after draining took
place.[68] In central Illinois, the output of corn multiplied in the
1880s at the time of tile draining.[69] By 1890, the corn crop represented 60% of the total value of all cereals produced in Illinois and
it maintained its ascendancy to the middle of the present century.
The acreage planted to corn was highest in the central prairie districts, where it accounted for between 40% and 50% of land in
crops, a higher proportion than that for the state as a whole.
Drained lands also produced the highest yields per acre.[70] In north
central Iowa, the pattern of change was faithfully repeated. The production of corn multiplied in the 1880s and 1890s on newly
drained land. It has been asserted that drainage undertakings were
largely paid for in corn.[71] Selling corn for cash provided immediate
returns without added costs or risks incurred in feeding livestock. It
was an essential element in the economy of tenant farmers, especially those paying shares of their crops. It also attracted owner-occupiers heavily encumbered with interest charges. Describing the
persistence of cash grain farming in north central Illinois, Arlin
Fentem concluded that "once agricultural patterns are established,
they prove extremely resistant to change," because "specializations
themselves confer an advantage which outweighs all others."[72] The
proportion of cropped land devoted to corn in Story County, Iowa,
rose from 24% before draining to 42% at the beginning of the
1950s.[73] While it is true of most lands dominated by Western culture that wherever corn can be grown, corn will be the leading crop
or at least the second crop, on wet prairies corn invariably ranks

68. Kaatz, The Black Swamp, 1955, 29.

69. Watterson, *McLean County, Illinois,* 1950, 72.

70. Andreas Grotewold, *Regional changes in corn production in the United States from
1909 to 1949*, Department of Geography Research Paper 40 (University of Chicago, Chicago
1955) 11–15.

71. Berry, Environment in north central Iowa, 1927, 294.

72. Arlin D. Fentem, Cash feed grain in the corn belt, Ph.D. diss., Geography, University
of Wisconsin 1974, 141.

73. Hewes and Frandson, Artificial drainage in Story County, 1952, 42.

first and its acreage may exceed the total of all other crops put together.[74]

In almost all newly drained areas oats came to be the second most important crop. In Illinois, the acreage of oats doubled between 1880 and 1890. For a brief period, drained areas in north and central Illinois became "one vast oatfield trespassing largely on the acreage of corn."[75] Production rose to a maximum in the second decade of the present century as the number of machines drawn by teams of horses increased. When horses were replaced by tractors in the 1920s, production fell, but the acreage in 1950 was still about double what it was before draining and in some areas it was the only crop of importance besides corn.

After 1881, returns from wheat sales failed to equal costs of production except on the best prairie soils, and wheat as a cash crop virtually disappeared. During the depths of depression in the 1890s and occasionally in boom years, corn, oats, and hay were diverted to livestock feeding. A substantial proportion of all crops were consumed in the counties in which they were grown, mostly by hogs. Fattening hogs was the most efficient and dependable method of converting corn into meat. Fattening cattle for meatpackers was a more speculative venture, highly remunerative when steers and corn were cheap and fatstock prices high. Hogs could be fattened on cattle droppings supplemented by rations of corn. The dominance of corn over all other crops was matched by the dominance of hogs over all other livestock. The greatest corn-producing areas were also the foremost producers of porkers.[76] During the last quarter of the nineteenth century, large national enterprises, including Swift, Armour, Kingan, and Fowler, rapidly expanded their slaughtering and packing operations.[77]

An immense change took place in the Midwest as a result of draining wet prairies. After a long period of painfully slow and hesitant change, preceded by many unsuccessful endeavors by speculators, railroad companies, and cattlemen, draining effected a spec-

74. Richard Hartshorne, A classification of the agricultural regions of Europe and North America on a uniform statistical basis, *Annals Assoc. Amer. Geog.* 25 (1935) 99–120.

75. Destler, Agricultural readjustment in Illinois, 1947, 105.

76. Earl Shaw, Swine production in the corn belt of the United States, *Econ. Geog.* 12 (1936) 364, 372.

77. Margaret Walsh, From pork merchant to meat packer: The Midwestern meat industry in the mid-nineteenth century, *Agric. Hist.* 56 (1982) 133–37.

tacular transformation and brought rich rewards. In Illinois, more than half the land in crops was artificially drained. Not only in Illinois but also in Iowa and Indiana the most productive lands were formerly waterlogged. Wetlands, once drained, enjoyed several clear advantages for arable farming not possessed by neighboring areas. Wet prairie soils were rich storehouses of plant foods. Webster soils were described by the Soil Survey Division of the USDA as "not surpassed by any other of equal size in the production of corn."[78] After draining, Clyde, Newton, Maumee, and Toledo soils were equally well adapted to corn-growing. Not only were these soils remarkably fertile; they were also less subject to erosion than soils on sloping surfaces. The flatness of wet prairies facilitated the efficient operation of farm machinery and made possible the division of farms into large, square fields. Production rapidly outstripped that on formerly timbered slopes and higher, drier ground on prairies themselves. In 1950, Leslie Hewes observed, "the durable flat prairie has come into its own."[79]

78. USDA, *Soils and men,* 1938, 1054.

79. Leslie Hewes, Some features of early woodland and prairie settlement in a central Iowa County, *Annals Assoc. Amer. Geog.* 40 (1950) 56.

7

Occupying, Draining, and Abandoning Northern Bogs and Swamps

Draining transformed wet prairies in Indiana and Illinois. A curse that seemed to have hung over them for half a century suddenly lifted. Crop yields shot up. Corn production multiplied and land values rose to new heights. Not only upland wet prairies but also riverine marshes were successfully reclaimed for agriculture. In the Black Swamp in Ohio and Kankakee Marsh in Indiana and Illinois, rivers were channelized and bountiful corn crops were harvested. Prospective purchasers began to approach wetlands everywhere with less fear and trepidation. Northern bogs and swamps were regarded more as a challenge than a threat. Caution was relaxed just where and when it was most needed.

Land in the northern lakes states was taken in a rush. The frontier was a live frontier. To the north lay British Canada. Trading posts and missions were situated deep in Indian territory. Military forts were outposts of American occupation and among the earliest American pioneers were prospectors for lead, copper, iron, and other minerals as well as lumber cruisers who came to measure the volume of white pine in the forests. These scouts were ahead of a front line of advancing miners and lumbermen. Mines were dug and sawmills built before treaties with Indians had been signed, before government surveyors had platted the land, before land offices had been opened for public sales. Vast areas of public domain were not offered at federal auctions; they were assigned by the federal government to newly formed states.

States grew at unprecedented speed as people surged in, anticipating cheap land, quick riches, and cities springing up in the wilderness. Between 1830 and 1840, Michigan's population rose from 31,639 to 212,267; Wisconsin's increased from 30,945 in 1840 to 305,391 in 1850; and Minnesota's population soared from 6,077 in 1850 to 172,023 in 1860 and continued to rise to 439,706 in

1870. Within twelve years of Michigan being admitted to the federal union, a railroad was constructed from Detroit to Lake Michigan and in 1852 the Michigan Central entered Chicago. The University of Wisconsin at Madison was founded in 1848, the year Wisconsin attained statehood. In St. Paul, a state capitol, a cathedral, scores of churches, banks, and hotels were erected within a few years of Minnesota becoming a state. The precocious growth of cities with dense concentrations of population and high-order central place functions contrasted with thin scatterings of people over the rest of the region.[1]

In the back country, railroads steamed through silent woods and over quaking bogs in advance of almost all settlers. On rivers, mills were set up before towns were laid out. Many pioneers lived in temporary accommodation. Trappers, prospectors, lumber cruisers, and surveyors were accustomed to sleeping under the stars or making their own rudimentary shelters in the depths of northern winters. Mining camps, lumber camps, and railroad construction camps provided temporary quarters for teams working long distances from their operational bases. In the middle years of the nineteenth century it was widely believed that mines, sawmills, and railroad stations would attract permanent settlements; that churches, schools, stores, homes, and farmsteads would be built; that stumplands would be cleared and marshes drained and converted into smiling cornfields. After a phase of sleeping rough and youthful roving, it was expected that people would settle down and build family farms.

At the end of the nineteenth century, inducements to drain northern peatlands were strong and growing stronger. In 1890, the director of the U.S. Census reported that the westward advance of the frontier of settlement had ended, and Frederick Jackson Turner, Wisconsin-born historian, examined the significance of that statement in a paper to the American Historical Association in 1893.[2] In

1. Ralph H. Brown, *Historical geography of the United States* (Harcourt Brace, New York 1948) 287, 331; Michael P. Conzen, The spatial effects of antecedent conditions upon urban growth, M.A. thesis, Geography, University of Wisconsin, Madison 1968; Conzen, The maturing urban system in the United States, 1840–1900, *Annals Assoc. Amer. Geog.* 67 (1977) 88–108.

2. Frederick Jackson Turner, The significance of the frontier in American history, *American Historical Association Annual Report* (1893) 199–227, reprinted in *The early writings of Frederick Jackson Turner*, with introduction by Fulmer Mood (University of Wisconsin Press, Madison 1938) 183–229.

Michigan, Wisconsin, and Minnesota almost all public lands had been disposed of and free 160-acre homesteads were no longer to be had. Henceforth, the land market was in private hands and vacant parcels were eagerly sought after.

A new opportunity for land purchasers arose as lumbermen felled or burned northern pine forests. If no buyers came forward, lumber companies simply abandoned stumplands and allowed them to revert to state ownership as tax delinquent. For a few years after lumbermen left, state and other local agencies as well as Chicago land companies and many private speculators engaged in promoting sales of stumplands. Those brief opportunities occurred at different times in different places, mostly between 1880 and 1910. After 1920, it was clear that most cutover, especially cutover wetlands, offered very poor prospects for successful agriculture.[3]

Railroads Advance into Northern Swamps and Bogs

The federal government gave northern lakes states two valuable bonuses to hasten the colonization of wetlands. Enormous land grants were awarded under the provisions of the Swamp Land Act and even larger grants were handed out to finance railroad construction. Grants of swamplands to northern states were much larger than those awarded to prairie states. Michigan alone received 5,655,816 acres, exceeding by over two million acres the sum total of 3,612,016 acres granted to Ohio, Indiana, Illinois, and Iowa put together. Minnesota's grant of 4,663,007 acres was a million acres larger and Wisconsin received 3,251,684 acres, barely 360,000 acres less than the combined total for the four prairie states. In 1861 and again in 1865, state governors proposed that half a million acres of Minnesota's swampland grant be applied to founding asylums for the blind, deaf, dumb, and insane, endowing three state normal schools, and erecting a new state prison.[4] In addition, swamplands were sold in order to build roads and bridges. Much larger areas of swampland were transferred to railroad companies. Between 1861 and the abolition of the practice in 1881, the State of

3. Gordon G. Whitney, *From coastal wilderness to fruited plain: A history of environmental change in temperate North America, from 1500 to the present* (Cambridge University Press, Cambridge 1994) 247.

4. Ben Palmer, *Swamp land drainage with special reference to Minnesota*, University of Minnesota Studies in Social Sciences 5 (Minneapolis 1915) 89–90, 92, cites Governor Ramsey's message of 26 January 1861 and Governor Miller's message of 4 January 1865.

Table 7.1 Swamp land grants to railroads in Minnesota, 1861–81

Year	Grantee	Acres
1861	Lake Superior and Mississippi RR Co. (St. Paul to Duluth; later Northern Pacific)	694,399
1861	Taylors Falls and Lake Superior RR Co. (Taylors Falls to Wyoming, Minnesota; later Northern Pacific)	91,830
1863	St. Paul and Chicago RR Co. (St. Paul to Winona; later Chicago, Milwaukee and St. Paul)	425,300
1865	Minneapolis and St. Cloud RR Co. (St. Cloud to Hinckley; later Great Northern)	425,664
1865	Southern Minnesota RR Co. (Fillimore County to western boundary of Minnesota)	–
1865	Minnesota Central RR Co. (Red Wing to Mankato; later Great Northern)	275,000
1875	Duluth and Iron Range RR Co. (Duluth to Brainerd; later Northern Pacific)	–
1881	Sioux Falls and Dakota RR Co. (Little Falls to western boundary of Minnesota; later Northern Pacific)	265,856
	Total	2,178,049

Source: Ben Palmer, *Swamp land drainage with special reference to Minnesota* (University of Minnesota, Minneapolis 1915) 90–94.

Minnesota granted over two million acres of swamplands to aid railroad construction (table 7.1). In 1861, the state awarded seven full sections of swampland for every mile of railroad completed. Later acts awarded from four to ten sections for every mile built, larger grants being given to lines that opened up uninhabited areas. No act honored a pledge to the federal government by requiring railroad companies to reclaim swamps.

In addition to grants of swamplands from states, railroad companies received vast areas of land direct from the federal land office. By 1923, railroads had received 2,595,133 acres of public land in Illinois; 4,929,849 acres in Iowa; 3,133,232 acres in Michigan; 3,649,869 acres in Wisconsin; and 8,035,578 acres in Minnesota.[5] Grants to railroads in the northern states were vastly larger than those in the prairie states and recipient companies were correspondingly more active in selling land, promoting settlement, and generating traffic. Such large tracts of land were granted to railroad compa-

5. Benjamin H. Hibbard, *A history of the public land policies* (Peter Smith, New York 1939) 264, cites General Land Office, *Report of the Commissioner, 1923,* 37–38.

nies because tracks were to be laid in areas where settlement had scarcely begun. Land values were low not only because of long distances from market centers but also because soils and climate were less favorable to agriculture than on the Grand Prairie in Illinois. Railroads had to push sales of land to raise money to build their lines. For some, capital dried up before projects were completed.[6] Other companies sold bonds at a discount in order to proceed with construction but found it difficult to recover sufficient from land sales to repay investors when bonds matured.[7] In 1881, the Minnesota legislature voted to redeem defaulting bonds.[8] After acrimonious debate about large private corporations receiving relief in preference to debt-laden small farmers, the legislature decided that no further land grants should be made to railroad companies.

Draining was seen by James J. Hill, president of the Great Northern Railroad, as a means of increasing land values and attracting settlers. In 1879, in the flat floodplain of the Red River of the north, about 45 miles of ditches were dug along the line of the railroad. They carried storm waters in spring, in time for sowing wheat; and more important "was the object lesson they furnished to the farmer of what service even small drains could be." The initiative taken by the railroad company aroused widespread interest. In June 1886, local newspapers invited people to take part in "a general mass convention," to be held at Crookston, "for the purpose of considering the subject of drainage, and to devise means for the accomplishment of a thoroughly effectual and general system of drainage for said section of country."[9] The Great Northern, which at that date owned over one million acres in the Red River valley, promised to support a large-scale drainage scheme, offered to contribute $5,000 toward a topographical survey, and nominated C. G. Elliott, former chief of drainage investigations for the USDA, as chief engineer. A convention again met at Crookston on 8 December

6. Julius Grodinsky, *Transcontinental railway strategy, 1869–1888: A study of businessmen* (University of Pennsylvania Press, Philadelphia 1962) 127.

7. Theodore C. Blegen, *Minnesota: A history of the state* (University of Minnesota Press, Minneapolis 1963) 252.

8. State of Minnesota, *Laws* (19 November 1881) chap. 71. [Railroad Adjustment Bonds Act].

9. Palmer, *Swamp land drainage*, 1915, 65, cites Minnesota Drainage Commission, *Report 1899;* and *Crookston Times,* 5 June 1886.

1886 to receive a preliminary survey and engineer's report recommending digging about 275 miles of ditches at an estimated cost of $750,000. The meeting elected a committee of five members, one from each of five counties, to prepare and present to the state legislature a general drainage law for the organization of drainage districts that would be given powers to assess benefits and damages and charge rates for constructing and maintaining ditches. The meeting also resolved to ask the governor of Minnesota for "a very liberal appropriation by the state to open up the obstructed river channels" in the district. In 1887, two laws implemented the principal proposals of the Crookston convention, and drainage works were launched with the aid of state bonds and further subventions from the Great Northern Railroad.

The lead taken by railroads in initiating ditching and mobilizing public interest in land reclamation was a distinctive feature of relations between corporate interests and states in the northern lakes region. In effect, the states of Michigan, Wisconsin, and Minnesota relied on railroads to promote settlement and economic development and thus expand the tax base. Railroads responded in the conviction that what was good for the Great Northern was good for the state of Minnesota. It is remarkable that an ad hoc convention held in Crookston, on the western boundary of the state, should take upon itself responsibility for drafting a general drainage law for the whole state, and that the state legislature should adopt the measure with minor revisions and provide financial backing for the project. Railroads pressed the state to fulfill its obligation to the federal government by authorizing the draining of wetlands, and in 1901 the governor of Minnesota agreed to introduce new drainage laws conferring wider powers on counties and judicial districts. He also set up a state drainage commission and appointed a state drainage engineer.

Lumbering in Northern Pineries

Railroads were drawn deeply into the northern forests by lumber companies whose aim was to reach districts remote from logging streams. Swamplands granted to railroad companies contained some dense stands of white pine. Lumbermen were interested almost exclusively in what they called "pineries" and they urged railroads to build lines that gave access to these tracts.[10] In the late 1870s, the Wisconsin Central Railroad completed a line from Neenah to Ashland on Lake Superior, midway between lumbering districts tributary to the Wisconsin River on the east and the Chippewa River, containing about a sixth of all pine timber in the lakes states, to the west. By 1892, over 400 million feet of lumber were cut from the logging district served by the railroad, and twelve sawmilling centers at the side of the tracks were producing boards for consignment to all parts of the West.[11] Branch lines were constructed to carry logs from distant woods to mills on main lines. As well as generating traffic, lumber companies supplied railroads with ties, telegraph poles, and wood for fuel. Wealthy lumbermen invested heavily in railroad stock and some acquired controlling interests in lines that sustained their operations. Frederick Weyerhaeuser owned a major share in the Chicago Great Western and a large holding in the Great Northern. His interests extended from lumber and railroads to agricultural land and urban property. John S. Pillsbury and W. D. Washburn both made fortunes in lumbering before setting out to become the world's leading flour millers.[12] The basis of their wealth was white pine.

White pines grew singly or in groups, soaring to great heights above a canopy of hardwoods. They frequently stood on mounds where roots of windthrown or burned-down trees had decayed. Most characteristically they occupied hummocks within or at the edges of wetlands, often in association with hemlocks, spruces, and

10. William Cronon, *Nature's metropolis: Chicago and the great west* (W. W. Norton, New York 1991) 152, comments on a prevailing materialist attitude toward forests as resources, trees as lumber, plants as commodities for trade, sources "of gain to the inhabitants," in a phrase written by a visitor to Manitowoc, Wisconsin, in 1852. Cronon notes: "When most nineteenth-century Americans saw a white pine, they could summarize their reaction with a single, compelling word: 'lumber.' No other tree was so highly prized." The foremost nineteenth-century historian of lumbering in the lakes region, George W. Hotchkiss, *History of the lumber and forest industry of the Northwest* (G. W. Hotchkiss, Chicago 1898) 752, declared: "No wood has found greater favor or entered more fully into supplying all those wants of man which could be found in the forest growths."

11. Charles E. Twining, The lumbering frontier, in Susan L. Flader (ed.), *The Great Lakes forest: An environmental and social history* (University of Minnesota Press, Minneapolis 1983) 132; Robert F. Fries, *Empire in pine: The story of lumbering in Wisconsin, 1830–1900* (State Hist. Soc. Wisconsin, Madison 1951) 86–87.

12. Agnes M. Larson, *History of the white pine industry in Minnesota* (University of Minnesota Press, Minneapolis 1949) 410–12.

firs.[13] The association of white pine with poorly drained soils presented both difficulties and opportunities. Many sites were so miry that it was possible to haul logs only when the ground was frozen solid.[14] Opportunities for profitable exploitation arose where states permitted swamplands to be purchased at prices below the government minimum and when it was prudent for swamps to be sold cheaply to lumber companies rather than let timber be stolen by trespassers.[15]

Frederick Jackson Turner's *Significance of the frontier in American history*, 1893, mentions neither wetlands nor lumbering.[16] Apart from drawing a broad contrast between forest and grassland, the essay pays little attention to differences in soil and vegetation that were subjects of intense scientific investigation by late-nineteenth-century geologists, botanists, agricultural writers, topographers, and local historians. More remarkable is the unexplained absence of a lumberman from Turner's "procession of civilization, marching single file" across the continent.[17] Turner was born at Portage, Wisconsin, in 1861. As a boy, he would have heard axes ringing and tall pines crashing in surrounding woods in winter and watched logs running downstream in spring. Lumbering on the Wisconsin River was at its height in the late 1880s when Turner began research on the fur trade in Wisconsin. In his home area, fur traders and trappers were followed by lumberjacks and log drivers who, in turn, were succeeded by ditchers and farmers. Many of Turner's contemporaries who worked as loggers during the winter, grew potatoes and oats and cut hay during the summer.[18] They were part-time lumberjacks, making farms.

Clearing trees and grubbing up stumps were steps toward cultivating fields, whether or not the felled logs and branches were useful or could be sold. Many nineteenth-century observers regarded woods as encumbrances on potential agricultural land and their re-

13. Eric A. Bourdo, Jr., The forest the settlers saw, in Flader (ed.), *Great Lakes forest*, 1983, 9.

14. Cronon, *Nature's metropolis*, 1991, 156.

15. Fries, *Empire in pine*, 1951, 168, 196.

16. Twining, Lumbering frontier, 1983, 128, describes the absence of the logger from Turner's account as "the greatest omission of all."

17. Turner, Significance of the frontier, 1893, in *Early writings*, 1938, 199.

18. Walker Wyman and Lee Prentice, *The lumberjack frontier: The life of a logger in the early days on Chippeway* (University of Nebraska Press, Lincoln 1969) 46–47, recalls early experiences of Louie Blanchard, who combined lumbering and farming on a seasonal basis.

moval "as a worthy sacrifice to the cause of civilization."[19] In 1868, a visitor to the upper peninsula of Michigan reported: "The pioneer is insensible to arguments touching the future supply; to him the forest is only fit to be exterminated, as it hinders his plough and obstructs his sunlight."[20] Similar sentiments were expressed in Wisconsin in the 1870s; and in Conrad Richter's novel, *The trees*, those that were cut down and burned in an Ohio swamp were described as "all worthless, good for nothing, cluttering up the black land."[21] The effort expended on clearing "seemed to consecrate the transfiguration" of wilderness into cultivated fields.[22] Where a pioneer, a logging crew, or a lumber company profited from sales of timber and wood, it was judged to be no more than a just reward for the sweat, courage, and enterprise that produced marketable lumber.[23]

The lumber industry was praised for performing a valuable service in supplying prairie farmers with materials for frame houses, barns, fenceposts, and fuel. As early as 1854, a Minnesota booster rejoiced in the contribution that northern pinewoods were making toward the settlement of prairies to the south. "We are ashamed," he wrote, "that we ever distrusted Providence, or suspected that our munificent Maker could have left two thousand miles of fertile prairies down the river, without an adequate supply of pine lumber at the source of the river, to make those plains habitable."[24] A sawmilling firm, Laird, Norton, at Winona, Minnesota, began shipping lumber down the Mississippi waterway to Iowa in the 1860s. Later, during the 1880s, it supplied ready-made materials for building houses and barns, and shipped them by rail to "line yards," which were located at the railside in Dakota territory.[25] Northern lumber furnished waterways with quays, booms, mills, warehouses,

19. Cronon, *Nature's metropolis*, 1991, 205.

20. C. H. Brigham, The lumber region of Michigan, *North Amer. Rev.* 107 (July 1868) 97–98.

21. Twining, Lumbering frontier, 1983, 128, citing Conrad Richter, *The trees* (A. A. Knopf, New York 1940).

22. Ibid., 129.

23. James Willard Hurst, The institutional environment of the logging era in Wisconsin, in Flader (ed.), *Great Lakes forest*, 1983, 145.

24. J. W. Bond, *Minnesota and its resources* (Redfield, New York 1854) 30, cited in Cronon, *Nature's metropolis*, 1991, 154.

25. John N. Vogel, *Great lakes lumber on the Great Plains: The Laird, Norton Company in South Dakota* (University of Iowa Press, Iowa City 1992).

and shipyards. It also provided railroads with constructional timber and fuel. In effect, lumbering was viewed widely as a means of creating agricultural land and generating wealth on a grand scale. It exploited nature for the benefit of society.

Lumbermen were not free agents in deciding whether to cut pine or save it. They acted within a framework of land laws that were designed to encourage freehold ownership of family farms. No provision was made for loggers to obtain temporary licenses to fell trees and move on. Lumbermen, lawyers, speculators, and many others in the Midwest valued, above all, individual liberty, initiative, enterprise, and independence secured by ownership of land in fee simple. Owners enjoyed freedom to dispose of natural endowments on their property, subject only to police powers and liability to pay taxes imposed by elected governments. Before 1880, planning and regulation of resource uses were not issues raised by politicians seeking election to legislative bodies or by electors. Those engaged in the lumbering industry were simply not free to opt for conservation. Whole communities depended on jobs in logging camps and sawmills and work was expected to continue until firms went bankrupt or the last marketable tree was cut down.[26]

Lumber companies, shareholders, and banks that lent money could not afford to hoard land or spread cutting programs over periods long enough to permit forests to regenerate. A threat of theft or trespass deterred lumber companies from buying tracts of forest before they were ready to begin logging.[27] The tax system rewarded lumbermen who clear-felled and disposed of cutover land as rapidly as possible. The longer they stayed, the higher the tax bill; the faster they removed timber, the sooner the value of land declined to that of cutover.[28] When timber had gone, taxes would have to be reduced or owners would quit. As early as 1878 in northern Minnesota, Laird, Norton lumber company decided it was commercially expedient to take a clean cut, cutting all merchantable timber on their land at the first logging. Where stumpland was worth holding for speculation, the company's agent was instructed to negotiate with local government for a "mere nominal tax," threatening to pull

26. Twining, Lumbering frontier, 1983, 133.
27. Fries, *Empire in pine,* 1951, 187–94.
28. Larson, *White pine industry,* 1949, 342–43.

out and default unless a reduction was conceded. In a sparsely settled area, officials were in a weak position to refuse a lower rate of tax if the alternative was to receive none at all.[29] Robert Fries concluded that "taxes were perhaps the expense that was most annoying to the timberland owner, especially if he did not control the election of the assessors."[30] When rivals began to compete for dwindling stands of timber, prices rose, cutting accelerated, and, because the process was difficult to reverse or slow down, great waste and destruction followed. Even small measures to prevent wanton destruction of forests were resented and resisted. Government agencies failed to stop theft and trespass and were reluctant or too weak to provide effective fire protection.[31]

Exploitation of the pinery was attended by two myths: first, that the resource was inexhaustible; second, that lumbering was a necessary prelude to more intensive use of the land by farmers. Writing in the 1940s and 1950s, Agnes Larson, Robert Fries, and other historians, as witnesses to extensive fresh scars left by fifty years of unrestrained clearing, discussed the myth of inexhaustibility. By the 1980s, many devastated areas were covered with second-growth forests and the notion of exhaustion could be approached from a more optimistic viewpoint. The most confident forecasts of unlimited stocks had been made in the mid-nineteenth century. In July 1852, a Wisconsin congressman, Ben Eastman, looking at lands drained by the upper Mississippi and its tributaries, declared: "there are interminable forests of pine, sufficient to supply all the wants of the citizens . . . for all time to come."[32] Also in 1852, James Goodhue, editor of the *Minnesota Pioneer*, surveying the St. Croix valley, wrote: "Centuries will hardly exhaust the pineries above."[33]

At the rates at which forests in Maine had been depleted, woods in the Great Lakes region should have lasted a very long time, but demand increased steeply when farms and towns on treeless prairies called for building materials, fences, railroad ties, and telegraph poles. The market area expanded enormously and levels of con-

29. Fred W. Kohlmeyer, *Timber roots: The Laird, Norton story, 1855–1905* (Winona Hist. Soc., Winona 1972) 139.

30. Fries, *Empire in pine*, 1951, 175.

31. Hurst, Institutional environment, 1983, 144–50.

32. *Congressional Globe*, 1851–52, cited in Twining, Lumbering frontier, 1983, 124.

33. Larson, *White pine industry*, 1949, 85.

sumption also rose. In the 1870s and 1880s, evidence for diminishing yields and shrinking stocks were disputed by spokesmen for the
lumber industry. The *Northwestern Lumberman* contested the validity of statistics and pessimistic economic projections made by various authorities, and even questioned estimates of the remaining
timber supply compiled for a special report on the nation's forests
in the census of 1880.[34] Easterners were suspected of trying to
damage Western enterprises. New Englanders, it was said, had been
content to profit from plundering their own forests but were now
trying to stop their successors from doing the same.[35] The state of
the market clearly indicated that no timber famine was impending.
Demand had not slackened nor were prices escalating. New sources
of supply in the South and Pacific Northwest were making good deficiencies from the Midwest, and lumber companies that had been
active in the lakes region moved their operations to these newly
opened areas of production.

By 1880, it was apparent to most observers that northern pineries
would not last forever. In the upper peninsula of Michigan,
white pine already faced extinction and many districts in Wisconsin
and Minnesota were sorely depleted. At this time, a second myth
captured popular imagination: that cutover tracts were destined to
become prosperous farmlands. Many people "took for granted that
the natural long-term value of the forest lay in being cleared for
farming."[36] Most lumber companies played little part in selling
stumplands but, at the end of the 1880s, a few realized that north
European immigrants were potential buyers for otherwise worthless
property. By 1899, a Milwaukee lumber merchant, J. L. Gates,
nicknamed "Stump Land" Gates, had acquired over 700,000 acres
of stumplands and was making "conversion of these lands a life
business."[37] In Minnesota, a few lumber companies turned themselves into land companies to dispose of cutover tracts. The American Immigration Company dealt in lands formerly owned by lumber
companies operating on the Chippewa and Mississippi rivers. The
Immigrant Land Company, organized in 1898, sold cutover above

34. Cronon, *Nature's metropolis,* 1991, 200.
35. Hurst, Institutional environment, 1983, 145.
36. Ibid., 141.
37. Fries, *Empire in pine,* 1951, 178; Michael Williams, *Americans and their forests: A
historical geography* (Cambridge University Press, Cambridge 1989) 235–36, discusses the
activities of J. L. Gates and his relations with other lumbermen.

Little Falls on the upper reaches of the Mississippi River. The American Colonization Company attracted foreign buyers to Chippewa valley stumplands through agents stationed in Finland, Sweden, Norway, Austria, and Russia.[38]

Railroad companies had well-organized agricultural departments, promoted agricultural extension services, established demonstration farms, maintained experimental plots, enlisted assistance from agricultural colleges, and offered practical help in blowing up stumps and draining swamps.[39] In 1914, the Northern Pacific circulated many leaflets advertising sales of railroad lands and listed 25 booklets addressed to applicants interested in particular localities. The Chicago, Milwaukee, St. Paul, and Pacific Railroad published a guide, *Wisconsin for home seekers,* and another brochure entitled *Many acres open to settlers;* and, in 1916, the railroad ran an excursion train to northern Wisconsin to demonstrate how cutover lands might be made ready for cultivation. The Great Northern provided cheap fares and hostel accommodation for settlers and their families on the way to new farms.[40] In northern Minnesota, some railroad land was sold very cheaply at the beginning of the 1920s. It was so wet that settlers were said to "cut their hay in hip boots and spread it on the stumps to dry."[41] In Michigan, individual speculators, in Wisconsin, large land companies, in Minnesota, railroad companies, and throughout the lakes region, a host of state and local government agencies were engaged in promoting agricultural settlement on land stripped of timber and hurriedly vacated.[42] Many

38. Larson, *White pine industry,* 1949, 405–6.

39. Roy V. Scott, Land use and American railroads in the twentieth century, *Agric. Hist.* 53 (1979) 683–703.

40. Ibid., 690–97; James B. Hedges, The colonization work of the Northern Pacific. *Mississippi Valley Hist. Rev.* 13 (1926) 311–42; Stanley N. Murray, Railroads and the agricultural development of the Red River valley of the north, 1870–1890, *Agric. Hist.* 31 (1957) 64–66.

41. Wallace Ashby, Problems of the new settler on reclaimed cutover land, *Agric. Engineering* 5 (February 1924) 27–29.

42. Disposal of cutover land and farming settlements are discussed in J. D. Black and L. C. Gray, *Land settlement and colonization in the Great Lakes states,* USDA Bulletin 1295 (GPO, Washington, D.C. 1925); W. A. Hartman and J. D. Black, *Economic aspects of land settlement in the cutover region of the Great Lakes states,* USDA Circular 160 (GPO, Washington, D.C. 1931); N. J. Schmaltz, The land nobody wanted: The dilemma of Michigan's cutover lands, *Michigan History* 67 (1983) 32–40; Gordon G. Whitney, An ecological history of the Great Lakes forest of Michigan, *Journ. of Ecology* 75 (1987) 667–84; L. Kane, Selling the cutover lands in Wisconsin, *Business History Review* 28 (1954) 236–47; J. I. Clark, *Farming the cutover: The settlement of northern Wisconsin* (Wisconsin State Hist. Soc.,

colonization companies sought only quick profits and lured suscep-
tible clients with glowing promises and enticing offers. "At no time
in American history," writes Hazel Reinhardt, "was the prospective
settler at so great a disadvantage."[43]

Attracting Settlers to Poorly Drained Cutover Lands

A decline in employment in lumbering threatened the survival of
remote settlements in northern counties, and reductions in freight
traffic dealt serious blows to northern railroads. State governments
directed special efforts to support established settlements and keep
open existing lines of communication. Their policy was to encourage immigration and promote agricultural colonization. It was
hoped that farmers would bring fresh life to beleaguered communities and inaugurate a new era of prosperity. All three northern states
established immigration bureaus and collaborated with land companies, railroads, and newspapers in maintaining agencies in New
York and, from time to time, in European cities.[44] All three states
organized propaganda campaigns to appeal to farmers. In 1895, the
Wisconsin legislature commissioned the state college of agriculture
"to prepare a bulletin or handbook describing the agricultural resources of Wisconsin, especially the newer and most thinly settled
districts, with reference to giving practical, helpful information to
the homeseeker." The primary aim was that "this book shall set
forth the advantages of the newer portions of this state for those
seeking homes on the land."[45] The dean of agriculture, William A.
Henry, was the author of *Northern Wisconsin: A handbook for the
homeseeker,* which was intended to inspire confidence, offer useful
advice, and inform prospective settlers how they might obtain assistance in pulling up stumps and draining swamps.[46] In Joseph
Schafer's opinion, the book was a reliable guide, "without question,

Madison 1956); Arlan C. Helgeson, *Farms in the cutover: Agricultural settlement in northern
Wisconsin* (Wisconsin State Hist. Soc., Madison 1962); Harold F. Peterson, Some colonization projects of the Northern Pacific Railroad, *Minnesota History* 10 (1929) 127–44.

43. Hazel H. Reinhardt, Social adjustments to a changing environment, in Flader (ed.),
Great Lakes forest, 1983, 209.

44. Fries, *Empire in pine,* 1951, 241–42.

45. James Willard Hurst, *Law and economic growth: The legal history of the lumber industry in Wisconsin, 1836–1915* (Belknap Press, Harvard University, Cambridge 1964) 595.

46. William A. Henry, *Northern Wisconsin: A handbook for the homeseeker* (Democrat
Printing Co., Madison 1896).

the most valuable single source of information in regard to northern Wisconsin at that time."[47] Wisconsin and other states loaned money to farmers to purchase land in the north and carry out improvements. They also subsidized counties to build roads and bridges and either invested in ditch-digging or backed the formation of drainage enterprises.

The most confident promotion was dedicated to the settlement of swamplands. A report prepared by the Minnesota State Drainage Commission was published in 1907. It recommended that "the reclamation of the state swamplands be continued on a more extensive scale and that a liberal annual appropriation be made for carrying on this work." The report presented the following conclusions:

First: The state's swamp lands can all be reclaimed and rendered productive at comparatively small cost.

Second: That when reclaimed these lands will be the most productive lands of the state.

Third: That in assisting in making Minnesota a greater state it is the duty of the state legislature to do at least its part in helping to develop the state's greatest natural resources.

Fourth: As a business proposition there is none better and none other before the people at the present time wherein the state can get as good returns for the money invested.

Fifth: The ever increasing population, the constant demand for farm lands and homes makes it imperative that the state watchfully guard her best interests and keep pace with the wonderful settlement and development now taking place in other western states. To do so it is necessary to make possible the settlement of these waste lands.[48]

Those conclusions were to be severely tested during the next twenty years. Costs of draining were to increase, peat soils were to prove less productive than expected, swamplands as natural resources were to be at least as much damaged as developed, drainage enterprises as businesses were to be badly hit by economic depression, and rural settlements were to suffer losses as well as gains in population. Meanwhile, in other lakes states, surveys of peat de-

47. Joseph Schafer, *A history of agriculture in Wisconsin*, Wisconsin Domesday Book 1 (State Hist. Soc. Wisconsin, Madison 1922) 140.

48. George A. Ralph, *Report on topographical and drainage survey of swamp and marshy lands owned by the state of Minnesota* (State Drainage Commission of Minnesota, Crookston 1907) 1.

posits carried out by geological and soil scientists in the early years of the twentieth century reported widespread progress in draining and advocated further rapid development of these areas "as additional territory for our people."[49] Frederick Newell, director of the U.S. Geological Survey Bureau of Reclamation, strongly urged that drainage be extended. In 1909, he wrote of swamplands in general that their soil "is extremely fertile and with effective systems of drainage the lands are capable not merely of supporting large and prosperous agricultural communities, but will be sources of strength to each commonwealth in which they are situated, instead of being, as now, breeding places of mosquitoes and other pests, centers of disease and a menace to land values in the neighborhood."[50] For many observers, the outlook was bright.

Swamps and Bogs in Central Wisconsin

Opinions expressed by scientists and technologists, endorsed by state governments, were likely to be received as authoritative by people to whom they were addressed. Many lumber companies, on the other hand, either had difficulty in persuading farmers to buy cutover, did not consider it worthwhile waiting for bids, or wished to discourage an influx of settlers who might impose fresh tax demands.[51] Government publicity held out hopes that poorly drained, cutover areas might be transformed into productive farms, but as early as 1898, a report by the Wisconsin State Forestry Commission warned that there was "no prospect that our denuded lands will be put to agricultural uses."[52] In the minds of both sellers and buyers, hopes that draining and cultivation would succeed struggled and conflicted with fears that debts would not be repaid out of low farm

49. Palmer, *Swamp land drainage,* 1915, 1; other reports in J. D. Towar, *Peat deposits of Michigan,* in *Mich. State Agric. Coll. Expt. Sta. Bull.* 181 (1900). Charles A. Davis, Peat: Its origin, uses, and distribution in Michigan, in State Board of Geol. Surv., *Report for 1906* (Lansing 1907) 279, warned of the risks of attempting to cultivate peat swamps on a large scale. Frederick W. Huels, *The peat resources of Wisconsin,* in *Wisc. Geol. Nat. Hist. Surv. Bull.* 45 (Madison 1915) 46, also issued a note of caution: "Peat has been used to some extent in connection with agricultural operations. But such use has not been extensive."

50. F. H. Newell, What may be accomplished by reclamation, *Annals Amer. Acad. Pol. and Soc. Sci.* 33 (1909) 658.

51. Hurst, *Law and economic growth,* 1964, 85; Helgeson, *Farms in the cutover,* 1962, 1.

52. *Report of the Forestry Commission of the State of Wisconsin* (Democrat Printing Co., Madison 1898) 18.

incomes. A history of draining and abandoning swamps and bogs in central Wisconsin throws light on these opposing attitudes. Swamps and bogs, whose soils were classed mostly as peat and muck, covered nearly 900,000 acres in seven central Wisconsin counties: Adams, Juneau, Monroe, Jackson, Clark, Portage, and Wood (fig. 7.1). Until the beginning of the twentieth century, they had few inhabitants and little economic development. Then, for a brief period from 1900 to 1920, pioneers scrambled to organize large drainage enterprises and establish farms. The attempt failed and by 1930, almost all the drainers had gone. The marshes were left almost as empty as they had been in the nineteenth century. After a few years, visitors came from cities to hunt, fish, and enjoy open space. At first they spent a week or two in the summers, but in the 1950s, some began to stay for longer periods and a few built permanent homes.

Exploiting Undrained Bogs and Swamp Forests

The swamps and lakes of central Wisconsin were valuable sources of beaver, muskrat, and mink for Indian trappers as well as French, English, and American fur traders.[53] By 1830, American lumbermen had started cutting the southernmost stands of pine along the Wisconsin River, and in 1831 a sawmill was built deep in Indian territory near Nekoosa.[54] Lumbering advanced rapidly up the Wisconsin River and its tributaries. After 1836, when the government obtained a cession of lands six miles wide along the river, sawmills sprang up in quick succession at Grand Rapids (now Wisconsin Rapids), then at Stevens Point, at Big Bull Falls (near Mosinee), at Wausau, and at Merrill. The advance which swept up other logging rivers, beginning with the Lemonweir, can be traced with the founding of settlements, some short-lived, at Mauston, New Lisbon,

53. F. J. Turner, The character and influence of the Indian trade in Wisconsin [1891], in *Early Writings,* 1938, 87–181; Jeanne Kay, Wisconsin Indian hunting patterns, 1634–1836, *Annals Assoc. Amer. Geog.* 69 (1979) 402–18. The following account of the historical geography of changing attitudes to land use problems in central Wisconsin is drawn from Hugh Prince, A marshland chronicle, 1830–1960: From artificial drainage to outdoor recreation in central Wisconsin, *Journ. Hist. Geog.* 21 (1995) 3–22.

54. Contemporary accounts of this and other ventures appear in James H. Lockwood, Early times and events in Wisconsin, *Wisc. Hist. Coll.* 2 (1856) 141, also 132–40; Ebenezer Childs, Recollections of Wisconsin since 1820, *Wisc. Hist. Coll.* 4 (1859) 175–77.

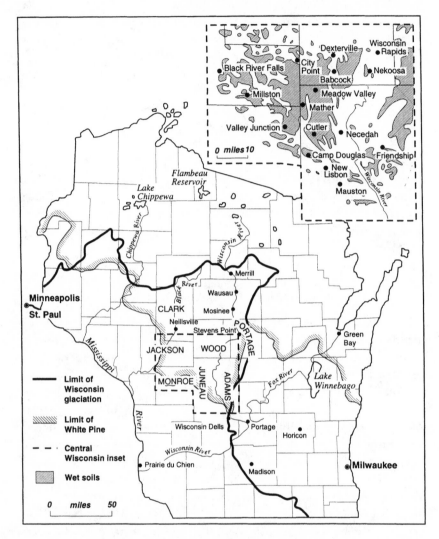

Figure 7.1 Location of wetlands in central Wisconsin

Knapp, Millston, and Black River Falls.[55] The swamps, which contained much tamarack, very little cedar, no hemlock, and not much
white pine, were passed over quickly.[56] Farmers began to cultivate

55. Albert G. Ellis, Upper Wisconsin country, *Wisc. Hist. Coll.* 3 (1857) 435–52; John G.
Gregory, *West central Wisconsin: A history* (Indianapolis 1933), vol. 1: 174–94.

56. T. C. Chamberlin, *Geology of Wisconsin: Survey of 1873–77* (Commissioners of
Public Printing, Madison 1877), vol. 2, pt. 3: 450.

Table 7.2 Area of swamp land grant and poorly drained soils in central Wisconsin

County	Swamp land grant (acres)	Poorly drained soils (acres)
Adams	70,157	99,264
Clark	30,189	54,516
Jackson	83,763	149,176
Juneau	155,358	191,296
Monroe	47,472	80,832
Portage	106,642	138,688
Wood	144,789	179,800
Total	638,370	893,572

Sources: Acreages of swamp land grants are set out in E. R. Jones, Keeping faith with the Swamp Land Fund, Seventh Report, Wisconsin State Drainage Association (Madison, March 1927) 1; Compilation of laws relating to the Swamp and Overflowed Land Fund of the State of Wisconsin (Madison 1882) 237. Acreages of poorly drained soils are compiled from soil surveys of counties in Wisc. Geol. Nat. Hist. Surv. Bull. (Madison, 1903–24).

cutover land near lumber camps, supplying hay and oats to feed teams of horses; potatoes, vegetables, meat, and dairy produce to feed lumbermen.[57] When lumbering moved north to denser pine-woods, farmers had to look further afield for markets.

In the 1850s farming settlement not only changed in character but expanded to a remarkable extent. During the decade 1850–60, population in the seven central counties of Wisconsin multiplied twenty-seven-fold. The expansion may be attributed first to a release of cheap swampland, second to an influx of immigrants from northern Europe, and third to the production of wheat as a cash crop. Under provisions of the Swamp Land Act of 1850, all swamp and flooded lands lying in the public domain were granted to the state on condition that proceeds from their sale were applied to reclamation (table 7.2). By 1857, the state had selected nearly a thousand square miles of swampland in central Wisconsin and this was sold to incoming settlers at less than a dollar an acre. In Juneau and Wood counties, some land was authorized to be sold for as little as 50 cents an acre. No mention was made of spending money from the fund on drainage works.[58]

57. Schafer, History of agriculture, 1922, 132; A. R. Whitson, Soil Survey of Adams County, in Wisc. Geol. Nat. Hist. Surv. Bull. 61D (Madison 1924) 15.

58. E. R. Jones, Keeping faith with the Swamp Land Fund, Seventh Report, Wisconsin State Drainage Association (Madison, March 1927) 1; Ellis, Upper Wisconsin, 1857, 446;

Cheap land attracted a stream of immigrants from northern Europe, many from Germany, Poland, Scandinavia, and Bohemia.[59] They reinforced a flow of English-speaking migrants who had been coming in considerable numbers from New York, Illinois, Michigan, and Ohio, and a few directly from Canada, England, and Ireland. In 1852, the state opened an Office of Emigration in New York to advise potential settlers, but the number of inquiries was far larger than expected. After three years the office was closed so as not to arouse further interest in immigration. Nearly 8% of Wisconsin's total population increase in the decade 1850–60 was concentrated in the seven central counties and in 1860, about one in every four inhabitants was foreign-born.[60]

What sustained farming settlement at this time was an expanding demand from industrial areas for wheat and a cheap means of shipping it from Milwaukee, Chicago, and other lakes ports to New York and across the Atlantic.[61] Wetlands were ill-suited to wheat growing, but in dry years farmers reaped bonanza harvests and also profited from scarcity prices. An exceptional drought occurred in 1859. While yields in other districts were disappointing, the crop from wetlands helped to raise Wisconsin to third place among America's wheat-producing states.[62]

Had it not been for the arrival of railways, the area of farmland might have expanded more slowly than it did. In the 1870s, lines were constructed northward to open the pineries.[63] Rail transport

---

Wisconsin Statutes 1867, chap. 51; Juneau County, Proceedings of County Board of Supervisors (1859), vol. 1: 145–50, 159; Adams County, Proceedings of County Board of Supervisors (23 November 1854), vol. 1 [State Historical Society Archives, Madison].

59. Hildegard B. Johnson, The location of German immigrants in the Middle West, *Annals Assoc. Amer. Geog.* 41 (1951) 32–36.

60. U.S. Census Office, *Eighth Census*, vol. 1, *Population of the United States, 1860* Washington, D.C. 1864), reported that the seven central counties of Wisconsin had a total population of 38,563 of whom 8,922 were foreign-born.

61. Schafer, *History of agriculture*, 1922, 81–96; John G. Thompson, The rise and decline of the wheat growing industry in Wisconsin, *Univ. Wisc. Bull.* 292 (1909).

62. A. P. Brigham, The development of wheat culture in North America, *Geog. Journ.* 35 (1910) 42–56. To the total of 15,657,458 bushels produced in Wisconsin, the seven central counties contributed 432,192 bushels in 1859.

63. Samuel Weidman, *Preliminary report on the soils and agricultural conditions of north central Wisconsin*, in *Wisc. Geol. Nat. Hist. Surv. Bull.* 11 (Madison 1903) 57, notes that the Wisconsin Central Railroad reached Stevens Point in 1871 and Prentice in Price County in 1873; the Wisconsin Valley Railroad, later acquired by the Chicago, Milwaukee and St. Paul Railroad, reached Grand Rapids in 1873, Wausau in 1874, and Merrill in 1881; the

made possible large-scale exploitation of such bulky commodities from the marshes as wire grass, used in the manufacture of matting, and sphagnum, used by nurserymen and florists. Speed of delivery was decisive for opening cranberry marshes in Wood, Juneau, and Jackson counties.[64] It was equally important for growers of other perishable products. Consignments of onions, cabbages, radishes, and strawberries were dispatched to Milwaukee, Minneapolis–St. Paul, and Chicago.[65] After 1880, railroads stimulated an expansion of dairying.

The new developments were important in preparing the way for an extensive conversion of wetlands into farms. At the end of the nineteenth century, less than half of the surface area of the seven counties had been improved.[66] In their unimproved state, bogs and swamps raised large numbers of prairie chickens, ducks, deer, and muskrat; in addition, many marshes were regularly cut for wild hay and some bogs were kept flooded for cranberries. They might have continued supporting a small number of hunters and gatherers and a few farmers indefinitely. As neighboring regions became more intensively utilized, the marshes were compared unfavorably as backward areas.

Draining Fever, 1900

By the end of the nineteenth century, lumbering had ceased in central Wisconsin and hopes for future prosperity were pinned on agriculture.[67] Vacant swampland was cheap, and now that the halcyon

---

Chicago, St. Paul, Minneapolis, and Omaha Railroad entered Neillsville in 1881.

64. Loyal Durand, Wisconsin cranberry industry, *Econ. Geog.* 18 (1942) 159–82, refers to the rapid expansion of cranberry growing in Wood County between 1873 and 1880; A. R. Whitson, W. J. Geib, L. R. Schoenmann, C. A. Leclair, O. E. Baker, and E. B. Watson, *Soil survey of Juneau County,* in *Wisc. Geol. Nat. Hist. Surv. Bull.* 38 (Madison 1914) 62; Weidman, *Preliminary report,* 1903, 47, 48, 58–62.

65. Whitson et al., *Soil survey Juneau,* 1914, 14. Distances by rail from Mauston were 128 miles to Milwaukee, 209 to Minneapolis–St. Paul, and 214 to Chicago.

66. The area of land in farms in the seven counties increased from 1,514,904 acres in 1880 to 2,528,697 acres in 1900. In 1880, only Monroe County had more than half its lands in farms; by 1900, farmland occupied more than half of all except Clark County.

67. Governor Smith's message to the state legislature, 31 December 1881, announced that a further 523,161 acres of swampland had been received from the federal government since 1855, bringing the total to 2,174,223 acres. The availability of land in central Wisconsin is referred to in passing in Arlan Helgeson, The promotion of agricultural settlement in northern Wisconsin, 1880–1925, Ph.D. diss., History, University of Wisconsin 1951. The state geologist, T. C. Chamberlin, *Geology of Wisconsin, Survey of 1873–79* (1883), vol. 1:

days of free land in the West were coming to an end, home-seekers began to search for unoccupied sections that had been passed by in the westward rush. The swamps looked promising. The surface was drier than in 1850, owing partly to increased runoff following tree felling, partly to the improvement of small patches that were farmed. Lumbering had prepared the ground for cultivation. It was cleared; some of it had been burnt down to the roots of the stumps; its surface had been enriched by a valuable layer of ashes; and the remaining stumps were easily freed from the loose peat. To those who had discovered the virtues of stiff, hard-to-drain mineral soils on wet prairies in Ohio, Indiana, Illinois, and Iowa the soil looked good. Most was pure black peat, richer in nitrogen than prairie soils, but lacking in other essential plant nutrients.[68] It was also stone-free, and so light that it required no plowing, only disking and rolling to make it ready to receive a crop. Other wet soils were admixtures of peat and sand or waterlogged fine sand of the Dunning series (fig. 7.2). Home-seekers needed no further inducement.

The rush was started by fire and spread like fire. In 1893, and again more extensively in 1894, fire swept across unoccupied cutover swamps in Juneau and Portage counties leaving a trail of ash behind. In 1895, one of the driest years on record in the marshes, when crops in Kansas failed because of drought, oats were sown on forty acres of burnt-over marsh near Valley Junction and an exceptional crop was harvested.[69] Dozens of farmers seized a chance to acquire derelict marshland on tax deed simply by paying outstanding taxes for it. The next year, 1896, was another exceptionally dry year and once more remarkable crops of oats and timothy were gathered.[70] Encouraged by their success, farmers and speculators sought more land.

A wet season occurred in 1897, then came three years from 1899 to 1901, each wetter than the last. Potash, left by the burning

---

684, forecast that "the highest utilization of the marshes will require artificial drainage."

68. Michael J. Goc, The Wisconsin dust bowl, *Wisconsin Magazine of History* 73 (1990) 166, reports that the presumed fertility of black peat brought settlers from Illinois. An old-timer recalled: "When they saw the black soil they thought it was just like home" but they soon came to realize "that's just about the poorest land on earth."

69. E. R. Jones, *Marsh problem old*, Report released 13 March 1928, 3 [Records of State Drainage Engineer, Madison]; A. R. Whitson, *Soils of Wisconsin*, in *Wisc. Geol. Nat. Hist. Surv. Bull.* 68 (Madison 1927) 133.

70. Whitson et al., *Soil survey Juneau*, 1914, 72.

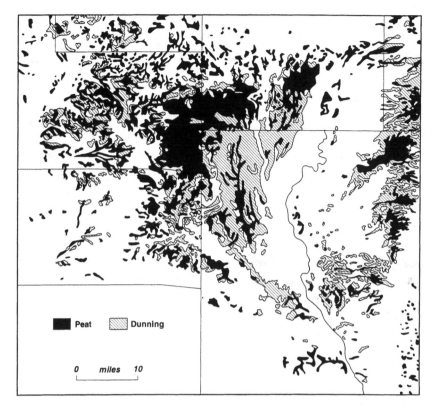

Figure 7.2 Wetland soils in central Wisconsin

Source: A. R. Whitson, Map of Soils of Wisconsin, Wisconsin Geological and Natural History Survey, 1926.

of peat, rapidly leached away. Oats failed and timothy died.[71] The only means of saving newly won fields was to provide them with ditches and drains. Families, including 75 from Illinois and Iowa, who were already settled on marshland along the Little Yellow River, had to secure financial assistance quickly to save their holdings.[72] But what was a desperate predicament for settlers offered speculators a golden opportunity to gain control over large blocks

71. Jones, *Marsh problem old,* 1928, 3.

72. Wisconsin Department of Agricultural Administration, Immigration Division Records, 1920–30, Proceedings of drainage district conferences, 1923–25 [State Historical Society Archives, Madison, Series 9/1/1, 3], examines histories of drainage undertakings from their foundation to the financial crises in the 1920s. For specific reference to the Little Yellow see Meeting, 9 April 1924, 37.

of threatened land by advancing loans or making outright purchases. It was confidently expected that the temporary distress
caused by one or two wet years would be recompensed by timely
investment in drainage works and that when land was improved it
could not fail to rise in value. County officials were pleased to co-
operate because they could now expect to collect taxes on lands
abandoned by lumbermen and they might soon be able to raise assessments on lands improved by draining.

In 1900 promoters organized, under Wisconsin law, 60,000
acres in the Little Yellow Drainage District almost overnight. In the
next three years no fewer than eight other districts were formed in
rapid succession, each promoter eager to sell his land before his rivals. By 1907, the area in organized drainage districts amounted to
at least 317,537 acres.[73] The first drainage districts were established
in the very center of central Wisconsin, occupying nearly all the
least promising wet land, situated on deep peat with a water table
almost at the surface. Settlers' first problems stemmed from the
poverty of the peat. They were problems of management: how to
check peat wastage, how to treat deficiencies of essential minerals,
how to avoid frost damage, how to control flooding, how to prevent
fire. Districts established at later dates in remote parts of Jackson,
Clark, and Wood counties suffered, in addition, from serious lack
of support from small investors. Far from large settlements, they
were unable to attract sufficient numbers of settlers with capital to
bear the costs of draining. Their difficulties were financial as well as
agricultural.

Early Difficulties of Peat Farming

In 1903, the first signs of trouble became apparent. Consulted about
further additions to Portage County Drainage District, the College of
Agriculture recommended that tests should be made to determine
the quality of the soil before any more districts were organized.
Some promoters rightly feared that such investigations would deter
potential buyers, but settlers struggling to wrest a living from their

---

73. A. R. Whitson and E. R. Jones, Drainage conditions of Wisconsin, *Univ. Wisc. Agric.
Expt. Sta. Bull.* 146 (Madison 1907) 22. It was reported that 215 miles of ditches had been
dug and the estimated cost of draining was about $2.68 per acre. These estimates were much
too low. Within twenty years, 344 miles of ditches had been dug and the cost had risen to
$6.83 per acre, more than two and a half times as much as the sum originally budgeted.

hard-won acres appealed to the college for help and advice. Everywhere the peat shrank as water drained out and the dry fibers began to decompose. Under an uninterrupted corn-grain-hay-hay rotation the rate of subsidence was about one foot in 35 years.[74] Soon the original ditches became too shallow to provide adequate outlets. Deepening them brought fresh problems: to maintain sufficient fall for effective discharge and to prevent the crumbling and collapse of sides cut down into sand below the peat.

When, after 1905, soils were tested, they were found to be markedly acid and seriously deficient in certain minerals essential for plant growth, notably potassium and phosphorus.[75] Two adjustments were called for. First, crops had to be chosen which were tolerant of acid conditions, since the cost of applying sufficient lime to make the soil mild enough for sensitive crops was prohibitive. Second, it was imperative to make good the deficiency of potassium and phosphorus by applying appropriate quantities of commercial fertilizers. The cost of initial treatment would be about two to four dollars per acre.

Frost was, and still is, one of the most frequent causes of crop failure, one of the least predictable and also least avoidable hazards of marsh farming. Several farms experienced frost two years out of three and in cold seasons lost one-tenth or more of their crops.[76] Yet the magnitude of the problem was not fully appreciated until recently, not because statistics were lacking, but because their significance was not properly understood. Why did farmers report the incidence of killing frosts in July while official statistics recorded the latest spring frost on 12 June 1903 and the earliest autumn frost on 8 August 1904?[77] Screen temperatures, five feet above the ground,

74. A. R. Albert, Status of organic soil use in Wisconsin, *Proc. Soil Sci. Soc. of Amer.* 10 (1945) 275–78, esp. 278.

75. H. W. Ullsperger, Report on Portage County Drainage District, 1916 [Records of State Drainage Engineer, Madison].

76. E. R. Jones and B. G. Packer, Drainage district farms in central Wisconsin, *Univ. Wisc. Agric. Expt. Sta. Bull.* 358 (Madison, October 1923) 6, 28.

77. U.S. Department of Agriculture, Weather Bureau, *Climatic summary of the United States* (Washington, D.C. 1930) sect. 48, 19, data for Meadow Valley, Juneau County, 1894–1930; Whitson and Jones, *Drainage conditions of Wisconsin*, 1907, 9, reported that in central Wisconsin in 1904 "a heavy frost in August killed corn, damaged sugar beet and potatoes." Before the record frosts of 1903 and 1904, Weidman, *Preliminary report*, 1903, 55, stated: "In general frosts are not earlier in the autumn than for the southern portions of the state, northern Illinois or Iowa though they are likely to occur later in the spring, but not to such an extent as to be at all discouraging to agriculture in most of its phases."

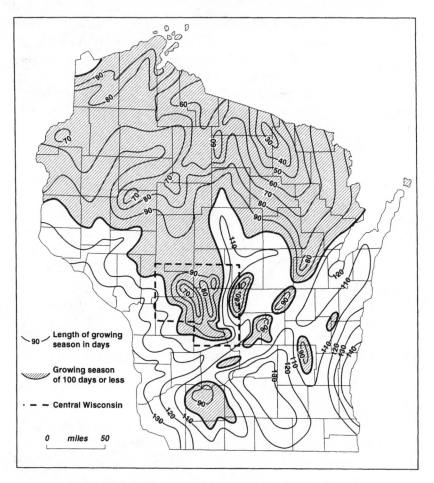

Figure 7.3 Normal annual growing season for cultivated
plants in Wisconsin

Source: Jen Yu Wang, The phytoclimate of Wisconsin, M.S. thesis, Meteorology,
University of Wisconsin 1955.

are almost always much higher than corresponding ground temperatures, five inches above the surface of the soil. The difference is yet
greater above peat soils than above mineral soils, because peat is a
poor conductor of heat[78] (fig. 7.3). Below the surface the temperature of peat changes little from day to day nor does it influence the

78. Jen Yu Wang, The phytoclimate of Wisconsin, M.S. Thesis, Meteorology, University
of Wisconsin 1955, 25.

temperature of air in contact with it. It releases little heat to raise the temperature of cold air above it, so that cold air descending into a peat lowland remains cold for a long period. Allowance must also be made for the fact that plants differ in their hardiness to frost. In most localities crops sensitive to frost, such as corn for grain, late potatoes, and soybeans for seed, were doomed to failure, but hay crops were generally hardy enough to survive and several truck crops could be raised in a short season of as little as 70 days. Draining, cultivation, and sanding reduced the incidence of frost.

Draining also increased the danger of fire. Unless quickly controlled, fire would ignite dry peat and burn down to the level of groundwater or bedrock, whichever was the higher. In Portage and Wood counties, some areas of peat had been completely burnt out before the First World War.[79] The remedy was worse than the disease as far as one season's crops were concerned, since the only way of preventing fire spreading once it was below the surface was to flood the area.

Another recurrent threat to drained land paradoxically was flooding, a menace that increased as the area protected from natural flooding was extended. Floodwaters mounted swiftly in height, particularly where they were prevented from ranging freely over floodplains. In a great flood of October 1911, part of Black River Falls was swept away and many acres of marshland in Jackson and Juneau counties were inundated.[80]

When the Wisconsin Soil Survey and College of Agriculture examined the peatlands, they recommended that farmers should concentrate their efforts on small tracts of better land. They repeatedly stated the benefit to be derived from putting in supplementary tile drainage and stressed the urgency of applying artificial fertilizers. They advised home-seekers to exercise caution and discrimination in selecting wetland, not to pay much for it, and to reserve enough capital to carry out further improvements. The state drainage engi-

79. Weidman, *Preliminary report,* 1903, 46, referring to the swamps of Wood and Portage counties in 1903, states: "Because of repeated fires but little of the peat in some swamps is left"; A. R. Whitson, W. J. Geib, T. J. Dunnewald, and Lewis P. Hanson, *Soil survey of Portage County,* in *Wisc. Geol. Nat. Hist. Surv. Bull.* 52C (Madison 1918) 63; A. R. Whitson, W. J. Geib, Guy Conrey, W. C. Boardman, and Clinton B. Post, *Soil survey of Wood County,* in *Wisc. Geol. Nat. Hist. Surv. Bull.* 52B (Madison 1918) 53.

80. Lawrence Martin, *The physical geography of Wisconsin,* in *Wisc. Geol. Nat. Hist. Surv. Bull.* 36 (Madison 1932) 360.

neer, E. R. Jones, attended personally to inquiries from potential
settlers from Chicago, Des Moines, and Iowa Falls, warning them
about conditions they were likely to encounter. He also corresponded with one or two promoters of drainage enterprises, urging
them to consolidate their gains, to cut ditches deep enough and
close enough to provide adequate outlets, to repair channels that
were choked with sand and vegetation, and to deepen those that
had become too shallow as a result of peat shrinkage.[81]

The College of Agriculture acted only in an advisory capacity. It
was powerless to prevent the formation of two vast, ill-considered
schemes in Clark and Jackson counties in 1916 and 1918. As pre-
dicted, both districts were in difficulties from the outset. They had
been unsuccessfully pioneered in previous decades and most of the
original settlers had left. The land company which launched the
projects claimed that it set itself to "relieve, if possible, the farmers
who are doomed to perpetual discouragement unless some big
ditches are made."[82] But the first need was for roads to make districts accessible. Ditches might follow roads; they could not profitably precede them. The prospective benefits of the Clark County
scheme were not sufficient, in the opinion of the state drainage engineer, to justify the cost of digging the ditches. Above all else, new
settlers had to be attracted in numbers sufficient to pay for improvements and this the district failed to accomplish. In 1919, a
new Drainage Law was passed conferring on the state drainage engineer powers to restrain ventures of this kind, but the damage had
already been done.

Collapse of Drainage Districts

The insecure financial structure of the weaker districts began to
crack before the First World War. In 1914, a soil survey report
from Juneau County observed that prospects for draining peat overlying sand were "very unpromising." Drainage district tax assess-

81. E. R. Jones, Letter to A. C. Willard, Necedah, re Little Yellow, 8 June 1915; E. R.
Jones, Letter to Nye Jordan, Mauston, re Cutler, 9 June 1915; E. R. Jones, Letter to Nye
Jordan, re overgrown ditches in Cutler, 5 March 1918; E. R. Jones, Statement to John F.
Baker, Attorney General, re Cutler affairs, 25 October 1920 [Records of State Drainage
Engineer, Madison].

82. John H. Gault, Letter to E. R. Jones re Clark County, 25 July 1914; further discussion
in E. R. Jones, Letter to James A. Phillips, Neillsville, 25 October 1915 [Records of State
Drainage Engineer, Madison].

ments were too high to yield profits.[83] In Wood County, it was reported that "drainage is universally unsatisfactory here," and in Portage County, some drainage districts, notably Dancy, made little progress in developing peat soils.[84] In 1917, a farmer in the notoriously mismanaged Beaver district, unable to pay the high taxes demanded, gave up his farm.[85] He was one of the first to be beaten by the struggle to meet the rising costs of draining, but many farmers were finding it increasingly difficult to maintain their tax payments while paying for supplementary tile drains and fertilizers, upon which survival of their operations depended. An additional burden fell upon farmers in Cranberry Creek district who had to fight a Supreme Court action to prevent cranberry growers interfering with the normal flow of water.[86] In a paper read to the State Drainage Association in March 1919, E. R. Jones observed that many of the existing projects were "unattractive because of their isolation." His sharpest criticism was directed at promoters who "have asked too much for their land and require initial payments that the settler cannot afford," adding that "in some cases they have misrepresented the state of drainage."[87]

Several speculators were having increasing difficulty in finding buyers for their unsold lands when B. G. Packer of the Wisconsin Board of Immigration publicly attacked their activities. In an article published in the *Milwaukee Journal* in January 1922 he is reported to have said that the activities of certain unlicensed traders in Chicago and other large cities "have been wicked, among them some operating near Mauston, Germantown, Nekoosa, New Lisbon, Necedah and south-east of Wisconsin Rapids."[88] The charge was unmistakably leveled at promoters of central Wisconsin drainage

83. Whitson et al., *Soil Survey Juneau,* 1914, 64.

84. Whitson et al., *Soil Survey Wood,* 1918, 61; Whitson et al., *Soil Survey Portage,* 1918, 59.

85. E. F. Bean, Letter to E. R. Jones re a farmer at City Point who gave $2,400 for an 80. His first assessment for drainage tax was in 1907 and it was to run until 1931. In 1916 he paid $101.79. In 1917 he was assessed for $140, which he did not pay and was obliged to give up his farm [Records of State Drainage Engineer, Madison].

86. *Cranberry Creek Drainage District* vs. *Elm Lake Cranberry Company and others,* Supreme Court of the State of Wisconsin, Autumn Term 1919.

87. E. R. Jones, A sane plan for marsh development, *Fourth Report, Wisconsin State Drainage Association Proceedings* (January 1919–December 1920) 48–54. The paper was read in March 1919.

88. Hewing a great new empire in northern Wisconsin, *Milwaukee Journal* (22 January 1922) pt. 6, 2.

districts, and as a result of ensuing bitter recriminations, a court action was brought against the commissioners of the Cutler district alleging that they had misrepresented the quality of the land to purchasers. It was most unfortunate that Cutler should have been singled out for attack, because the agricultural value of its land was higher than that of other districts. Much of it was underlain by silt, its ditches had a good fall, and some of them were in good repair. On the other hand, the plaintiffs had a strong argument against the organizers of the district. It was undeniably a speculative venture. A large tract of land had been bought by an agent of a Chicago lumber company and sold to the Peddie Land Company, which organized the district in 1911. The company still held about 60% of the land and were promoting it much more vigorously than their competitors. Settlers had been charged $40 or $50 per acre for it, a high price, but certainly not unreasonable, for land that was adequately drained. The verdict returned by a jury made up of local farmers at 1:30 AM on 8 May 1922 upheld the charge of fraud against the commissioners and $4,200 damages were awarded to the plaintiffs. After the publicity the case attracted, Cutler affairs went from bad to worse and the Peddie company went bankrupt.[89]

As the financial situation deteriorated the state drainage engineer stated clearly the problems facing the depressed districts and encouraged settlers to make the best of a bad job.[90] He collaborated with B. G. Packer in preparing a survey of drainage district farms in central Wisconsin, interviewing 87 farmers in the company of S. H. McCrory, chief investigator of the USDA's Bureau of Drainage.[91] A general impression was that projects had been overambitious and that farmers had attempted to reclaim more land than they could afford. The results of the survey were published in October 1923. In November, a conference was called by the State Department of Agricultural Administration to inquire into evidences of indebtedness of drainage districts.[92] Among the first districts to be inves-

---

89. Cutler File [Records of State Drainage Engineer, Madison].

90. E. R. Jones and O. R. Zeasman, An outlet drain for every farm, *Univ. Wisc. Agric. Expt. Sta. Bull.* 351 (Madison, December 1922); Jones and Zeasman, Drain wet fields, *Univ. Wisc. Agric. Expt. Sta. Bull.* 365 (Madison, June 1924).

91. Jones and Packer, Drainage district farms, 1923, 1–48.

92. Wisconsin Department of Agricultural Administration, Inquiry into indebtedness of drainage districts, 1923 [State Historical Society Archives, Madison].

tigated was Kert Creek, whose development was admitted by B. M. Vaughn, the attorney representing it, to be "probably as slow as any district in central Wisconsin."[93] Insufficient drains were provided in the beginning and many of the original settlers left; one who started on a large scale was unable to complete his project and retired broken in health; another who began to reclaim 200 acres committed suicide. New ditches were dug after the First World War at great expense, but not more than nine or ten settlers remained to benefit from the improved drainage and bear the increased costs of providing it. By June 1924, only one settler was left and the outstanding debt amounted to $132,688, the third-largest debt among fifteen districts.[94] In April 1924, the conference examined conditions in the rapidly failing Cutler district and then considered Little Yellow, the most heavily indebted of all districts. B. M. Vaughn was of the opinion that pioneers who had come from Illinois and Iowa did not know how to work these lands and was hopeful of attempts to attract Danes and Dutch "who are raised on lands of similar character and know how to farm them." In 1924, no more than 10,000 out of 60,000 acres were being farmed and only 40 farmers remained[95] (fig. 7.4). In June 1924, the debts of Clark and Jackson County districts exceeded their capital costs (table 7.3). Four districts organized by the beginning of 1903 still owed $518,818, a sum greater than the amount they had originally budgeted to spend.[96] Eight out of 15 districts had paid more than half their debts, but only Dandy Creek, Portage, and Lemonweir owed nothing. The total debt outstanding amounted to $1,096,477.

What hope remained in 1924 of paying the debts of the districts was extinguished by the deepening agricultural depression. Prices for almost all farm produce fell, but oats and hay, the leading crops of swampland farms, were severely hit by shrinking demand resulting from a decline in the number of horses as motor vehicles took the place of draft animals. In 1924, some districts, such as Remington, sought to issue refunding bonds to enable them to continue to

93. Ibid., Meeting 23 November 1923, 11.

94. E. R. Jones, Central Wisconsin Drainage Districts 1924, Reports and Memoranda [Records of State Drainage Engineer, Madison].

95. Wisconsin Department of Agricultural Administration, Meeting 9 April 1924, 37–42.

96. Jones, Central Wisconsin Drainage Districts 1924, Reports and Memoranda [Records of State Drainage Engineer, Madison].

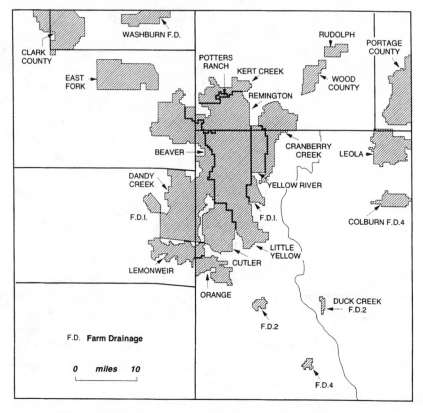

Figure 7.4 Drainage enterprises in central Wisconsin, 1924

Named enterprises are drainage districts; numbered county farm drainages are abbreviated F.D. *Source:* E. R. Jones, Central Wisconsin Drainage Districts 1924, Reports and Memoranda [Records of State Drainage Engineer, Madison].

meet their interest payments for another five or six years.[97] Others, such as Beaver, brought Wisconsin bonds into disrepute by defaulting.[98] By 1927, over 180,000 acres in central Wisconsin drainage districts were subject to tax deed and there was little prospect of the

97. E. R. Jones, Letter to Remington bondholder re refunding bonds, 2 April 1924 [Records of State Drainage Engineer, Madison].

98. Percival, Brooks and Coffin, Bankers, Chicago, Letter to Beaver Commissioners, 3 July 1921, regretting the second occasion on which Beaver had defaulted on its bonds, warning that such action "will only serve to place every drainage bond in Wisconsin under a cloud" [Records of State Drainage Engineer, Madison].

Table 7.3 Drainage districts in central Wisconsin

Founded date	Drainage district	Area (acres)	Wet acres	Ditch miles	Farms (no.)	Cost ($)	Debt ($)
1900	Little Yellow	60,000	48,000	40	40	412,648	250,000
1901	Beaver	33,480	27,600	50	25	176,303	72,903
1902	Dandy Creek	33,920	23,000	54	75	122,988	0
1902	Remington	25,920	21,820	50	25	145,552	63,227
1902	Portage County	52,731	43,452	100	25	469,102	0
1903	Kert Creek	8,867	8,204	50	1	143,988	132,688
1903	Dancy	36,082	22,951	41	20	315,775	81,478
1903	Lemonweir	15,000	11,000	30	–	48,400	0
1903	Cranberry Creek	19,139	13,785	67	25	188,481	114,839
1904	Juneau County	14,960	14,000	24	–	33,572	352
1906	Leola	17,438	15,000	54	20	121,000	19,000
1911	Wood County	6,000	1,120	20	15	40,000	35,000
1912	Cutler	20,745	20,745	49	27	104,990	69,990
1916	Clark County	98,000	32,000	73	20	152,500	165,000
1918	Jackson County	25,760	13,000	20	20	85,000	92,000
Total		468,042	315,677	722	338	2,560,299	1,096,477

Note: Table excludes Buena Vista and Orange drainage districts, for which no information is available.
Source: Survey of central Wisconsin drainage districts, 1 June 1924 [Records of State Drainage Engineer, Madison].

trend being arrested let alone reversed.[99] Before 1927, almost all tax delinquent land belonged to absentee owners. Resident farmers paid their taxes regularly until the Wisconsin Supreme Court decided, in 1927, in favor of bondholders in the Dancy district, that those who were still paying drainage taxes were liable to an additional assessment to make up the deficit caused by the delinquency of others.[100] The award was a hollow gesture because resident landowners left rather than pay an additional imposition. Farmers in other districts were alarmed and ceased to pay their taxes. The mischief spread further, because penalties accumulating on unpaid taxes soon exceeded the market value of the land.[101]

In 1930, drainage districts in Wisconsin, compared with those in six other Midwestern states, had a larger proportion of land in arrears on capital repayments and interest charges on bonded indebtedness.[102] In the seven central counties, the area of farmland provided with drainage was little more than half the acreage recorded ten years earlier.[103] Remote counties, such as Clark and Jackson, which had very little drained land in 1920, lost nearly all of it during the decade. Counties with extensive drainage enterprises, such as Juneau and Wood, suffered heavy losses but still had considerable areas of drained land in 1930[104] (fig. 7.5). Maps of 1937 show that many square miles were left without a single occupied farm (fig. 7.6). Hardly any abandoned farms were reoccupied; they remained empty until they rotted away, were burnt down, or were flooded by new lakes. During the 1930s, most drainage districts in cen-

---

99. E. R. Jones, Keeping faith with the Swamp Land Fund, *Seventh Report, Wisconsin State Drainage Association* (Madison, March 1927) 9.

100. Jones, *Marsh problem old,* 1928, 6.

101. E. R. Jones, Preliminary plan for putting permanent prosperity into the drainage districts and counties of central Wisconsin, submitted to the Attorney General, 1924 [Records of State Drainage Engineer, Madison].

102. U.S. Bureau of Census, *Fifteenth census of the United States, 1930,* vol. 4, *Drainage of agricultural lands,* 335: Wisconsin 37%, compared with Minnesota 16%, Michigan 7%, Iowa 5%, Illinois 3%, Indiana 3%, Ohio 1%.

103. U.S. Bureau of Census, *Fourteenth census of the United States, Agriculture,* vol. 7, *Irrigation and drainage,* for earlier acreages.

104. Leslie Hewes, Drained land in the U.S., in the light of the Drainage Census, *Professional Geographer* 5 (1953) 6–12, discusses the value and limitations of the census data. Goc, Wisconsin dust bowl, 1990, 169, discusses drainage assessment arrears. In 1932, in Wood County, 92,000 acres of wetlands were tax delinquent; and in 1933, in Juneau County, only 12% of drainage taxes were paid.

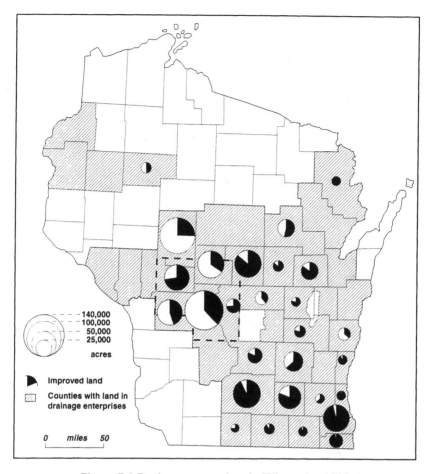

140,000
100,000
50,000
25,000

acres

◗ Improved land

▨ Counties with land in
drainage enterprises

0 miles 50

Figure 7.5 Drainage enterprises in Wisconsin, 1930

Source: U.S. Bureau of Census, *Fifteenth Census of the United States, 1930,* vol. 4, *Drainage of Agricultural Lands* (Washington, D.C. 1932).The area of each circle represents the total area of land in drainage enterprises. The black portion of a circle represents improved land within drainage enterprises.

tral Wisconsin were either dissolved or reduced in size. Only small areas were still recorded as "improved land" in the 1940 census of drainage (fig. 7.7). As farmers departed, an opportunity for restoring swamps to their former uses seemed to have returned. Cranberry growers openly welcomed the collapse of drainage districts and looked forward to regaining control of a few streams for their own

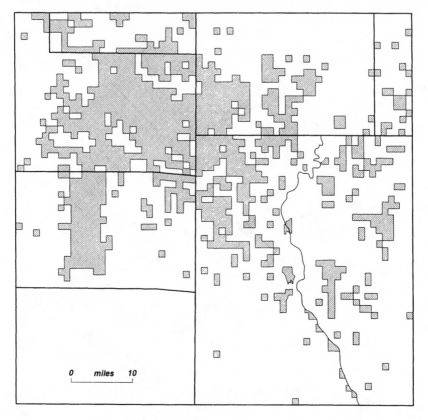

Figure 7.6 Sections within which no farm stood in central Wisconsin, 1937

Source: County maps prepared by the State Highway Commission of Wisconsin in the 1950s. Data for occupied and vacant farms are correct as for 1937.

purposes.[105] Meanwhile most of the swamps and bogs became as empty as they had been in 1890.

Conserving Swamps and Bogs in Central Wisconsin

Land that ceased to pay taxes was forfeited to the county, state, or federal government. In addition, the federal government acquired

105. C. L. Lewis, Presidential address, *Thirty-ninth Annual Meeting of Wisconsin State Cranberry Growers' Association* (Wisconsin Rapids, 9 December 1925) 7, declares: "Where ten years ago the sentiment of the country was strongly pro-drainage, we now have a strong anti-drainage sentiment. It is a very opportune time to change laws that are detrimental to our welfare."

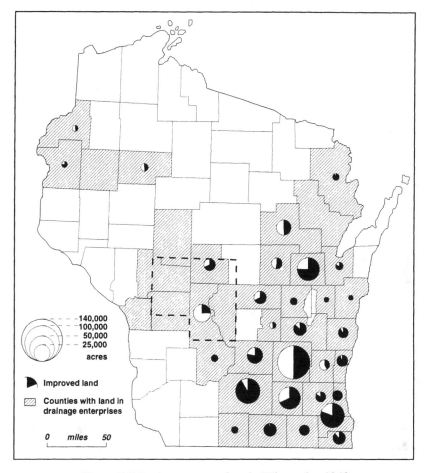

Figure 7.7 Drainage enterprises in Wisconsin, 1940

Source: U.S. Bureau of Census, *Sixteenth Census of the United States, 1940: Drainage of Agricultural Lands* (Washington, D.C. 1942).The area of each circle represents the total area of land in drainage enterprises. The black portion of a circle represents improved land within drainage enterprises.

land for the purpose of retiring isolated and unremunerative farms from agriculture and purchased other parcels to consolidate public holdings. It was not the intention of public authorities to create an empire in the swamps but eventually, by default, they found themselves in possession of almost all wetlands in central Wisconsin. In managing their estate, the primary objective was to conserve its natural resources: to promote regeneration of forests, to improve wildlife habitats, to control sources of water supply, to improve recreational facilities.

The State of Wisconsin had legislated on conservation issues over a long period.[106] It was among the first states to set up a forestry commission and appoint a game warden. Federal agencies also played an important role in reviving interest in forest and wildlife conservation. Aldo Leopold, assistant director of the U.S. Forest Service Products Laboratory, worked closely with William Aberg of the Izaak Walton League in drafting legislation that led, in 1933, to the establishment of a new Wisconsin Conservation Department. Christine Thomas describes 1927 to 1966 as "golden years" for conservation in Wisconsin. Federal laws set up new national forests, regulated timber exploitation, financed a Civilian Conservation Corps that in 1933 ran 14 summer camps, each manned by 200 recruits, to fight forest fires and help raise tree crops.[107] The state deployed further resources to manage its own forests and game reserves.

Over 250,000 out of two million acres of county forest land in Wisconsin in 1948 lay in the seven central counties, much of it in remote swamps in Clark and Jackson counties.[108] The objectives of forestry and wildlife conservation were in some respects incompatible, but over much of central Wisconsin the aims of both were pursued simultaneously. The traditional object of wildlife conservation was to protect and preserve stocks of game, above all, deer. A bold attempt to check the increase of the deer herd before it destroyed all the remaining palatable browse was made in 1936–37, when Adams, Jackson, Monroe, and Wood counties were declared open for hunting. In 1947, detailed observations of the effect of deer damage on half a million acres of commercial forest in central Wisconsin concluded that it was doubtful whether, "even with the foresters' efforts to favor it, the white pine can come back in any force."[109] The threat to the forest, like the threat to the deer herd, was that its greatest loss was suffered by its youngest members.

106. Christine L. Thomas, One hundred and twenty years of citizen involvement with the Wisconsin Natural Resources Board, *Environmental History Review* 15 (1991) 65, 68.
107. Ibid., 72, 73.
108. State Conservation Commission of Wisconsin, *Twenty-first biennial report* (Madison 1949) 40. Clark County had 128,097 acres and Jackson County 107,849 acres.
109. Stanley G. DeBoer, *The deer damage to forest reproduction survey*, Wisc. Cons. Dept. Publ. 340 (Madison 1947) 7.

The storage of water, management of trout streams, and erosion control measures in the central counties were harmoniously coordinated. The greatest change that occurred after the collapse of the drainage districts was the enlargement of the area of open water. Artificial reservoirs and flowages created about four-fifths of the 60,000 acres covered by lakes in 1950, the largest of them occupying much of the floodplain of the Wisconsin River, ponded behind two dams.[110] Between the 1920s and 1950, the Consolidated Water Power and Paper Company bought large tracts of former farmland, schools, and villages to create Petenwell Flowage and Castle Rock Flowage in the Wisconsin River valley (fig. 7.8). The impounded water was used to generate hydropower and run a sawmill. Archaeological and historical investigations have indicated that farms along the edges of the reservoirs were small and poorly equipped, lacked electricity or public water supplies, and changed hands fairly frequently.[111] In many former drainage ditches and tributary streams the Wisconsin Conservation Department was able to establish good habitats for trout, some 800 miles of waterways in the central counties being opened for fishing.[112] This was a great improvement on conditions that existed before drainage when water was thick with mud and flowed sluggishly.

Following the dissolution of drainage enterprises in central Wisconsin, attitudes to wetland conservation changed fundamentally. In the "decade of the drainage dream," 1910–20, conservationists, most of whom were trained as soil scientists or engineers, presumed that if the agricultural productivity of wetland could be improved by artificial drainage, then drainage should be regarded as beneficial.[113] Acknowledgment of forestry, wildlife, and water conservation as alternative goals revolutionized attitudes. In 1948, Aldo Leopold, most radical of all the new ecologists, proclaimed: "The ultimate value in these marshes is wildness." Experience had

110. Wisconsin Conservation Department, *Wisconsin lakes*, Publ. 218–51 (Madison 1951).

111. Mark S. Cassell, *A pale blank area: Archaeology and context of agrarian development in Wisconsin's central sands, 1860–1940* (Mississippi Valley Archaeology Center, University of Wisconsin–La Crosse 1995) 3, 13–19, 29–31.

112. Wisconsin Conservation Department, *Wisconsin trout streams*, Publ. 213–51 (Madison 1951).

113. The phrase "drainage dream" is from Aldo Leopold, *A Sand County almanac* ([Oxford University Press, New York 1949]; Ballantine Books, New York 1990) 11.

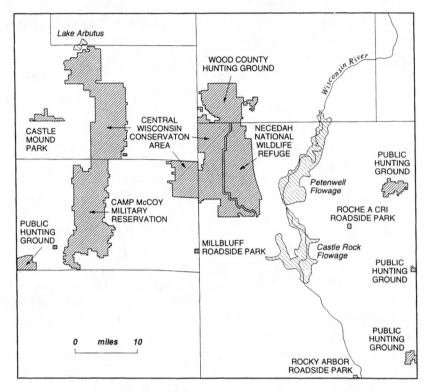

Figure 7.8 National and state reserves in central Wisconsin, 1953

Source: County maps of State Highway Commission of Wisconsin, 1953.

taught him that lasting solitude was not to be expected, let alone enforced: "All conservation of wildness is self-defeating, for to cherish we must see and fondle, and when enough have seen and fondled, there is no wildness left to cherish."[114] Development and free access were prevented in some public reserves. The Necedah National Wildlife Refuge, occupying a large tract of Juneau County along the Little Yellow River; conservation areas in Jackson, Juneau, and Monroe counties; and an extensive military training ground in Monroe County excluded members of the public. Bird sanctuaries and other reserves served minority interests. A majority of people wanted space to trek, canoe, hunt, fish, and build summer cabins.

114. Ibid., 107–8.

Within a few years, Leopold's wilderness dream had vanished as completely as the earlier drainage dream. More and more people sought to escape for longer periods from crowded cities, to enjoy the shade of young trees and cool waters that occupied areas once covered by open bogs and marshes. Those who had built shacks in the 1930s to spend summer vacations camping, in the 1950s began to turn them into better-equipped homes, connecting with electric power and telephone lines, installing kitchens and bathrooms, building garages larger than the original cabins to house not only one or two automobiles but also a power boat and later, when homes had been made habitable for winter, to provide additional space for snowmobiles.[115] The key to these developments was transportation by private car and the construction of paved, all-weather roads.

State and county authorities yielded to popular pressure to provide facilities for outdoor recreation. Derelict swamps in southwest Wood County were opened as public hunting grounds in the 1930s, together with smaller areas in Adams and Monroe counties. The State Conservation Department administered four parks in places of outstanding scenic beauty, and newly dammed lakes on the Wisconsin River offered visitors a variety of attractions. General taxes and license revenue were used to build new roads, furnish roadside picnic places, and lay out parking lots. Private developers also played an important part in building facilities for visitors and long-term residents.

Attitudes to Abandoned Wetlands in the Northern Lakes Region

The problem of how to manage swamps and bogs from which native Americans had been expelled, which lumbermen had deserted, or where farmers were left struggling against fearful odds, continued to perplex and torment people in the northern lakes region well into the twentieth century. Federal and state governments commissioned inquiries and attempted to formulate policies to deal with economic decline and social distress. Reports of government surveys provide evidence for contemporary thinking and prevailing attitudes. Many observers found it difficult to accept that poor drainage, poor soils, and short growing seasons prevented northern

---

115. John Fraser Hart, Resort areas in Wisconsin, *Geog. Rev.* 74 (1984) 193.

wetlands from being used more and more intensively. A hoped-for progression from hunting and gathering, through lumbering and other forms of destructive exploitation, to settled cultivation failed to materialize. The point at which development was arrested came when settlers attempted to drain poorly drained stumplands and convert them into farms. For a brief period from the late 1890s to about 1920, northern wetlands were feverishly sought after by eager pioneers. Prospective farmers were canvassed by unscrupulous salesmen, given bad advice by government agencies, and borrowed money on the strength of false promises. Robert Fries concludes: "The most depressing result of the lumber industry, and perhaps the most needless of all, was the plight of the settler on submarginal cutover lands."[116]

An early warning had been given in 1907 by Charles A. Davis that in northern Michigan, "it would be unsafe to try cultivating peat swamps on a large scale."[117] In 1910, Wisconsin's state forester forecast that no more than one-fifth of the state forest reserve would be profitable to farm.[118] In 1932, an economic survey of forests and cutover in Wisconsin confirmed that estimate.[119] Michigan's land economic survey, carried out in the 1920s, showed that in 16 northern counties barely one-fourth of the land offered good or fair potential for farm development.[120] In northern Wisconsin and northeastern Minnesota, surveys found that three-fourths of the surface was best suited to forestry.[121] Throughout the region

116. Fries, *Empire in pine*, 1951, 254.

117. Charles A. Davis, *Peat: Essays on its origin, uses, and distribution in Michigan,* Report State Board of Geol. Survey Michigan for 1906 (East Lansing 1907), 279; another cautionary note was sounded in Huels, *Peat resources of Wisconsin,* 1915, 46.

118. *Report of the state forester of Wisconsin, 1909–10* (Madison 1910) 56; at this period, prospects for farming peatlands were regarded as particularly doubtful. E. S. Delwiche, *Opportunities for profitable farming in northern Wisconsin,* in *Univ. Wisc. Agric. Expt. Sta. Bull.* 196 (Madison 1910); A. R. Whitson and F. J. Sievers, *The development of marsh soils,* in *Univ. Wisc. Agric. Expt. Sta. Bull.* 205 (Madison 1911). Ray Hughes Whitbeck, *The geography and industries of Wisconsin,* in *Wisc. Biol. Nat. Hist. Surv. Bull.* 26 (Madison 1913) 45, expresses an ambivalent attitude: "Some of these marshes are worthless or worse than worthless for the farmer. Some of them have been drained and where the land is muck [containing an admixture of mineral material] and not peat, it is often exceedingly productive."

119. Committee on Land Use and Forestry, *Forest land use in Wisconsin* (Madison 1932) 37.

120. H. Titus, *The land nobody wanted,* in *Mich. Agric. Expt. Sta. Bull.* 332 (1945); C. P. Barnes, Land resource inventory in Michigan, *Econ. Geog.* 5 (1929) 22–35; Norman J. Schmaltz, Michigan's land economic survey, *Agric. Hist.* 52 (1978) 229–46.

121. Raleigh Barlowe, Changing land use and policies: The lake states, in Flader (ed.),

a realization slowly dawned that farmers would not follow lumbermen on cutover land and draining would not make poor soils productive. By the late 1920s, it was clear that the bulk of the cutover, including most peat swamp, "was wholly unsuitable to farming" and small areas that had been reclaimed were barely capable of supporting their occupiers.[122]

A few exceptional enterprises succeeded in occupying wetlands for growing special crops such as mint, strawberries, lettuces, onions, and celery. In 1912, in bogs in south and east Michigan, cultivation of onions and celery had already "proved highly profitable."[123] In Luce County, one company drained a large tract of marsh and planted celery, but in 1920 only 25 acres of celery were grown and the enterprise failed.[124] In 1925, the state began to retire these remote, frosty lands from agriculture to protect them as wildlife refuges or forest reserves. In northern Wisconsin, the university College of Agriculture continued to encourage farmers to settle on poorly drained cutover until the late 1920s.[125] Soil survey reports, meanwhile, issued clear warnings about deficiencies of phosphorus and potash in peat soils and drew attention to hazards of frost and fire for growing crops. Under provisions of a new drainage law in 1919, the state drainage engineer was authorized to advise counties to postpone approval to possibly unsound applications for new drainage districts. By 1923, ten schemes had been postponed, including five in Chippewa County, where proposed costs exceeded likely benefits. Asking whether it was wise or economically worthwhile to extend farming on peatlands while millions of acres of other lands were still unreclaimed, E. R. Jones declared: "The answer is YES, where there is need of more tillable acres surrounding upland farms dipping down into the marsh. It is YES, where a sound and decently enforceable utilization plan is made a part of the drainage scheme. But it is emphatically NO, in all

Great Lakes forest, 1983, 166.

122. Hurst, *Law and economic growth*, 1964, 435.

123. Frank Leverett, *Surface geology and agricultural conditions of the southern peninsula of Michigan*, in *Mich. Geol. and Biol. Surv. Bull.* 9 (Lansing 1912) 85.

124. George F. Deasy, Agriculture in Luce County, Michigan, 1880 to 1930, *Agric. Hist.* 24 (1950) 37.

125. Hurst, *Law and economic growth*, 1964, 595; Hurst, Institutional environment, 1983, 153.

other cases."[126] In 1925, Wisconsin drainage district law was amended to prevent setting up any new county or judicial district; new projects could be established only under farm drainage laws. While cultivation of hay, oats, and potatoes ceased to be profitable in northern Wisconsin, small areas of peat continued to produce onions and celery in the southeast.[127]

In Minnesota, following highly successful ventures draining wet prairies in the south and west between 1890 and 1910, state and county governments welcomed proposals to drain large tracts of peatlands in the north.[128] In 1905, state drainage laws were revised to simplify procedures for forming drainage districts and to permit ditches to be extended "over larger and larger areas without regard to the artificial boundary lines of towns, counties or judicial districts."[129] In 1908, Congress passed the Volstead Act, enabling drainage districts to levy contributions from lands owned by federal agencies, including unpatented homesteads. In 1909, the Minnesota legislature made the state liable to pay assessments on undeveloped state-owned lands lying in drainage districts. These measures induced a few promoters to organize enormous districts, compelling public bodies to make substantial payments toward the costs.[130] Between 1907 and 1913, about six million acres were included in new drainage enterprises in Minnesota. In Roseau County alone, districts covered about 350,000 acres.[131]

Returns from agriculture were disappointing and many farmers were unable to repay loans, pay annual imposts for maintaining

126. E. R. Jones, Land drainage in Wisconsin, Press statement for State Department of Engineering and College of Agriculture, 1923 [Records of State Drainage Engineer, Madison].

127. Wisconsin Conservation Department, Progress report on drainage, Press release 15 March 1950 [Records of State Drainage Engineer, Madison].

128. Thomas J. Baerwald, Forces at work on the landscape, in Clifford Clark (ed.), *Minnesota in a century of change: The state and its people since 1900* (Minn. Hist. Soc. Press, St. Paul 1989) 48; John R. Borchert, *Minnesota's changing geography* (University of Minnesota Press, Minneapolis 1959) 49–50; Minnesota Department of Natural Resources, *Minnesota peatlands* (St. Paul 1978).

129. In Palmer, *Swamp land drainage*, 1915, 71; legislation and drainage procedures in Minnesota down to 1913 are reviewed comprehensively, 59–87.

130. Samuel Trask Dana, John H. Allison, and Russell N. Cunningham, *Minnesota lands: Ownership, use, and management of forest and related lands* (American Forestry Association, Washington, D.C. 1960) 25.

131. David L. Nass, The rural experience, in Clark, *Minnesota*, 1989, 130.

ditches, or bear increasing demands for local property taxes. Penalties and interest charges accumulated until outstanding dues were greater than the value of the land. Tax delinquent lands were forfeited to counties to such an extent that some counties were unable to meet their obligations to finance roads, schools, and other services and were, in effect, bankrupt. To protect the creditworthiness of Minnesota stock, the state was forced to rescue counties by guaranteeing redemption of outstanding drainage district bonds. In return, delinquent lands passed into state ownership. The financial crisis was deepest in seven northern counties whose lands were taken by the state to form a 1,651,000-acre consolidated conservation area.[132] The drainage of this vast area began in 1909 with encouragement from the Great Northern Railroad and funds raised by Beltrami County. After ditches were dug, extensive fires in 1910 and 1918 burned desiccated peat. Few farmers settled and those who stayed could not make a living. As late as 1927, when the project faced financial ruin, state engineers were still urging the county to spend more money to buy dredging machines to scour hundreds of miles of choked ditches. Not until 1929 did Minnesota's legislature and supreme court acknowledge that the government itself was responsible for pressing Beltrami and other counties into constructing drainage ditches that they could not afford.[133] In that year, over 1,170,000 acres in Beltrami, Lake of the Woods, and Koochiching counties were established as the Red Lake Game Preserve "to protect and to propagate wild life, to prevent forest fires, to develop forests, and for the preservation and development of rare and distinctive species of flora native in such areas."[134] Behind this declaration of intent to protect wildlife and develop forests lay a far more urgent purpose of preventing county governments from defaulting on drainage bonds, thereby threatening "the general credit of the State of Minnesota and all its political subdivisions."[135] In 1931, the state conservation area was given an additional 404,800 acres in

132. Marjorie H. Ahrens, A contribution to the history of land administration in Minnesota: The origins of the Red Lake Game Preserve, M.A. thesis, Geography, University of Minnesota 1987, 6. The scheme included lands in seven counties: Aitkin, Beltrami, Koochiching, Lake of the Woods, Mahnomen, Marshall, and Roseau.

133. Ibid., 62–76.

134. State of Minnesota, *Laws,* 1929, chap. 258

135. Ahrens, Red Lake Game Preserve, 1987, 8.

Aitkin, Roseau, and Mahnomen counties, and in 1933 it acquired another 75,800 acres in Marshall County.[136] By 1934, Minnesota had spent $4,000,000 undoing the work of draining: building dams to plug drainage outlets, planting trees on deserted fields, and assisting stranded settlers to move.[137]

From 1925 onward, as a result of widespread insolvency of drainage enterprises and forfeiture of tax delinquent lands, the federal government, the states of Michigan, Wisconsin, and Minnesota, and other public authorities came to possess large areas of land (fig. 7.9). By 1946, over 60% of land in ten counties in northeastern Minnesota and Schoolcraft County, Michigan, was publicly owned. In another 23 northern counties between 40% and 60% of the land area was in public ownership.[138] An immediate task for federal and state governments was to provide financial aid and recurrent grants to local authorities for repairing roads, running schools, and maintaining local services. Over a long term, plans for future management of public wetlands called for new political and economic measures, curtailing rights of private owners, reducing powers and responsibilities of local government, and reforming tax systems. Alternative goals of recreational use or forestry might be pursued but conflicts of interest needed to be resolved.

The course strewn with fewest snags led to recreational use. In 1930, K. C. McMurry reported a new appreciation of recreational opportunities afforded by wild lands in northern Michigan and similar districts in Wisconsin and Minnesota. During the next twenty years, automobiles and paved roads opened Luce County, Michigan, and other northern resorts to deer hunting, partridge shooting, and summer vacationing.[139] Large areas were designated public hunting grounds, game preserves, wildlife refuges, or conservation areas. From their inception, revenue was collected from sales of hunting licenses, fishing permits, duck stamps, and franchises for

136. State of Minnesota, *Laws,* 1931, chap. 407; 1933, chap. 402.

137. F. R. Kenney and W. L. McAtee, The problem: Drained areas and wildlife habitats, in U.S. Department of Agriculture, *Soils and men: Yearbook of agriculture, 1938* (GPO, Washington, D.C. 1938) 78, 80, 83.

138. Raleigh Barlowe, *Public landownership in the lake states,* in *Mich. Agric. Expt. Sta. Spec. Bull.* 351 (1948).

139. K. C. McMurry, The use of land for recreation, *Annals Assoc. Amer. Geog.* 20 (1930) 7–20; George F. Deasy, The tourist industry in a "northwoods" county, *Econ. Geog.* 25 (1949) 240–59.

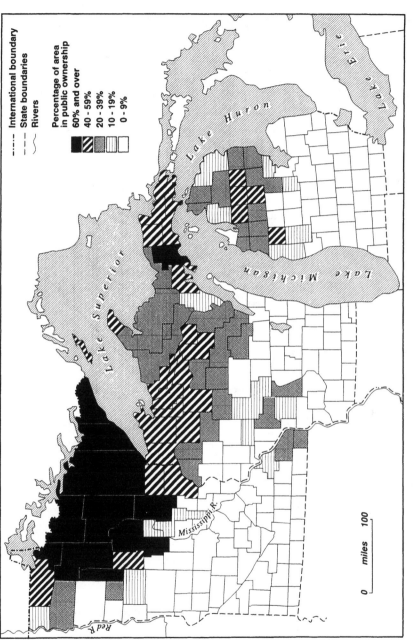

Figure 7.9 Land in public ownership in northern lakes states, 1946

Percentages of total surface areas of counties owned by public bodies. *Sources:* Raleigh Barlowe, *Public landownership in the lakes states,* in *Mich. Agric. Expt. Sta. Spec. Bull.* 351 (1948); Barlowe, Changing land use and policies: The lakes states, in Susan L. Flader (ed.), *The Great Lakes forest* (University of Minnesota Press, Minneapolis 1983) 168.

catering. Over a long term, as ditches filled up, were dammed by beavers, or were closed by artificial dams, new lakes spread out, waterfowl returned, and fish stocks multiplied.[140] A prerequisite for successful long-term wildlife management was that wetlands were not to be disturbed by draining. "To the extent that they are drained or cleared for farming, wild life habitat will be destroyed or reduced."[141] Some leisure activities, such as building vacation homes and powerboat cruising, competed directly with wildlife conservation.

A harder path led to forest management. During the first years, forestry yielded no income but incurred heavy expenditure. Initial outlays on planting, weeding, and fire prevention were not recovered until thinnings for pulpwood and poles were taken. In the 1930s, no entrepreneur would invest in a venture that offered little prospect of profit during the first twenty years. Private owners had to be tempted to engage in commercial timber production by government subsidies and special tax concessions. Lands growing tree crops were exempt from property taxes but were assessed for a severance tax at the time of cutting, based on the value of lumber sold.[142] National, state, and county forests were far more extensive than private forests. In 1945, over two-thirds of the surface of all northern peatlands was occupied by publicly owned forests.

Decisions about policies toward wetlands reflected changes in attitudes held by politicians and informed citizens. Increasing productivity of ill-drained soils by artificial draining was no longer regarded as the highest aim of landownership. Conservation of wildlife, forest, soil, and water resources was considered an important benefit. In sensitive wetland environments, individual freedom to dispose of natural resources was recognized to be in conflict with wider public concerns. People not resident in wetlands claimed interests in protecting sites of special scientific importance and ensuring sustainable use of soil, water, and vegetation. The USDA, whose original duty was to promote profitable agricultural development,

140. Oscar B. Jesness, Reynolds I. Nowell, and associates, *A program for land use in northern Minnesota: A type study in land utilization* (University of Minnesota Press, Minneapolis 1935) 52.

141. Marion Clawson, R. Burnell Held, Charles Stoddard, *Land for the future* (Resources for the Future, Baltimore 1960) 429.

142. Barlowe, Changing land use and policies, 1983, 166.

was, in 1935, put in charge of a newly established Soil Conservation Service. The Service conducted a crusade to stop soil erosion and urged restraint in draining peatlands. Discussing problems of draining peatlands in northern Minnesota, the *Yearbook of Agriculture, 1938* remarks: "The grass-covered bogs had better be left to serve as wire-grass meadows, or drained just enough to allow the cutting of wild hay, while the bogs with merchantable timber should be kept under proper forest management and all others left undisturbed."[143] This short passage is taken from the context of a study not originally commissioned by the USDA. The secretary of agriculture, Henry Wallace, may or may not have endorsed the opinion expressed, but many contemporaries, both in and out of government service, shared the author's general belief that over an "immense peat acreage, the profit of reclamation is to be regarded as extremely doubtful, even under the most skilled supervision and with every resource and facility for conducting the work economically."[144] Less than thirty years earlier, Frederick Newell, director of the U.S. Geological Survey Bureau of Reclamation, claimed that almost all artificially drained swamplands were capable of supporting large and prosperous agricultural communities. This represents a great change in outlook but not a total change. The goals of wetland management in the 1930s and 1940s were still productive. Deer and ducks were for hunting, trout were for fishing, and forests for making pulp and furniture. Recreation and tourism were commercial activities. Wetlands as natural landscapes were commodities to be marketed.

143. John R. Haswell, Drainage in the humid regions, in USDA, *Soils and men,* 1938, 727, quoting F. J. Alway, Agricultural value and reclamation of Minnesota peat soils, *Minn. Univ. Bull.* 188 (1920).

144. Ibid.

8

Utilizing and Conserving
Wet Prairies since 1930

By 1930, tile draining had rendered wet prairies less frequently wet. Conditions that had previously caused perennial or seasonal waterlogging were likely to recur when exceptionally heavy rains fell in summer or drains were blocked with roots or sediment. In most years, artificially drained lands were cultivated in the same manner as naturally well drained lands. For a considerable time after tiles had been laid, soils retained some wetland characteristics, and these distinctive markers indicated sites of former wet prairies.[1] Leslie Hewes considered that "the classification, description and mapping of the soil constitute the most valid means of approximating the extent of wet prairie which early pioneers encountered."[2] Areas of artificially drained farmland recorded in the 1930 census occupied the same localities as Clyde silty clay loam, Webster silt loam, Fargo clay loam, and related types of soil. Hewes noted that acreages of drained land in Illinois and Iowa were larger than areas of corresponding soil types in Minnesota and the Dakotas: overrepresenting "originally wet land on the south and southeast," underrepresenting it "to the north and northwest."[3] In Minnesota and the Dakotas, soil surveys designated as poorly drained larger tracts than in Illinois and Iowa, where surveys drew finer distinctions between well- or poorly drained soils. In the Red River valley, large drainage districts included much unimproved land, whereas smaller enterprises on the Grand Prairie in Illinois and within the Mankato lobe of glaciation in north central Iowa contained little or no unimproved land.

In most wet prairies, the area of artificially drained land ex-

1. U.S. Soil Conservation Service, *Soil taxonomy: A basic system of soil classification for making and interpreting soil surveys,* USSCS Agricultural Handbook 436 (Soil Conservation Service, Washington, D.C. 1975).

2. Leslie Hewes, The northern wet prairie of the United States: Nature, sources of information, and extent, *Annals Assoc. Amer. Geog.* 41 (1951) 307.

3. Ibid., 317.

panded steadily from 1880 to 1900, then very rapidly from 1900 to 1920, and slowed down in the 1920s.[4] Many farmers, as well as historians and geographers, made exclusive claims for the completion of ditching and tile draining at an early date (fig. 8.1). John Fraser Hart, who visited farms throughout the Midwest in the 1950s, was told by farmers that draining had been completed in the early years of the twentieth century, to prepare the ground for cultivation.[5] John Borchert also described draining wet prairies in southern Minnesota as a pioneer activity, commenting that it "was a huge task. Now it is mostly finished."[6] Draining ceased in the late 1920s but the process was neither ended nor mostly finished: it was resumed vigorously over large areas between 1945 and 1975. Michael Williams's summary that "much of the wet prairie draining is in the past" focused on the fundamental significance of early draining as a process for transforming wetlands into corn belt farms.[7] As draining advanced northward and westward, pockets of wet prairie and small lakes were mopped up.[8]

A uniform farming landscape emerged. In the late 1920s, O. E. Baker's classic description of the corn belt as a homogeneous region drew no distinction between former wetlands and dry lands.[9] All were highly cultivated; almost all were arable. Corn was the leading crop, grown in rotation with oats, hay, and wheat. John C. Weaver described the four-crop corn-oats-hay-wheat rotation that prevailed in 1929 as "the basic crop combination, the parent type from which most of the other associations are no more than variants."[10] Artificially drained land was, if anything, more intensively cultivated, more completely under the plow, and more productive in corn than other lands in the corn belt.

4. Mary R. McCorvie and Christopher L. Lant, Drainage district formation and the loss of Midwestern wetlands, 1850–1930, *Agric. Hist.* 67 (1993) 32–34.

5. John Fraser Hart, *The land that feeds us* (W. W. Norton, New York 1991) 130–31.

6. John R. Borchert, *Minnesota's changing geography* (University of Minnesota Press, Minneapolis 1959) 49–50.

7. Michael Williams, The human use of wetlands, *Progress in Human Geography* 15 (1991) 10.

8. John C. Hudson, *Making the corn belt: A geographical history of Middle-western agriculture* (Indiana University Press, Bloomington 1994) 140, 152–55.

9. O. E. Baker, Agricultural regions of North America, part 4: The corn belt, *Econ. Geog.* 3 (1927) 447–65.

10. John C. Weaver, Crop-combination regions for 1919 and 1929 in the Middle West, *Geog. Rev.* 44 (1954) 560–72, quotation from 565.

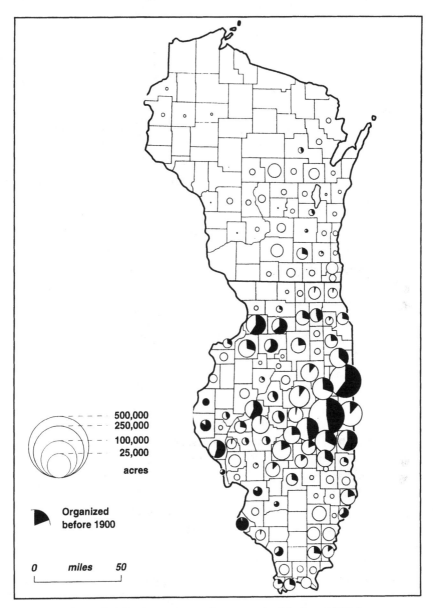

500,000
250,000
100,000
25,000

acres

Organized
before 1900

0   miles   50

Figure 8.1 Drainage enterprises organized before 1900 in
Illinois and Wisconsin

Areas of circles are proportional to total areas, in thousands of acres, by counties, of
land in drainage enterprises in 1950. Black portions represent areas organized in
drainage enterprises before 1900. *Source:* U.S. Bureau of Census, *Census of agriculture, 1950,* vol. 4, *Drainage of agricultural lands* (Washington, D.C. 1952).

Between 1930 and 1940, expansion of drained farmland ceased. The depression halted capital investment. In wet prairies in southern Minnesota and Iowa, in contrast to peatlands in the north, drainage districts managed to pay their debts and redeem their bonds. Little drained land became tax delinquent and farmers were not thrown into bankruptcy because of unpaid drainage debts. On the other hand, no new drainage districts were organized and plans for farm draining were postponed. In the 1930s, mortgage foreclosures in Iowa were more frequent on farms that had not been tile drained than on highly productive artificially drained lands.[11] Farmers responded to falling prices by reducing acreages growing wheat, except in the Red River valley and in Nebraska and Kansas. Timothy and rye grass were replaced as hay crops by clover and alfalfa, and increasing amounts of silage were made instead of hay. Artificially drained areas concentrated more and more on cash grain production. They led the way toward specialized farming within the corn belt.[12]

In the 1940s, as the economy recovered, a new phase of draining began. Some early tile lines needed repair and replacement; ditches that were silted or overgrown with trees and shrubs needed clearing, dredging, and deepening. After making good a backlog of neglected maintenance work, landowners embarked upon new schemes. Some permanently waterlogged sloughs and potholes were newly reclaimed and subsurface drainage was extended into stiff, hard-to-drain clays and silty clay loams. Many farmers had only hazy recollections of draining carried out in the period from 1945 onward. The building and rebuilding of barns, grain storage bins, and new silos, and the acquisition of giant tractors, combines, and other powerful machines remained fresh in the memory and conspicuous in the landscape. There was less to show for the removal of fences and laying of drain tiles, but clear evidence of widespread draining activity between 1945 and 1975 was recorded in official statistics.[13] A small number of landowners took an initiative in es-

11. Janel Curry-Roper and John Bowles, Local factors in changing land-tenure patterns, *Geog. Rev.* 81 (1991) 447–48, discusses the uneven incidence of foreclosure and corporate ownership.

12. John Fraser Hart, The Middle West, in Hart (ed.), *Regions of the United States* (Harper and Row, New York 1972) 258–82, esp. 267–68.

13. W. E. Frayer, T. J. Monahan, D. C. Bowden, and F. A. Graybill, *Status and trends of wetlands and deepwater habitats in the coterminous United States, 1950s to 1970s* (Depart-

tablishing or reactivating drainage districts or farm drainage schemes, and a majority of farmers joined in order to protect and possibly enhance the value of their land. Most were compliant, passive participants. The strongest reason for opting out was doubt about whether increased output would yield sufficient profit to cover the costs of draining.

Wildlife Conservation and Draining

From the 1930s farmers were subject to an increasing volume of federal regulations and subsidies and these measures played an ever greater part in decisions made on the ground. In response to falling world crop prices, the federal government intervened to lessen American farmers' dependence on exporting to overseas markets. Attempts were made to induce growers to set aside land that had been producing surplus crops.[14] The Agricultural Adjustment Act of 1933 made funds available to pay farmers to reduce acreages of corn and other crops. At the same time, observers who watched dark clouds blowing from the dust bowl and muddy waters flowing from badly gullied hillsides in the South were alerted to widespread damage caused by soil erosion. They could no longer remain confident that flat prairie surfaces were immune from wastage. Following the breaking of prairie sod and exposure of bare soils to rainstorms, hailstorms, and windstorms, deep layers of topsoil were washed away. A standard text on conservation described the onset of erosion: "Because the prairie soils of the midwest contained much organic matter and had gentle slopes, they resisted erosion longer than did the forest soils of the east. In time, however, even the rich prairie soils lost their fresh organic matter. The soil became compact; runoff and erosion followed."[15] It was estimated that Iowa farmers not practicing sound conservation techniques were losing two bushels of rich black topsoil for every bushel of corn pro-

ment of Forest and Wood Science, Colorado State University, Fort Collins 1983).

14. D. E. Ervin, Some lessons about the political-economic effects of set-aside: The United States' experience, in British Crop Protection Council, *Set aside: proceedings of a symposium held at Cambridge University, 15–18 September 1992*, BCPC Monograph 50 (Farnham, Surrey 1992) 3–12; Clive Potter, Paul Burnham, Angela Edwards, Ruth Gasson, and Bryn Green, *The diversion of land: Conservation in a period of farming contraction* (Routledge, London 1991) 44–59.

15. A. F. Gustafson, C. H. Guise, W. J. Hamilton, and H. Ries, *Conservation in the United States* (Comstock Publishing, Ithaca 1939) 108–9.

duced.[16] In little more than a century, probably half of Iowa's topsoil was wasted.[17] In 1938, an Agricultural Conservation Program was set up to make payments and grants to owners of land who carried out approved soil and water conservation schemes. It was intended to retire eroded or flooded land into a soil bank. Another set of measures had an opposite effect: the payment of subsidies for leveling, liming, and field draining raised production and brought new land into cultivation. Farmers were torn between claiming grants to retire land from cultivation and applying for subsidies to drain land to increase production. Adding to the confusion, an entirely new objective was introduced to protect wet spots for breeding and feeding waterfowl.

Concern was first expressed about losses of wetlands essential for waterfowl production in southern Minnesota in the late nineteenth century. Increasingly intense agricultural use of prairie soils eliminated many brood marshes in southern and western parts of the state during the first half of the twentieth century. In September 1951, under the slogan "Save Minnesota's Wetlands," the state Bureau of Wildlife Development launched a program to acquire small wetland areas threatened by agricultural drainage.[18] Purchases were funded with federal aid provided by the Pittman-Robertson and Dingall-Johnson acts, supplemented by state revenues derived from the sale of game and fish licenses. By the end of June 1964, the state had acquired 112,000 acres under this program, bringing the total of state-owned land whose primary function was wildlife management up to 1,006,000 acres.[19] Small areas of high-priced land were acquired in wet prairies. Purchases of this kind in

16. Amory Lovins, L. Hunter, and Marty Bender, Energy in agriculture, in W. Jackson, W. Berry, and B. Colman (eds.), *Meeting the expectations of the land* (North Point Press, San Francisco 1984) 81; G. A. Larson, G. Roloff, and W. E. Larson, A new approach to marginal agricultural land classification, *Journ. Soil and Water Conservation* 43 (1987) 103–5.

17. Joe Paddock, Nancy Paddock, and Carol Bly, *Soil and survival* (Sierra Club Books, San Francisco 1986) 7. Stuart Chase, *Rich land, poor land: A study of waste in the natural resources of America* (AMS Press, New York 1969) 92, cites a special report of the Iowa state planning commission's natural resources board published about 1935 which concluded: "Since the prairie cover was first disturbed, Iowa has lost approximately 550,000 tons of good surface soil per square mile." Forty percent of the area of the state had lost from 50 to 75% of its surface soil. Iowa lakes were filling with silt.

18. Minnesota Outdoor Recreation Resources Commission, *Acquisition of wildlife land in Minnesota (Wetland Program)* (State Capitol, St. Paul, Minnesota 1965) 3.

19. Ibid., 5.

TABLE 8.1 Wildlife management areas in four wet prairie counties
in south central Minnesota.

County	Acres	Dollars
Blue Earth	710	32,583
Brown	723	48,825
Le Sueur	1,613	63,672
Nicollet	22	234

Note: Areas purchased by the State of Minnesota Department of Conservation, Game and Fish Division, up to 31 December 1964.
Source: Minnesota Outdoor Recreation Resources Commission, *Acquisition of wildlife land in Minnesota (Wetland Program)* (State Capitol, St. Paul 1965) 9, 10.

four south central counties in Minnesota are listed in table 8.1. In addition to protecting waterfowl habitats, some marshes preserved native prairie-type vegetation and associated fauna that had ceased to exist over most of the surface of the prairie counties. These were regarded as valuable for educational and scientific as well as recreational purposes.

In April 1965, a report by the Minnesota Outdoor Recreation Resources Commission on the acquisition of wetlands for wildlife stated that "most of Minnesota's present-day drainage is in the flat to gently rolling grassland region of the state where soils are inherently more fertile. This is also the region where most of the remaining wetlands are rated high in waterfowl value. High soil fertility and high wildlife production seem to go hand in hand where wetlands are concerned. Widespread drainage, of course, can upset this direct agriculture-waterfowl relation. Since the best agricultural lands are the ones receiving the most drainage, waterfowl habitat on such lands often becomes locally scarce."[20] In prairie counties in southern and western Minnesota, wetlands and associated meadows had become rare in the 1940s, and during a twelve-year period from 1943 to 1954 it was estimated that a further 350,000 prairie potholes were drained. In the early 1960s, a concentration of new drainage projects in the south central part of the state, including Brown and Nicollet counties, had "largely eliminated waterfowl habitat from this region."[21] By 1964, southern Minnesota had "already been drained to the point where its waterfowl production

20. Ibid., 22.
21. Ibid., 41.

capabilities [were] insignificant."[22] This happened in spite of strenuous efforts by federal and state agencies to prevent further draining.

In 1962, Congress passed Public Law 87–732, imposing restrictions on the use of federal assistance for drainage projects affecting wetlands. The law required farmers applying for federal aid to satisfy the Fish and Wildlife Service that the project would not harm waterfowl conservation. In practice, the law did not deter farmers. The Fish and Wildlife Service had to raise objections within 90 days of receiving an application. If the service did not grant permission it either had to buy the land affected or obtain an easement from the owner within twelve months. The service did not have sufficient funds to compensate owners for highly valued agricultural land. A more effective restraint was imposed by administrators of the Agricultural Conservation Program, who refused to award cost-sharing subsidies to proposals for ditching or tile draining that planned to bring new land into cultivation. It was presumed that landowners who wished to drain fresh wetlands would do so entirely at their own expense.

Powerful economic incentives to drain persisted in the 1960s. Jon Goldstein's study of 13 drainage projects carried out with the help of the Soil Conservation Service in Blue Earth County, Minnesota, in 1963 reported that the full cost of laying tiles was between $123 and $228 per acre, and these systems were expected to remain effective for 20 to 50 years.[23] An analysis of benefits gained in returns from arable farming indicated that investment in draining seasonally waterlogged soils was profitable without a drainage project grant from the Agricultural Stabilization and Conservation Service or without a low-interest loan from the Farm Home Administration. It might even have been just marginally profitable without USDA price support for leading cash crops, principally, corn and soybeans. With crop subsidies, a farmer's capital investment could be repaid in a very short period of between three and seven years. On those terms, draining was a bargain that few farmers would refuse unless they had more urgent calls for capital expendi-

22. Ibid.

23. Jon H. Goldstein, *Competition for wetlands in the Midwest: An economic analysis* (Resources for the Future, Washington, D.C. 1971) 10–44. In a preface, ix, John Krutilla strongly asserted the prior claims of farmers in competing for land: "Wetlands represent potential nuisances at best and substantial increases in costs of agricultural production at worst for farmers on whose land they occur."

ture. No advantage was to be gained from not draining or setting aside temporary or seasonally waterlogged land. On the other hand, improvements to permanently wet land would not normally repay investment in tile drainage at levels of farm prices prevailing in 1963. The reason why draining permanent wetland was unprofitable was that digging outlet ditches, removing wetland plants, and preparing soils for cultivation were costly investments, and farmers themselves had to bear the costs of breaking new ground. Only where nuisance costs were high might modest expenditure be reckoned worthwhile.

In the 1960s, farmers were subject to strong external pressures pushing them in different directions. Political messages and public opinion urged them to protect wildlife habitats and conserve soils. Economic forces drove them to farm more intensively, to raise output, cut production costs, invest in new techniques, to speed up a technological treadmill.[24] Social pressures weakened ties that held together family farms, drawing husbands, wives, and children into nonfarm employment, disrupting plans for intergenerational succession.

Corn Belt Farmers on a Technological Treadmill

From the earliest period of draining, farmers on wet prairies were thoroughly commercialized, buying manufactured goods such as drain tiles, sodbusting plows, and barbed wire, selling almost all produce from the farm to processors such as millers and meatpackers.[25] Technological advances demanded increasing capital investment and caused downward pressure on prices. Heightened sensi-

24. A theory of a "treadmill of technology" was put forward by agricultural economist Willard Cochrane, who argued that a majority of farmers were compelled to adopt technological innovations with little reward, simply to stay in business. Willard W. Cochrane, *Farm prices* (University of Minnesota Press, Minneapolis 1958). Among later discussions of the theory is Brian Page and Richard Walker, From settlement to Fordism: The agro-industrial revolution in the American Midwest, *Econ. Geog.* 67 (1991) 281–315, observing (292): "Farmers were compelled to be improvers by the logic of the market; rising productivity and total output created a strong downward pressure on prices, thereby propelling farmers further into the market (and into debt) to secure better and better equipment and breeding stock." A wide-ranging review of recent literature on the topic is R. J. C. Munton, T. Marsden, and S. Whatmore, Technological change in a period of agricultural adjustment, in P. Lowe, T. K. Marsden, and S. J. Whatmore (eds.), *Technological change and the rural environment* (Fulton, London 1990) 104–26, who remark (104) that farmers have been "continuously obliged to adopt cost-cutting technologies to offset falls in real prices."

25. Page and Walker, From settlement to Fordism, 1991, 291–93.

tivity to profit margins upset the stability and uniformity of a regular corn belt four-crop rotation. John C. Weaver's analysis of changing patterns of cropland use from 1929 to 1949 indicated that a dynamic variability of crop combinations was "most notably characteristic of the short-growing-season north and subhumid west."[26] By 1949, a "maximum diversity in crop associations for the entire Middle West was found to exist in the northwestern reaches of the 'corn belt,' in southwestern Minnesota and northeastern South Dakota," culminating in an area where seven crops held a significant acreage.[27] At the opposite end of the scale, a monoculture of corn or an alternating succession of corn and oats was gaining ground in northern Illinois, western Iowa, and eastern Nebraska. These areas specialized in cash grain farming based on artificially drained land.[28] Hay declined in rank order among crop acreages. In southern Illinois, it was displaced by soybeans, a new arrival in the late 1940s. Weaver remarks upon an anomalous increase in the acreage of oats in many counties where farm horses had almost disappeared between 1939 and 1949: "Somewhere in the cost accounting machinery there must be a failure to attain a completely realistic appraisal of the situation."[29] In the period from 1945 to 1975 price mechanisms rectified most anomalies in cropping and stocking.

An acceleration in the rate of change followed the introduction of hybrid corn in 1933, although the full impact of the innovation began to register only in the 1950s. Hybrid corn enabled farmers to double or even triple output from prairie soils, but artificial fertilizers enabled them to obtain the highest yields; expenditure on artificial fertilizers multiplied sixfold between 1949 and 1969.[30] The most potent addition to a corn-grower's stock of chemicals was anhydrous ammonia, a gas injected into the soil to boost the supply of nitrogen. To benefit from high-yield strains of seed and heavy dressings of fertilizers, it was necessary to protect growing crops by con-

26. Weaver, Crop-combination regions for 1919 and 1929, 1954, 563.

27. John C. Weaver, Crop-combination regions in the Middle West, *Geog. Rev.* 44 (1954) 190.

28. Arlin D. Fentem, Cash feed grain in the corn belt., Ph.D. diss., Geography, University of Wisconsin 1974.

29. John C. Weaver, Changing patterns of cropland use in the Middle West, *Econ. Geog.* 30 (1954) 23.

30. John Fraser Hart, Change in the corn belt, *Geog. Rev.* 76 (1986) 51–72, esp. 72.

trolling insects and diseases with chemical pesticides, and by suppressing weeds with herbicides. To gather larger harvests and spread larger quantities of chemicals, more powerful, sophisticated, and costly machines were brought into use. To get the most out of new equipment, the area of farms had to be enlarged.

Between 1949 and 1982, the average size of farm in the corn belt nearly doubled, farms in Illinois, Iowa, and southern Minnesota expanding faster than those in Ohio and Indiana. In 1973, the average Midwestern corn farm was about 263 acres, but an economic study concluded that 800 acres was the most efficient size for a single-handed unit.[31] Farm enlargement proceeded mainly by owner-occupiers adding rented fields to their holdings. Young farmers customarily rented land from their parents or other relatives.[32] An additional source of rented land was offered by neighbors who gave up farming.[33] Part-owned farms increased in number and in size at the expense of both wholly owner-occupied and tenant farms. In 1949, the average size of part-owner operations was 205 acres; in 1987, it was 463 acres.[34] The remarkable expansion failed to close the gap between actual farm size and that dictated by optimum efficiency. There was insufficient land to meet demand from all competitors. In 1978, landholdings in the Midwest were

31. Warren Bailey, *The one-man farm* (USDA Economic Research Service, Washington, D.C. 1973).

32. From 1915 to 1923, in Blackhawk and Clay counties, Iowa, between a half and a quarter of all tenants were related to their landlords. In Blackhawk County, tenants averaged eight years younger than owner-operators; and landlords, mostly retired farmers, were seven years older than owner-operators. George H. von Tungeln, *A rural social survey of Orange Township, Blackhawk County, Iowa,* in *Iowa Agric. Expt. Sta. Bull.* 184 (Ames 1918); von Tungeln, *A rural social survey of Lone Tree Township, Clay County, Iowa,* in *Iowa Agric. Expt. Sta. Bull.* 193 (Ames 1920); von Tungeln, E. L. Kirkpatrick, C. R. Hoffer, and J. F. Thaden, *The social aspects of rural life and farm tenantry, Cedar County, Iowa,* in *Iowa Agric. Expt. Sta. Bull.* 217 (Ames 1923); von Tungeln and Harry L. Eells, *Rural social survey of Hudson, Orange, and Jesup consolidated school districts, Blackhawk and Buchanan counties, Iowa,* in *Iowa Agric. Expt. Sta. Bull.* 224 (Ames 1924).

33. A surprising conclusion of an examination of land transfers in seven prairie townships in Fayette and Benton counties, Iowa; and Kane County, Illinois, in the heyday of family succession between 1870 and 1950, was that 50% of farmers quit or sold out to nonfamily purchasers: 32% were renters who left without owning land and 18% were owners who left without handing over to a family heir. Mark Friedberger, The farm family and the inheritance process: Evidence from the corn belt, 1870–1950, *Agric. Hist.* 57 (1983) 1–13, ref. on 7.

34. John Fraser Hart, Part-ownership and farm enlargement in the Midwest, *Annals Assoc. Amer. Geog.* 81 (1991) 69.

more widely distributed among small or medium-sized private owners than in any other part of the United States. The proportion of
land held by the largest private owners was lowest in the western
tier of Midwest states.[35] In western states, not only was privately
owned land concentrated in the hands of a few individuals and large
corporations, but also the federal government and its agencies
owned more land than in eastern states (table 8.2). Until the end of
the 1970s, a struggle to enlarge farms on former wet prairies was
not resolved by widespread takeovers by giant enterprises. Farms
over 250 acres increased in size by means of part-ownership; those
between 50 and 200 acres diminished in size and numbers.[36] From
1925 to 1982, the total number of farms in Iowa declined by nearly
one-half, from 213,390 to 115,413 (table 8.3). In the last eighteen
years of that period, from 1964 to 1982, the number of large farms
(over 500 acres) in Iowa increased by 118%, the number of middle-
sized farms (50 to 500 acres) decreased by 41%, and the number of
small farms (less than 50 acres) increased by 31%. A revival of small
farms is explained by a growth in part-time farming by people engaged in businesses or employed away from the farm.[37] In every
part of the corn belt, some farmhouses were pulled down, their sites
marked by shelterbelts and driveways. Many others were taken over
by commuters as dormitory residences or vacation homes. Some
newcomers rented buildings to neighboring farmers for storage of
machinery or as accommodation for livestock[38] (fig. 8.2).

Government funding helped to keep the technological treadmill
turning. In the 1960s and 1970s, an increasing proportion of farmers' incomes was made up of price support payments. At the end of
the 1970s, the profitability of cash grain production largely derived
from price support. The subsidized profits of farming kept the price
of land buoyant and induced corporate investors (including pension
funds and investment trusts as well as private individuals earning
high off-farm incomes) who were seeking tax shelters to bid for
farmland in competition with family farmers.[39] At the same time,

35. Charles Geisler, Ownership: An overview, *Rural Sociology* 58 (1993) 532–46, data
tabulated 535.

36. Hart, Part-ownership and farm enlargement, 1991, 66–79.

37. Paul Lasley, The crisis in Iowa, in Gary Comstock (ed.), *Is there a moral obligation to
save the family farm?* (Iowa State University Press, Ames 1987) 99–101.

38. Hart, Change in the corn belt., 1986, 66.

39. Terry Marsden, Richard Munton, Sarah Whatmore, and Jo Little, Towards a political

federal and state agricultural advisory services were promoting technological innovations by recommending that farmers buy labor-saving plant and machines. Government funds for research and development, channeled through colleges of agriculture and research institutes, developed products and processes that substantially benefited manufacturers.[40] As a result, manufacturers increased their share of profits earned in the chain of food production and became ardent defenders of productivist agricultural policies pursued by successive administrations.

A direct consequence of technological progress was that farmers needed to borrow larger and larger sums of money. Farmers who bought more powerful machines also needed more land in order to gain maximum benefit from their labor-saving capacity. These farmers bid up rents and prices for all land, whether well drained or ill drained, and money borrowed for machinery, rents, and draining increased their debt-to-asset ratios. Prices of tractors and combines kept rising and, in 1985, "an electronic monitor for a corn planter was more expensive than the price of equipping an entire farm in 1950."[41] In real terms, prices of equipment increased faster than concurrent rates of inflation. Machinery manufacturers advanced credit facilities, fertilizer suppliers deferred payments, banks extended mortgages on land purchases, public lending agencies, including the Farm Home Administration, financed improvements such as drainage schemes. Industrialized agriculture led farmers into debt. A confident outlook in the 1970s is expressed by Hiram Drache: "With an asset to debt ratio of about five to one, the progressive farmers are going to continue to use their financial leverage for future expansion, all opposition notwithstanding."[42]

---

economy of capitalist agriculture: A British perspective, *International Journ. Urban and Regional Research* 4 (1986) 513–15; Gregg Easterbrook, Making sense of agriculture: A revisionist look at farm policy; and Luther Tweeten, Has the family farm been treated unjustly? Both in Comstock (ed.), *Obligation to save the family farm,* 1987.

40. Munton, Marsden, and Whatmore, Technological change in a period of agricultural adjustment, 1990, 111–14; F. H. Buttel and L. Busch, The public agricultural research system at the crossroads, *Agric. Hist.* 62 (1988) 292–312; Don F. Hadwiger, *The politics of agricultural research* (University of Nebraska Press, Lincoln 1982).

41. Hart, Change in the corn belt, 1986, quotation on 53. Carol Bly, *Letters from the country* (Harper and Row, New York 1981), discusses the impact of mechanization on rural life in "Getting tired," 8–13, and other essays.

42. Hiram M. Drache, Midwest agriculture: Changing with technology, *Agric. Hist.* 50 (1976) 290.

TABLE 8.2 Percentage of land held by largest 5% of private owners, 1978, and by federal government, 1979, in rank order

State	Land held by largest 5% (%)	Land held by federal gov't. (%)	State	Land held by largest 5% (%)	Land held by federal gov't. (%)
Minnesota	31	7	South Carolina	67	6
Iowa	34	–	West Virginia	67	6
North Dakota	37	5	North Carolina	69	6
Vermont	37	4	Virginia	69	9
Missouri	39	4	Texas	71	2
Kansas	43	1	Alabama	72	3
Nebraska	47	1	Arkansas	72	10
Indiana	49	2	Georgia	73	6
Wisconsin	53	5	New Hampshire	79	12
Kentucky	53	5	New York	80	1
Rhode Island	56	1	Louisiana	80	4
Illinois	57	1	Colorado	83	36
Connecticut	57	–	Utah	84	66
South Dakota	58	7	Arizona	85	45
Tennessee	58	6	Idaho	86	64
Ohio	59	1	Maine	87	–
Maryland	59	3	Washington	87	29
Delaware	60	3	California	87	44
Oklahoma	61	3	Nevada	89	86
Mississippi	62	5	Florida	90	10
Massachusetts	63	1	New Mexico	90	34
New Jersey	63	2	Wyoming	90	48
Pennsylvania	64	2	Oregon	90	52
Montana	64	30			
Michigan	65	9	Total U.S.	75	22

Note: States in which the federal government held a high percentage of land were among those where landownership was concentrated in few hands.
Sources: J. A. Lewis, *Landownership in the United States*, Economics, Statistics and Cooperatives Service Staff Report 80-10 (U.S. Department of Agriculture, Washington, D.C. 1980); General Accounting Office, The federal drive to acquire private lands should be reassessed, Report CED–80–14, (GAO, Washington, D.C. 1979).

Another result of industrialization was increasing specialization in production. A strength of early farming on drained prairie soils had been its diversity and versatility. When corn prices were high, farmers sold corn off the farm as grain; when corn prices were low, they used it to feed hogs or fatten beef or both. In the 1950s and 1960s, rearing hogs and feeding beef became highly industrialized operations requiring massive investments in new buildings, silage production, slurry disposal, and restructuring of crop rotations.

TABLE 8.3 Change in number of Iowa farms by size category, 1925–82

Acres	No. of farms 1925	No. of farms 1982	Change (no.)	Change (%)
Under 50	25,718	20,232	−5,486	−21.3
50–500	185,779	77,218	−108,561	−58.4
Over 500	1,893	17,963	+16,070	+848.9
Total	213,390	115,413	−97,977	−45.9

Source: U.S. Bureau of Census, Iowa census of agriculture, vol. 1, pt. 15 (GPO, Washington, D.C. 1982).

Specialization on cash crop farming concentrated more and more heavily on producing corn and soybeans to the exclusion of most other crops. The acreage growing corn increased continuously from the late nineteenth century, and in the 1980s occupied as much as 40% of many cash grain farms. In 1950, corn yields averaged about 50 bushels an acre; by 1980 they had risen to 125 bushels.[43] The meteoric rise of soybeans, "the wonder crop of the Corn Belt," began in the 1950s. It was eagerly welcomed as a cash crop to take a place in the rotation instead of hay and oats, which had never been important money earners.[44] In the 1980s, soybeans ranked as the second crop after corn throughout the prairies but were especially important in south central Minnesota, north central Iowa, and the Grand Prairie in Illinois.[45]

An inevitable result of producing more meat, more corn, and more soybeans was to flood the market and force down prices. Attempts by the federal government to reduce surplus production by diverting millions of acres of cropland into a soil bank failed to prevent rising yields of corn and soybeans from raising total output from land remaining in cultivation. The soil bank was abolished in 1973; acreages in corn and soybeans again expanded and yields continued to rise.[46] An elaborate system of price support served to encourage exports. In 1980, the United States exported 34.5% of its

43. Hart, Change in the corn belt, 1986, 53.
44. John Fraser Hart, The look of the land (Prentice-Hall, Englewood Cliffs, N. J. 1975) 141.
45. Hart, Change in the corn belt, 1986, 62.
46. Hudson, Making the corn belt, 1994, 199.

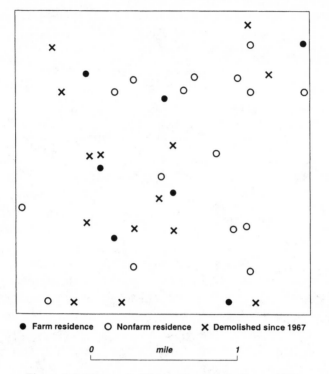

● Farm residence   ○ Nonfarm residence   ✕ Demolished since 1967

0   mile   1

Figure 8.2 Houses in Tipton County, Indiana, 1982

The four-mile-square area had 37 houses standing in 1967. In 1982, 14 had been demolished, 16 were occupied by nonfarm families, and only 7 were still residences of farm families. *Source:* John Fraser Hart, Change in the corn belt, *Geog. Rev.* 76 (1986) 68.

corn crop and 40.4% of the soybean crop.[47] In 1981, the world market was unable to absorb any more produce and prices tumbled until they sank below costs of production.

Financial Crisis and Draining

In the 1980s, investment in farming was cut back as returns on capital diminished and turned into losses. At the same time, the federal government inaugurated a tight money policy to curb inflation and reduce government expenditure. Interest rates were raised to re-

47. Hart, Change in the corn belt, 1986, 68; Harry D. Fornari, U.S. grain exports: A bicentennial overview, in Vivian Wiser (ed.), *Two centuries of American agriculture* (Agric. Hist. Soc., Washington, D.C. 1976) 137–50.

strain borrowing. Banks were forced to pass on higher charges to their customers while the value of land, on which loans were secured, followed a downward spiral.

The first financial crisis that faced a farmer was to pay higher interest rates and keep up capital repayments out of reduced income. A second crisis occurred when land values fell below the level necessary to cover outstanding debts. Even if lenders did not call in loans at this point, they either refused to advance credit needed to buy seeds, fertilizers, fuel, and household goods until the next harvest or they increased interest charges or tried to obtain additional collateral. A farmer in this dilemma reported: "Every time I went in [to the bank] they needed more security, and I had nothing left but the land, and I wouldn't give them title to that. Well, that's when they started raising these [interest] rates . . . Our net worth was going down with this land price going down and machinery prices going down, livestock prices going down. Naturally, our net worth is going right down along with it. And our net worth, we could go back, and I can show you . . . bank statements of about four years ago. They had me worth over $800,000, and there's no way in heck that I was worth that—no way—but this was when they were trying to make things look good."[48]

Many farmers felt betrayed by banks and government lending agencies who encouraged them to borrow more than they ought to have borrowed and in some cases pressed them to spend more than they needed. An aggrieved couple recalled that in 1979, "When we went to Farmers Home Administration, we wanted to buy a new tractor, thirty beef cows and just pay a few bills. They wanted us to go build and build and buy a whole bunch more cattle. If we didn't do it, they said they wouldn't give us the loan." A short-term loan was arranged and the following spring the supervisor again insisted that they borrow over $200,000 to build a dairy barn and two big silos, to which they replied: "We won't do it. It just won't cashflow. There's no sense all of us going out there night and day and not even getting enough out of it to eat." The supervisor then "really got angry, and from that moment on, he was just on us constantly." They ended up filing for bankruptcy in 1983.[49] No fewer than 20

48. Paul C. Rosenblatt, *Farming is in our blood: Family farms in economic crisis* (Iowa State University Press, Ames 1990) 8.

49. Dianna Hunter, *Breaking hard ground: Stories of the Minnesota Farm Advocates* (Holy

of the 24 farm families interviewed by Paul Rosenblatt blamed a
lending agency or its officers for their economic troubles. In general,
the opinion of respondents was that "no matter who makes the decision, them or you, it's still your problem, and that doesn't sit real
well with me . . . As far as I'm concerned, when I signed that loan
agreement, we signed a business agreement. And if I'm going to
have some tough times, [they should] have some tough times."[50]
Some thought there ought to have been provisions for writing down
debts if the value of land depreciated or rescheduling repayments if
interest rates shot up.[51]

The conflict between lenders and borrowers flared into violence. In September 1983, at a farm on the prairie in western
Minnesota, a bankrupt farmer and his 18-year-old son ambushed
and shot a banker and his loan officer who had come to arrange a
foreclosure sale. The case attracted nationwide publicity and set in
train a flurry of death threats and copycat holdups.[52] Two years
later, on a winter morning, seven miles from Iowa City, a distraught
63-year-old farmer shot his wife, drove into town, shot the president
of a bank which had been pressing for repayment of loans amounting to nearly $800,000, drove back to the farm of a neighbor with
whom he had a land dispute, shot him, attempted to kill his wife
and daughter, then returned home and shot himself. Later that af-
ternoon, while police were removing the bodies of the morning's
victims, an anonymous caller telephoned Iowa State Bank and said:
"Tell Marty, he's next." On the front page of the *New York Times*,
the events of 9 December 1985 were headlined as "Death on the
Iowa Prairie: Four New Victims of Economy."[53]

Much of the farmers' anger was directed against themselves in
heavy drinking and suicide. A testimony of farm suicides in the mid-
1980s in Iowa is recounted by Paul Hendrickson.[54] In 1983, the

Cow! Press, Duluth 1990) 90.

50. Rosenblatt, Farming is in our blood, 1990, 14.

51. Neil E. Harl, The financial crisis in the United States, in Comstock (ed.), *Obligation
to save the family farm*, 1987, 112–29, esp. 126.

52. Andrew Malcolm, *Final harvest: An American tragedy* (Signet Books, New York
1987) xiv, 228–29.

53. Osha Gray Davidson, *Broken heartland: The rise of America's rural ghetto*
(Doubleday, New York 1990) 89–93. "Marty" was apparently a reference to Iowa State
Bank's agricultural loan officer.

54. Paul Hendrickson, Those who are no longer with us, in Comstock (ed.), *Obligation to
save the family farm*, 1987, 47–53.

suicide rate among Iowa farmers was 46 per 100,000 compared with a national rate of 29 per 100,000 for all adult men. By 1987, suicides in Iowa reached the highest total since the 1930s. Farmers' families also suffered from violence, abuse, and other symptoms of stress. Social workers observed that farm children were hiding their favorite toys for fear that they would be taken away by the bank, just as their parents' tractors and cars were repossessed. In 1986, a survey of adolescents in rural Minnesota reported that three out of every hundred had attempted suicide in the previous month, a figure 15 times higher than the national average.[55]

Distress was particularly acute in prairie districts in Iowa and Minnesota where protests and demonstrations were organized. In Minnesota in 1985, hundreds of farmers took nonviolent action to prevent foreclosure farm sales, but the state responded with a show of force. The spread of unrest was checked by the opposition or neutrality of many merchants, shopkeepers, and farming neighbors who were owed money and found themselves in the same predicament as bankers and the state.[56] From 1985 to 1987, the crisis escalated to ruin retailers, banks, and professional services in small towns, and manufacturers and suppliers of farm equipment in distant industrial centers.[57] For farmers, the crisis reached its lowest depth in 1986. From 1987, recovery was slow and faltering, checked by crop failures caused by drought in 1988 and flood in 1993. The Iowa Farm Finance Survey, conducted by Iowa State University and Iowa Agricultural Statistics, monitored the profitability, liquidity, and solvency of over 1,000 Iowa farm families from 1984 onward. The proportion of farms in financially stressed groups fell from 31% in 1986 to 15% in 1991 but rose again to 22% in 1993 (table 8.4).[58]

How did draining fit into this turbulent context? Some farmers continued to spend money on draining, but the extent of the activity

55. Davidson, *Broken heartland*, 1990, 94–97.

56. Hunter, *Breaking hard ground*, 1990, 107–8, 27–31.

57. Davidson, *Broken heartland*, 1990, 47–68, observed the repercussions of this widening depression from the standpoint of Mechanicsville, Iowa..

58. Robert W. Jolly, *1993 Iowa Farm Finance Survey* (Iowa State University Extension Service, Ames 1993) 1, maintains that financially stressed farms classed as "weak" can survive if operating changes are made and asset or debt restructuring occurs. They are vulnerable to income losses or asset value declines. This group includes farms with large losses and high equity as well as those with positive earnings and low equity. Farms classed as "severe" are unlikely to survive unless earnings or debt positions are significantly improved.

TABLE 8.4 Financial status of Iowa farms in 1986 and 1991

	1986	1991
Average farm size (acres)	301	438
Average age of farmer (years)	49	58
Return on assets (%)	−1.6	4.8
Cost of debt (%)	11.5	10.6
Debt to asset (%)	27.2	18.8
Profit margin (%)	−6.5	19.3

Return on assets computation: (net farm income + interest paid − family living expenses) ÷ value of farm assets. Interpretation: the ratio estimates pretax earnings per dollar of investment. It can be used as an index of profitability that is independent of the way in which the farm is financed. Changes in asset values can cause the ratio to fluctuate. Family living expenses are used as a proxy for unpaid labor and management.

Cost of debt computation: interest paid ÷ total liabilities. Interpretation: this ratio measures interest charges as a percentage of debt.

Debt to asset computation: total liabilities ÷ total assets. Interpretation: this ratio measures the indebtedness of a farm in percentage terms. Changes in asset values will influence the ratio.

Profit margin computation: (net farm income + interest paid − family living expenses) ÷ total gross income. Interpretation: this ratio measures the pretax return to the total capital stock per dollar of sales. It is an index of average profitability, before financing, for a unit of production.

Source: Robert W. Jolly, *1993 Iowa Farm Finance Survey* (Iowa State University Extension Service, Ames 1993) 9, 13, app. A.

is difficult to measure. Because dependable data on the location of county drainage ditches and the extent of drained land were lacking in 1978, county boards in four south central Minnesota counties cooperated with the University of Minnesota Water Resources Research Center and the federal Office of Water Research and Technology to fund an extensive mapping program. A team led by Henry W. Quade from the Department of Biological Sciences at Mankato State University calculated that the four counties had nearly half their total area drained by public drainage systems (table 8.5). About two-thirds as many ditches were dug and almost as large an area of land was drained in the thirty years from 1950 to 1979 as in the peak period from 1900 to 1929 (table 8.6). The survey indicates that draining was still active in the late 1970s.[59] In the upper Midwest as a whole, it was estimated that about 29% of cropland was located on wet soils.[60]

59. Henry W. Quade et al., *The nature and effects of county drainage ditches in south central Minnesota,* in *Water Resources Research Center Bulletin* 105 (University of Minnesota, Minneapolis 1980).

60. R. T. Diedrick, The agricultural value of wet soils in the upper Midwest, in B. Richardson (ed.), *Selected proceedings of the Midwest conference on wetland values and*

Draining was an expensive operation, yet it is unclear how it was financed. None of the 24 heavily indebted farm couples who spoke to Paul Rosenblatt and his assistants between 1986 and 1988 cited draining as a cause of their economic difficulties. Nor were any of the 31 stories reported by Dianna Hunter and the Minnesota Farm Advocates concerned with debts incurred in draining. Records kept by the Farm Home Administration might contain information on loans and repayments for drainage projects but government lending agencies were directed to discourage farmers from reclaiming wetlands.[61] Banks and business associates would have preferred to finance schemes that yielded profits directly and rapidly, such as buildings, equipment, and stock for new livestock enterprises. Draining did not produce large returns in the first or second year; it benefited productivity over a long period of time.

Investment in draining might have been inhibited for other reasons. A tenant would not spend his own money improving a landlord's property unless the landlord were his father or a relative from whom he would eventually inherit an improved estate. Many farms that were enlarged by adding land rented from corporate institutions would not be drained at the occupier's expense. Similarly, detached portions would be less likely to be drained than ring fence farms, and fragmented holdings were also becoming more numerous. Capitalization increased competitiveness and diminished cooperation. Many farmers borrowed money to buy equipment that would make them more efficient and less dependent on neighbors and contractors. Draining, on the other hand, demanded cooperation with neighbors in setting up a drainage district and digging an outfall ditch. Even within the boundaries of a single farm, it was rarely possible to drain land without discharging water into a ditch that crossed someone else's land.

Would it have been possible for individual farmers to have saved sufficient money out of the profits of their enterprises or from off-farm earnings? It seems unlikely because draining was normally carried out at the beginning of a farming career. It was a long-term investment that brought rewards after many years in occupation. It was not worthwhile if the future of a farm was in jeopardy. A profile

---

management (Freshwater Society, St. Paul, Minn. 1981) 97–106; data relate to Indiana, Illinois, Iowa, Michigan, Minnesota, North Dakota, South Dakota, and Wisconsin.
   61. Information received from Graham A. Tobin, 24 March 1995.

TABLE 8.5 Drained land in four south central Minnesota counties

County	c.1860[a] % swamp	c.1860 % lake	1930[b] % drained	1971[c] % drained	1978[d] % drained
Blue Earth	5.1	3.3	21.4	50.4	39.9
Le Sueur	9.9	5.8	6.9	43.5	46.7
Nicollet	7.3	5.4	14.6	59.4	58.9
Brown	2.6	1.8	9.5	48.2	45.9
All four counties	5.8	3.8	13.9	50.2	46.7

Note: Drained area is expressed as percentage of total surface area.

[a] Original Land Survey Plats and Notes, c.1860, in Henry W. Quade et al., *The nature and effects of county drainage ditches in south central Minnesota*, in *Water Resources Research Center Bulletin* 105 (University of Minnesota, Minneapolis 1980).

[b] U.S. Bureau of Census, *Fifteenth Census of the United States, 1930*, vol. 4, *Drainage of agricultural land* (Washington, D.C. 1932) 185.

[c] U.S. Geological Survey, Drainage survey by counties for Minnesota, 1971–72, unpublished report, in Quade et al., *Drainage ditches*, 1980.

[d] Quade et al., *Drainage ditches*, 1980, 6, 11–18.

of potential drainers closely resembled that of the youngish, healthy, well-educated, enterprising, dedicated farmers interviewed by Rosenblatt. Rosenblatt's respondents had farming in their blood but little money in their pockets. Might capital have been subscribed by relatives, partners, or shareholders in family corporations? That possibility must be examined in the light of the debate over family farms.

Tragedy for Farming Families

Tragedy is deeply embedded in the Midwestern family farming system. The notion of a homestead as a free gift, or nearly free gift, from the U.S. public domain has been a poisoned chalice handed from generation to generation. A total stranger seeking a working farm would have had to raise an enormous sum of money to start up, to buy sufficient land, buildings, and equipment. The return on the investment, heavily discounting the value of labor put in by the farmer and his family, would have been very much smaller and less certain than from most other businesses. What allowed the system to function at all was the inheritance of more or less free land. If the land had been encumbered with debt or burdened with taxes the system would have broken down. In order to prevent an operating farm being divided at death among wife and nonfarming children,

TABLE 8.6 Drainage ditch construction by decade
in south central Minnesota, 1880–1979

	Ditches (no.)	Ditches (%)	Drained (acres)	Drained (% total)
No info	11	4.1	–	–
1880–89	4	1.5	7,142	1.2
1890–99	5	1.9	10,208	1.7
1900–09	35	13.0	94,976	17.2
1910–19	70	26.0	112,000	20.3
1920–29	36	13.4	56,909	10.3
1930–39	1	0.4	525	0.9
1940–49	11	4.1	30,554	5.5
1950–59	34	12.6	128,602	23.2
1960–69	41	15.2	83,706	15.1
1970–79	21	7.8	27,149	4.9
Total	269	100.0	551,771	100.3

Source: Quade et al., *Drainage ditches,* 1980, table 13, p. 30; table 15, pp. 33–34.

an estate might be incorporated and shares allotted to heirs. Some farms were converted into corporations to reduce liability to taxes, and indebted or bankrupt farms, taken over by banks or other lenders, might be run as separate corporate entities.[62] Reorganization changed the role and status of a farmer. A president or chief executive is not the same as a father or head of household.

As long as a holding remained intact and free to be passed on, a family was bound to the soil. The owner and his wife went on working until their child or children could take over. Children were brought up and trained to succeed their parents. Their future was preempted. Continuity of family farming was most secure where patrilineal bonds between father and son were strongest, where a father took more account of opinions of his heir than his wife in making decisions about property and investment. Sonya Salamon has drawn comparisons between adherence to family succession in

62. Caroline Tauxe, The myth of the family farm: An essay on hegemonic process in American society, *Dialectical Anthropology* 17 (1992) 291–318, esp. 307; Curry-Roper and Bowles, Local factors in land-tenure patterns, 1991, 447–48. Hart, *The land that feeds us,* 1991, 375, challenges the notion that large corporate farms are gobbling up small family farms. In the eastern United States, including the Midwest, they are a legal device for keeping farms in the family, "created solely to avoid having to sell off a chunk of the farm for death and estate taxes." In 1982, according to the Census of Agriculture, about 90% of farms organized as corporations were held by farming families.

seven long-established communities in central Illinois: three of German descent, three of Yankee descent, and one of mixed German and Yankee origins. In the German communities, nearly half the family names have continued for over a century, whereas only one-fifth of Yankee family names have survived. In one of the German communities, 23% of households report some form of father-and-son operating agreements, whereas only 9% of Yankee households have such working arrangements. Because Yankee parents feel little obligation to consult their children about planning and management, "generational inequalities often make the successor role demeaning for sons."[63]

Since the 1960s, most Midwestern communities have moved toward a free, entrepreneurial Yankee style of family relationship. A flexible intergenerational link in families of Dutch origin in Marion County, Iowa, succeeded in keeping only one farm in the hands of its founding family and, in the fourth generation, that farm passed out of direct patrilineal descent from an uncle to a nephew. It is inferred that family succession is maintained in times of economic stability, "but in times of economic stress it is disrupted by external forces."[64] Only a very tightly knit family will have been able to hold an heir since the crisis in 1981. One couple farming on the black soil of a Minnesota prairie felt good about "getting our arms around those little grandchildren and knowing we had to fight for the farm to keep them here, instead of shipping them off to the Twin Cities or who knows where. If we wanted to keep our son and family in farming we had to hang on to this land. That's what put the determination in us."[65] Murray Straus has found that sons choosing to follow a farming career tended to come from high-income, owner-operated farms, had few siblings, enjoyed close relationships with their parents, and upheld traditional rural values. The price paid for their dependent status was lack of initiative, lack of self-reliance, deficiencies often magnified by shyness and mediocre educational achievements.[66]

63. Sonya Salamon, Culture and agricultural land tenure, *Rural Sociology* 58 (1993) 580–98, quotation from 591.

64. Curry-Roper and Bowles, Local factors in land-tenure patterns, 1991, 454.

65. Hunter, *Breaking hard ground*, 1990, 80.

66. Murray A. Straus, Societal needs and personal characteristics in the choice of farm, blue collar, and white collar occupations by farmers' sons, *Rural Sociology* 29 (1964) 423–25.

A loosening of family ties and loss of confidence in the future for farming conspired to weaken attachment to family farms. An event that broke a succession brought tragic consequences. Accidental death, chronic illness, physical injury, mental breakdown, divorce, remarriage, or simply renouncing a commitment to farm (which in less constricted families would have been regarded as exercising a right to choose freely an independent way of life) were likely to lead to a loss of farm, home, kith, and kin. In a highly competitive environment, external events such as floods, droughts, storms, and crop or livestock diseases might precipitate disaster. Jane Smiley constructed the plot of her novel, *A Thousand Acres*, around a decision by a domineering old farmer to hand over his Iowa prairie farm to his three daughters. The youngest daughter, who practices law with her husband in Des Moines, declines an offered share and is promptly disinherited. As the tale unfolds, childhood beatings and sexual abuse of the older daughters are repaid in hatred and violence toward the old man, and the daughters' marriages end in suicide for one husband and divorce for the other.[67] A scenario that contrived to avoid catastrophe was described in Robert Waller's *Bridges of Madison County,* in which an Italian wife of an Iowa farmer has a passionate love affair with a roving photographer. The affair ends abruptly after five days so that the wife might resume her marriage, bring up the children, and save the farm. In the outcome, the farmer dies, the children leave the land, and the wife alone survives at the farmhouse. Leaving aside the implausibility of a magically terminated romantic interlude, the novel drew a familiar picture of a woman's loneliness on a Midwest farm and the sacrifices she was called upon to make to keep home and family together.[68]

Draining, because of its long deferred rewards, underscored the covenant that tied a family to the land. It also provided to the observant eye a sign of good farming. In addition to "clean fields, neatly painted buildings, breakfast at six, no debts," an infallible indicator of good management was "no standing water." The tile system on Jane Smiley's thousand-acre farm "drained fields that were nearly as level as a table . . . The old watercourses, such as they were, had been filled in and plowed through, so the tile lines drained into

67. Jane Smiley, *A thousand acres* (Flamingo, London 1992).
68. Robert Waller, *The bridges of Madison County* (Warner, New York 1992).

drainage wells. These wells, thrusting downward some three hundred feet, still dot the township, and there were seven around the
peripheries of our farm. A good farmer was a man who so organized his work that the drainage well catchment basins were
cleaned out every spring and the grates painted black every two
years."[69] The author described a most peculiar, if not impossible,
form of outfall, discharging water down extraordinarily deep shafts,
presumably into a porous substratum. Over most of the Midwest,
outfalls were conducted very gradually along open ditches into tributaries of major streams, and these ditches were maintained by collaboration among neighbors. Collaboration and trust were out of
character with Smiley's portrait of the farmer as a competitive,
rugged individualist who cleaned out his own drains at regular intervals and sank wells at the edges of his own property. A drainage
system bounded by lines of private property, draining into private
wells, imposed a formidable obstacle to transfers of ownership. It
would have been difficult to subdivide or sell a piece of land whose
tile lines drained into a soakaway or outlet belonging to a private
owner other than the prospective buyer. It would also have been difficult to add a piece of land that did not drain into an outlet belonging to the same property. Smiley mentions a pond "down toward
Mel's corner. The pond, an ancient pothole that predated the farm,
was impressively large" to the eye of a child. In the 1960s, the
farmer drained it and "took out the trees and stumps around it so
he could work that field more efficiently."[70] The draining would
have enhanced the good appearance of the farm and bound the field
firmly to the patrimony.

A competitive struggle for possession of land on the part of successful farmers seeking to expand their holdings and struggles by
indebted farmers to avoid expenditures not yielding immediate returns were two powerful disincentives to draining. Disputes between
father and heir apparent arose when father sought to sell a parcel of
land in order to buy another home or an annuity for retirement.
Conversely, investment in draining was contested as favoring an heir
and depriving older members of a family of their savings. A farmer
who was about to retire had a stronger incentive to enter into a set-
aside agreement than a successor who wanted to maximize produc-

69. Smiley, *Thousand acres*, 1992, 45.
70. Ibid., 85, 205–6.

tivity. During a recession, a successor would strive to obtain part-time employment and other means of supplementing farm income by renting campsites, providing transport services, contracting for highway maintenance, or engaging in building construction. Aggressive individualism severely curtailed the scope for collaborative ventures in draining.

As rural areas became dormitories for commuters, growing numbers of nonfarming neighbors increased pressures on farmers to refrain from damaging waterfowl habitats, spraying beyond the edges of fields, dumping slurry in watercourses, or burning stubble and straw. Newcomers would not tolerate the smell of pigs and complained about foul water in ditches and streams. In the early 1970s, residents in south central Minnesota, as well as water engineers and planners, perceived "water pollution" and "water resource planning" to be more important problems than "wet agricultural fields."[71] In 1986, a poll conducted by the *Des Moines Sunday Register* showed that 78% of Iowans and over half of all farmers wanted to restrict the use of agricultural chemicals to protect groundwater from contamination.[72] A scientific review indicated that insecticides sprayed from the air were being carried into prairie potholes, raising concentrations to levels dangerously high for waterfowl.[73] Concern was expressed over damage caused to the environment by artificial drainage through discharging sediment and nutrient loads into outflowing streams. Some historical geographers joined the chorus of complaint. In 1993, Mary McCorvie and Christopher Lant described the draining of Midwestern wetlands as "a massive environmental insult."[74]

Under a barrage of hostile publicity, farmers were driven to consider alternatives to draining their land and other ways of earning a

71. Robert Moline, *The citizen and water attitudes in southern Minnesota*, Office of Water Research and Technology Project B-042-Minnesota (Minneapolis 1974).

72. Duane Nellis, Agricultural externalities and the environment in the United States, in I. R. Bowler, C. R. Bryant, and M. D. Nellis (eds.), *Contemporary rural systems in transition: Agriculture and environment* (CAB International, Wallingford 1992), vol. 1: 133.

73. C. E. Grue, M. W. Time, G. A. Swanson, S. M. Borthwick, and L. R. Deweese, *Agricultural chemicals and the quality of the prairie pothole wetlands for adult and juvenile waterfowl: What are the concerns?* National Symposium on Protection of Wetlands from Agricultural Impacts (Colorado State University, Fort Collins 1988); Edwin Clark II, The United States, in Kerry Turner and Tom Jones (eds.), *Wetlands: Market and intervention failures* (Earthscan, London 1991) 39–72, esp. 44.

74. McCorvie and Lant, Drainage district formation, 1993, 22.

livelihood.[75] Many farm operators already earned more from non-farm occupations than from selling crops and livestock. Data collected in 1990 by the Agricultural Economics and Land Ownership Survey of the Census of Agriculture led Gene Wunderlich to conclude that "the classic concept of a farm continues to wither. Farm households encompass a diversity of enterprises. Farm resources are owned by absentees. Farm families enjoy a wide range of employments."[76] What remained was a tract of land whose use was changing. Farm operators began to think of themselves in roles other than farmers, as salesmen, teachers, lawyers, even conservationists. But however much their self-images changed, most continued to cherish an ideal of family farming. In 1993, Iowa Farm and Rural Life Poll received responses from 2,390 agricultural producers in answer to a questionnaire about perceived threats to rural America. "Loss of family farms" topped the poll by a substantial margin, 76% of respondents rating it the most severe threat to the future of rural America.[77]

Landowners' Attitudes to Draining since 1980

In the 1980s, government policy effectively put a stop to subsidizing drainage and offered inducements to landowners to restore wetlands. Senator George Mitchell optimistically claimed that tax reforms and discriminatory farm support measures "removed incentives that had encouraged wetlands destruction."[78] The 1985 farm bill, the Food Security Act, initiated a Conservation Reserve Program to take land out of cultivation, thereby reducing productive capacity and protecting erodible soils. The bill also contained a

75. John Fraser Hart devotes the preface of *The land that feeds us*, 1991, 11, to trying to understand the bitterness behind the question "Why do you city folks hate us farmers so much?"

76. Gene Wunderlich, The land question: Are there answers? *Rural Sociology* 58 (1993) 547–59, quotation on 557.

77. Paul Lasley, *Iowa farm and rural life poll: 1993 summary report* (Iowa State University Extension Service, Ames 1993) 8. Among other "severe threats" respondents voted "closing of small businesses" 67%, "increased use of illegal drugs" 61%, "lack of jobs" 60%, "changes in traditional values" 56%, "decline in the American work ethic" 55%, "changes in traditional family structure" 52%, and, trailing at the bottom of the poll, "decline in the quality of the environment" 14%.

78. George J. Mitchell, foreword to Jon A. Kusler and Mary E. Kentula, *Wetland creation and restoration: The status of the science* (Island Press, Washington, D.C. 1990) ix.

Swampbuster provision, denying commodity benefits to anyone growing crops on land reclaimed from swamps after 23 December 1985. In return, the USDA set up a Wetlands Reserve Program, a voluntary easement, enabling landowners to restore and protect wetlands. Under this scheme, the government agreed to pay a rental for land withdrawn from commercial cropping and offered grants of up to 50% toward the costs of laying retired land down to grass or planting it with trees.[79] The 1990 farm bill, the Food, Agriculture, Conservation, and Trade Act, introduced an even more radical Environmental Conservation Acreage Reserve Program to provide additional protection for wildlife and waterfowl habitats, improve water quality, and enhance aesthetic, educational, and scientific values of wetlands.[80] The measure aimed at enrolling an additional forty to fifty million acres in the reserve by 1995. In the first year of the program, over 1,700,000 acres were enrolled and it was reported that "there is a strong regional shift away from the Great Plains to the Corn Belt and Lake States, areas where water-caused erosion is perceived to cause more serious damages."[81] The higher value of land in the corn belt, albeit damaged by erosion, was reflected in the average rental compensation going up to $57 per acre and grants for restoring wetlands being raised to 75% of costs.

Facts about the status of draining in the Midwest since 1980 are sparse and vague. In northern and central Illinois expenditure on drainage has not been reported in farm business records since 1982.[82] On the other hand, reports from the U.S. Fish and Wildlife Service suggest that it continued in the 1980s.[83] In 1985, Congress directed the secretary of the interior to conduct an inquiry into the impact that federal programs had and were likely to have on the conversion, degradation, and conservation of wetlands in different regions in the United States. A first volume of the report dealt with

79. Potter et al., *Diversion of land,* 1991, 55.

80. U.S. Department of Agriculture, *What the conservation provisions of the Food, Agriculture, Conservation, and Trade Act mean to you* (Washington, November 1991) 1–3.

81. Ervin, Political-economic effects of set-aside, 1992, 6.

82. D. F. Wilken et al., *Fifty-eighth annual summary of Illinois farm business records* (University of Illinois, College of Agriculture, Urbana 1983).

83. Thomas E. Dahl, *Wetlands losses in the United States, 1780s to 1980s* (U.S. Fish and Wildlife Service, Washington, D.C. 1990) 5, 9, 10; W. E. Frayer, *Status and trends of wetlands and deepwater habitats in the coterminous United States, 1970s to 1980s* (Michigan Technological University, Houghton 1991) 3, 18, 19.

the two most important regions: the lower Mississippi alluvial plain
and the prairie pothole region in the upper Midwest.[84] It concluded
that federal agricultural programs had significantly increased the
profitability of drainage in the prairies, that price and income
supports provided the strongest incentives to drain wet prairies, that
low-interest loans and cost-sharing schemes for carrying out
drainage offered further encouragement to expand the area of farmland. In addition, tax allowances and stream channelization works
induced individuals to undertake draining at their own expense. In
1988, on wet prairies, farm owners were "financially better off after
drainage than before and they almost always benefit more from
agriculture programs after draining their land than if the land had
remained undrained." Private drainage projects were still "profitable, even in the absence of government incentives."[85] Efforts by
federal agencies to stop further draining were not successful.
Swampbuster provisions of the Food Security Act were at variance
with the primary objectives of the USDA. Controls exercised by the
U.S. Army Corps of Engineers under Section 404 of the Clean
Water Act of 1977 were regarded as "not very effective in Wisconsin, having reduced wetland losses by only 15 per cent, in contrast to estimates from across the country ranging from 24 per cent
to 50 per cent."[86] The Corps of Engineers had little power over
drainage of wetlands in the Midwest because it had agreed that the
Natural Resources Conservation Service (formerly the Soil Conservation Service) should be responsible for agricultural land, including
wetlands on farms. The Clean Water Act was not drafted as a measure for wetland protection, and until judicial decisions were taken
in the late 1970s, Section 404 could not be invoked to prevent agricultural drainage.

Farmers either ignored issues of draining and wetland conservation or continued to believe that land not in cultivation was "less
productive than it ought to be" and that "the only good wetland is a
drained one."[87] A climatologist at a college of agriculture who

84. U.S. Secretary of the Interior, *The impact of federal programs on wetlands*
(Washington, October 1988), vol. 1: 2, 4.

85. Ibid., 96.

86. C. R. Owen and H. M. Jacobs, Wetland protection as land-use planning: The impact
of Section 404 in Wisconsin, U.S.A., *Environmental Management* 16 (1992) 352.

87. Robert T. Moline, Cultural modification of wet prairie landscapes, in Thomas J.
Baerwald and Karen L. Harrington (eds.), *A.A.G. '86 Twin Cities field trip guide* (Assoc.

owned a farm in northern Iowa said in 1990 that he knew his land was tiled. The tiles were supposed to last forty or fifty years but he had no clear idea when they had been put in. He was a little surprised that anyone should even be asking about drainage. After further inquiry he learned that tiles cost 70 to 90 cents a foot and were spaced at intervals of 70 to 100 feet. His estimates worked out at $400 to $500 an acre, but neighbors told him that his figure was too low. He did not expect to have to replace his drainage system. It was a once-in-a-lifetime operation, not to be repeated and, once done, out of sight, unlike buildings or fences which remained visible and could be seen to deteriorate. A geographer who farmed in northern Illinois spent $20,000 in the 1980s replacing blocked drains, and another part-time farmer in south central Indiana asked for an estimate for draining but could not afford the cost.[88] In southern Minnesota tiling was still going on in 1990 but it was difficult to find out the extent of the activity. It did not enter official records. A farmer in Waseca County said that his dealings with the county drainage board were very informal. The county drainage inspector doubled as soil conservation officer. When the ditch was overgrown with trees, farmers asked for them to be pulled out, tenders were invited, a contract drawn up, and costs shared on a "good neighborly" basis. Tiling was a private expense, except that no new lines of tile were allowed to be laid on land not already cultivated.[89]

Evidence for converting drained land back into wetland is extremely scanty. In 1986, lakes on wet prairies in southern Minnesota were fringed with sites for day visitor recreation and strings of summer homes. Along the shores of lakes Washington and Jefferson, hundreds of permanent residences had been built since 1965.[90] Nutrient-rich runoff from surrounding fields and seepage from septic tanks made the water unpleasant for swimming and sailing but wetland restoration helped to abate the nuisance and brought an added attraction of wildlife. One large landowner in southern Minnesota who was a keen hunter and bird-watcher recognized an opportunity for extending breeding grounds and feeding

Amer. Geog., Minneapolis 1986) 196.

88. Conversations with John Fraser Hart at University of Minnesota on 3 November, and 3 and 6 December 1990.

89. James Zimmermann, telephone conversation on 3 December 1990.

90. Moline, Wet prairie landscapes, 1986, 201–2.

places for ducks and developed his sloughs as an outdoor recreation resource. A landscape architect, Joan Nassauer, asserted that "the look of land" might "be critical to farmers' decisions to participate" in Swampbuster provisions of the 1985 farm bill and might lead "to economic development benefits realized by rural communities."[91] On the basis of consultations with residents in Olmsted County, Minnesota, she designed visually pleasing schemes for wetland restoration, retaining an appearance of neatness and care associated with well-farmed landscapes.[92] Small farmers had less room for adjustment or rural ornament. They had to carry fixed overhead costs for buildings and machinery to be paid for out of returns from a reduced productive acreage. Many were hesitant to talk about plans for the future.

Wet Prairies and Floods in the Upper Mississippi River Basin, 1993

In addition to economic pressures exerted by market forces, technological treadmill, surplus production, financial crisis, and social ties to family farms and local traditions, owners of wet prairies were subjected to severe natural hazards, occasionally drought, as in 1988; more frequently floods, as in 1913, 1927, 1937, 1951, 1955, 1972, and 1993, when nonwetlands reverted temporarily to something like their original state as wet prairies. The damage and destruction caused by these events led landowners to reconsider whether it was worthwhile keeping vulnerable land under cultivation or whether they ought to set it aside and restore it to wetland.

Residents on wet prairies were no strangers to mud and waterlogging. Mud returned almost every spring and was a recurrent theme in Garrison Keillor's Lake Wobegon stories. Pioneers "had known mud in New England, of course, but there was no mud like what they found in Minnesota. The town sat in a swamp from April until June, and then again in July, and often in August, too, while September brought some more, and one October, three days of rain made them a lake that promptly froze over for six months."[93] In 1993, most soils throughout the upper Mississippi River basin were

91. Joan Iverson Nassauer, Agricultural policy and aesthetic objectives, *Journ. Soil and Water Conservation* 44 (1989) 384.

92. Nassauer, The aesthetics of horticulture: Neatness as a form of care, *HortScience* 23 (1988) 973–77.

93. Garrison Keillor, *Lake Wobegon days* (Penguin, Harmondsworth 1986) 46.

saturated before the onset of exceptionally heavy summer rains. Record downpours fell on different localities between mid-June and the end of July. Higher than average rains continued to fall in August and floods did not recede until mid-September. At the end of September, thunderstorms again deposited seven inches of rain in eastern Missouri, causing another burst of flash floods.[94] Heavy rains fell repeatedly on the same region to the west of the Mississippi, keeping the ground waterlogged week after week. Most of the rainfall quickly ran off. A remarkable feature of the 1993 floods was the rapidity with which storm waters rushed from distant parts of the watershed into streams and swelled into giant crests. Peak discharges in main stem rivers were observed within hours of heavy rainstorms.

Most of this rain fell on uplands and spread widely over flat prairie surfaces. One such temporary lake, covering nearly 10,000 square miles of uplands in northern Iowa and southwestern Minnesota, was recorded on a satellite image on 14 July 1993 (fig. 8.3). At the height of the floods, when all eyes were fixed on gaping breaches in levees along the Mississippi and Missouri and forecasters were awaiting anxiously the fate of beleaguered cities at the rivers' edges, an astute reporter acknowledged that "most of the inundation affected thinly populated farmland."[95] The full extent of land that was at one time or another under water, amounting to 36,000 square miles, far exceeded the combined areas of floodplains of the great rivers and their tributaries. Over much of the uplands water did not stand for long in fields, farmyards, or highways. It ran rapidly into ditches carrying loads of corn stalks, agricultural chemicals, and sediment. John F. Sullivan, water quality specialist for the State of Wisconsin's Department of Natural Resources, commented: "There is no question—sediment is the real threat to the upper Mississippi today."[96] A data sheet issued by the federal

94. Kenneth E. Kinkel, A hydroclimatological assessment of the rainfall, in Stanley A. Changnon (ed.), *The great flood of 1993: Causes, impacts, and responses* (Westview Press, Boulder 1996) 52–67. Other general accounts of the 1993 floods include Graham Tobin and Burrell E. Montz, *The great Midwestern floods of 1993* (Harcourt Brace, Orlando 1994); Hugh Prince, Floods in the upper Mississippi River basin, 1993: Newspapers, official views, and forgotten farmlands, *Area* 27 (1995) 118–26.

95. G. J. Church, The worst is supposed to be over this weekend, *Time* (26 July 1993) 31.

96. W. S. Ellis, The Mississippi: River under siege, *Water: National Geographic Special Edition* (November 1993), quotes Sullivan on 98.

Soil Conservation Service in October 1993 reported that 60,000 acres, mostly flooded bottomlands, were covered to depths of more than two feet with sand and silt. On uplands, drainage ditches were filled with sediments, and other agricultural infrastructures were damaged or destroyed by scour and deposition. It was estimated that removal of sediment and debris from ditches would cost at least $10 million.[97] Record volumes of floodwater did not dilute releases of pesticides, herbicides, and other toxic materials. Concentrations of two herbicides, atrazine and cyanazine, in some samples taken from the Mississippi River, exceeded permitted safety limits for drinking water.[98] Effects of discharges of agricultural chemicals and deposition of sediments have yet to be scientifically examined.

By the end of the summer in 1993, the president declared 523 counties in the upper Mississippi valley "flood disaster counties." Floodwater did not cover the entire area of those counties, but all suffered some flood damages (fig. 8.4). Preliminary estimates of total damages ranged from $12 billion to $16 billion, losses to agriculture accounting for half that sum. By June 1994, the federal government had contributed over $5.4 billion toward emergency assistance and restoration, about half of which was given to agriculture. Almost all aid to agriculture was compensation for crop losses. Most crop losses and the largest share of USDA compensation payments went to farmers whose fields were too wet to plant or were planted too late for a good harvest. Mary Swander, a correspondent to the *New York Times Magazine*, writing from the heart of the flooded region in Iowa, described the protracted wait for a dry spell. In mid-July, "*everywhere* the ground was wet, sodden and soggy, with just a few dry spots surrounded by damp fields or swirling water. Wetlands that had been drained for decades and used to grow corn returned to swamps."[99] Many who succeeded in getting corn and soybeans to germinate on wet soils suffered losses later in the season through low temperatures and early frosts.

97. U.S. Department of Agriculture, Soil Conservation Service, Data sheet (Columbia, Missouri, October 1993).

98. Donald A. Goolsby, William A. Battaglin, and E. Michael Thurman, *Occurrence and transport of agricultural chemicals in the Mississippi River basin, July through August 1993*, Circular 1120-C (U.S. Geological Survey, Washington, D.C. 1993).

99. Mary Swander, Iowa, colored blue, *New York Times Magazine* (19 September 1993) 36.

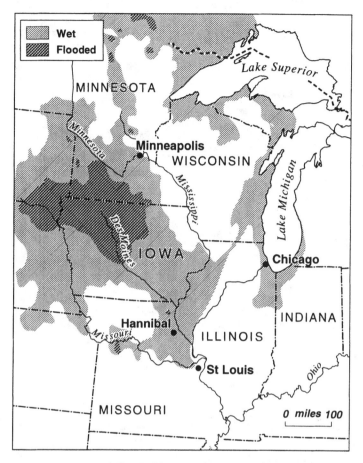

Figure 8.3 Wet and flooded land in the Midwest on 14 July 1993

A satellite image recorded a soil moisture index: "flooded" areas were under water; "wet" areas were saturated. A sheet of water covered about 10,000 square miles in northern Iowa, southwestern Minnesota, and southeastern South Dakota. *Source:* National Oceanographic and Atmospheric Administration, AVHRR SSM/I Image, 14 July 1993.

Wet prairie counties in the Dakotas, Minnesota, and Iowa received 70% of the USDA's flood disaster assistance and 80% of crop insurance payments. Minnesota had the largest acreage of harvest failure. Iowa had the highest losses in crop yields and incomes. Missouri had the largest percentage of cropland flooded. Table 8.7 shows payments for crop losses made from federal flood relief funds

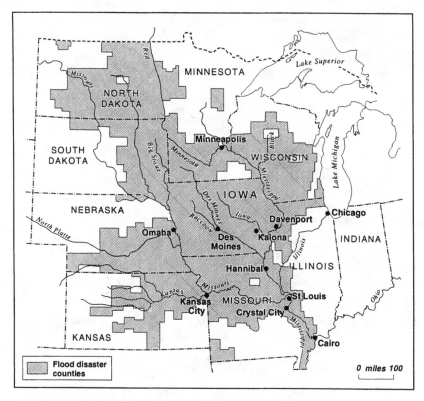

Figure 8.4 Flood disaster counties in the upper Mississippi basin, 1993

In a nine-state region, 523 counties, including all counties in Iowa, were declared federal disaster areas. *Source:* Stanley A. Changnon, Losers and winners: A summary of the flood's impacts, in Changnon (ed.), *The great flood of 1993*, 1996, 290.

to 523 counties designated disaster areas in nine states up to 13 June 1994. Payments under the Crop Insurance Program (CIP), administered by the Agricultural Stabilization and Conservation Service (ASCS), were related to amounts of coverage farmers had subscribed to. In the disaster areas as a whole, 56.7% of eligible croplands were insured. Crops in North Dakota, Kansas, and Iowa were more extensively covered than crops in Wisconsin or Missouri. Payments under the program were highest in Minnesota and Iowa and lowest in Missouri. Disaster assistance provided relief at a somewhat lower rate to landowners who had not enrolled in the crop insurance scheme. In December 1993, a total of $219 mil-

TABLE 8.7 Participation in Crop Insurance Program, payments for crop losses in 1993 floods, and deficiency payments for 1992 crops

	% of acres participating in CIP 1993	CIP payment 1993–94 ($ million)	Disaster assistance 1993–94 ($ million)	Total loss payments 1993–94 ($ million)	Deficiency payments 1992 ($ million)
Illinois	44.4	26	55	81	114
Iowa	64.9	283	396	679	453
Kansas	76.4	41	71	112	118
Minnesota	52.4	354	481	835	204
Missouri	24.0	28	129	157	98
Nebraska	56.1	49	86	135	239
N. Dakota	93.4	140	119	259	105
S. Dakota	47.0	54	163	217	87
Wisconsin	11.3	47	141	188	76
Total	56.7	1,022	1,641	2,663	1,494

Sources: U.S. Department of Agriculture, Federal Crop Insurance Corporation, *Midwest flood/xcess moisture loss projections for 1 Dec 1993*; USDA Flood Information Center, *USDA emergency assistance paid to flood states*, 13 June 1994; Nancy S. Philippi, *Revisiting flood control: An examination of federal control policy in light of the 1993 flood event on the upper Mississippi River* (Wetlands Research, Chicago 1994) 15, 18, 20.

lion in disaster assistance had been paid.[100] By June 1994, the total had risen to $1.6 billion, bringing the total amount of compensation paid for 1993 flood crop losses to $2.7 billion. The size of these emergency relief payments acted as a disincentive to participate in the Crop Insurance Program. They also compared favorably with deficiency payments paid to farmers when prices for crops fell below a stipulated minimum. In 1992, a bumper harvest was accompanied by a fall in prices that was recompensed by high levels of deficiency payments. Nebraska, Illinois, and Kansas received greater compensation for high production in 1992 than for flood losses in 1993. In normal years, government price support payments boosted crop production and did not encourage farmers to think about conserving wetlands.

The 1993 floods upset prevailing attitudes. Local residents,

100. Critical discussions of crop insurance compensation payments and flood disaster relief are outlined in Nancy S. Philippi, *Revisiting flood control: An examination of federal flood control policy in light of the 1993 flood event on the upper Mississippi River* (Wetlands Research, Chicago, May 1994); Philippi, *Spending federal flood control dollars: Three case studies of the 1993 Mississippi River floods* (Wetlands Research, Chicago 1995).

weary after weeks struggling against water and mud, disheartened by
the stinking mess left behind, angry about the haphazard way in
which financial relief was handed out, demoralized by confusion,
lack of coordination, lack of planning, lack of direction among different agencies, were ready to listen to new ideas. Many wanted to
free themselves from the strain of trying to raise crops in such a haz-
ardous environment. Aspirations of residents began to converge
with projects taking shape in the minds of government officials. The
first response by a spokesman for the U.S. Army Corps of Engineers
to questions about future flood protection in the upper Mississippi
River basin was to declare the intention of the Corps to rebuild all
federal levees and bring most nonfederal levees up to the same high
standards of construction. They aimed to "make the system whole,
as it was before the flood."[101] Four months later, Major General
Stanley Genega, director of civil works, backed away from a
promise of total restitution and assured the public: "There's no intent on the part of the Corps of Engineers to line the Mississippi
with concrete."[102] In June 1994, an interagency floodplain management review committee outlined a new strategy for flood damage
reduction offering less than complete protection to all lands liable to
flood. The committee recommended that additional areas of bottomlands should be left open to occasional flooding, that property
in such washlands should be bought out and people exposed to
floods should be offered new homes on the uplands.[103] They would
not be protected by higher levees nor would they be eligible for
federal emergency aid.

A policy of buyouts marked a fundamental departure from earlier measures of disaster relief. A National Flood Insurance Act introduced in 1968 required property owners taking part in the program to elevate structures above anticipated 100-year flood levels,
by building on stilts or raised platforms. Section 1362 of the act enabled the Federal Emergency Management Agency to purchase

101. Now they rebuild the levees, *Washington Post* (10 August 1993) 3A.

102. Alan Mairson, The great flood of '93, *National Geographic* 185 (January 1994) 73.

103. Interagency Floodplain Management Review Committee, Scientific Assessment
and Strategy Team, *A blueprint for change: Science for floodplain management in the 21st
century* (SAST, Washington, D.C. 1994), followed by a directive, U.S. Executive Office of
the President, *Sharing the challenge: Floodplain management into the 21st century*
(Washington, D.C. 1994); detailed proposals were outlined in U.S. Army Corps of Engineers,
Floodplain management assessment (Washington, D.C. 1995).

flood-damaged property and offer owners an opportunity to relocate in areas not prone to flooding, although funds were limited to $5 million for the whole United States. In 1993, the Robert T. Stafford Disaster Relief and Emergency Assistance Act instituted a Hazard Mitigation Grant Program, up to 75% of whose costs were to be funded by the Federal Emergency Management Agency. New measures were taken to acquire properties in floodplains and assist in the removal and relocation of individuals and businesses. States and local communities were authorized to take possession of the purchased properties and manage them as public open spaces.[104] Federal and state governments presented residents with a clear choice between accepting a once-for-all payment to forgo further claims to compensation and taking responsibility for insuring themselves against risks they might face in future. In depressional wetlands on uplands as well as in floodplains, buyout appealed to many stricken landowners. An Emergency Wetlands Reserve Program, administered by the federal Soil Conservation Service, received $15 million under the Flood Disaster Relief Bill in 1993 and a further supplement of $85 million in the spring of 1994. Under this program, 25,000 acres were signed up for immediate restoration to wetlands.[105] Some landowners were glad to be released from economic pressures in pursuit of productivist goals and some were glad to escape from social obligations to hand on family farms. A few local communities welcomed opportunities to manage restored wetlands for amenity and recreational purposes.

Opening parts of floodplains to serve as washlands or reservoirs to hold floodwaters led to renewed interest in the role of wetlands in the flow of surface and groundwater in upper reaches of the catchment. A long debate was revived over whether and how far wet prairies or potholes helped in retaining storm waters and reducing floods. Paul Adamus and L. T. Stockwell defined "flood storage" as a "process by which peak flows (from runoff, surface flow, groundwater interflow and discharge, and precipitation) enter a wetland basin and are delayed (slowed) in their downslope journey."

104. Federal Emergency Management Agency, Federal programs for property acquisition, relocation, and elevation (FEMA leaflet, Chicago, 20 December 1993).

105. Committee on Characterization of Wetlands, William M. Lewis (chair), *Wetlands: Characteristics and boundaries* (National Research Council, Washington, D.C. 1995) 160; personal communication in 1995 from Billy Teels, Wetlands Staff Leader, Natural Resources Conservation Service (formerly Soil Conservation Service).

Another process, "flood desynchronization," occurred when peak flows were delayed simultaneously by numerous basins within a watershed and water was later discharged in a nonsimultaneous manner.[106] Richard Novitzki's study of the influence of wetlands in Wisconsin on floods, stream flow, and sediment yields has shown that even small remnants of undrained wetland may soak up and retard outflow to the point where wetland is totally inundated. Wetlands also retain sediment, intercepting not less than 80% of the sediment entering them.[107] A more recent study has found that for rare floods occurring at intervals of more than 100 years, loss of wetlands significantly increased the rate of discharge in peak stream flow.[108]

Quite clearly, the speed at which water was discharged into streams was increased by artificial drainage. A survey has shown that in Illinois, where almost all wet prairies have been drained for agriculture, the highest proportion of stream channelization was in the drained areas of the Grand Prairie, having 30% of its streams channelized, and in the Kankakee Marsh, having 40% of its streams channelized.[109] In Iowa, where less than 1% of the surface is now occupied by wetlands, not only was a high proportion of streams channelized but a more or less rectilinear network of drainage ditches supplemented the preexisting stream pattern.[110] These

106. P. R. Adamus and L. T. Stockwell, *A method for wetland functional assessment*, 2 vols., U.S. Department of Transport, Federal Highway Administration Report FHWA-IP-82-23 (Washington, D.C. 1983), vol. 1: 12.

107. R. P. Novitzki, Hydrologic characteristics of Wisconsin's wetlands and their influ - ence on floods, stream flow, and sediment, in P. E. Greeson, J. R. Clark, and J. E. Clark (eds.), *Wetland function and values: The state of our understanding* (American Water Resources Association, Minneapolis 1979) 377–88.

108. H. Ogawa and J. W. Male, Simulating the flood mitigation role of wetlands, *Journ. Water Resources Planning and Management* 12 (1986) 114–28.

109. Rosanna L. Mattingly, Edwin E. Herricks, and Douglas M. Johnston, Channelization and levee construction in Illinois: Review and implications for management, *Environmental Management* 17 (1993) 785.

110. Leslie Hewes and P. E. Frandson, Occupying the wet prairie: The role of artificial drainage in Story County, Iowa, *Annals Assoc. Amer. Geog.* 42 (1952) 24–50; Natural Resources Defense Council, *Land use controls in the United States: A handbook on the legal rights of citizens* (Dial Press, New York 1977) 33, reported that the U.S. Soil Conservation Service planned and paid a large share of the costs of stream channelization and ditching in the Midwest. B. W. Menzel, Agricultural management practices and the integrity of inter-stream biological habitat, in F. W. Schaller and G. W. Bailey (eds.), *Agricultural manage - ment and water quality* (American Water Resources Association, Minneapolis 1978), esti-mated that the total length of streams in Iowa was halved by channelizations.

straightened channels not only discharged water more rapidly but also carried greater loads of sediment. In a postscript added to a White House paper on environmental policy on 24 August 1993, some lessons were drawn from the floods, acknowledging the importance of wetlands higher in the watershed that might serve to absorb rain and snowmelt, releasing water slowly, thereby reducing the severity of flooding downstream. "We must be cautious," it added, "not to repeat policies and practices which may have added to the destruction caused by these floods. One way to assist landowners while alleviating some flood risks was through funding wetlands restoration and acquisition programs targeted to help those in flood-ravaged areas."[111]

In 1993, much of the upper Mississippi River basin was saturated with water at the surface and at groundwater level from early spring. The buffering capacity of lakes and wetlands was largely filled up and the only storage available was in deep depressions in the prairie pothole region of the Dakotas, Minnesota, and Iowa. Hydrologic models of four watersheds representative of distinctive terrain units were analyzed by the Scientific Assessment and Strategy Team in 1994. In four watersheds that were modeled, the maximum reduction for a deeply incised pothole in uplands was 23% of a one-year event, 11% of a 25-year event, and 10% of a 100-year event. In a shallow depression in an upland, restored wetland reduced peak discharge by 9% of a one-year event, 7% of a 25-year event, and 5% of a 100-year event. The models indicated that a combination of land treatment and wetland restoration would reduce runoff between 12% and 18% for floods occurring less than once in 25 years. Wetlands functioned in much the same way as upland lakes or reservoirs in small floods but their effect in larger events has not been fully analyzed. In the 1993 flood, upland wetlands seem to have had little effect because their storage capacity was exceeded before the onset of summer storms.[112]

Whether depressional wetlands on uplands exerted a significant or insignificant influence on the 1993 floods, government policy directed toward restoring wet prairies would certainly help to reduce

111. White House, Office on Environmental Policy, *Protecting America's wetlands: A fair, flexible, and effective approach* (GPO, Washington, D.C. 1993) 26.

112. Gerald E. Galloway, The Mississippi basin flood of 1993, Paper prepared for Workshop on reducing the vulnerability of river basin energy, agriculture, and transportation systems to floods, Foz do Iguacu, Brazil (29 November 1995) 14.

the frequency and magnitude of less than 25-year events. Some landowners signed up for buyouts. Others converted farmland into wetland for renting to hunters, hikers, and weekend visitors. Others built vacation and retirement homes at the edges of restored marshes and lakes.[113] They hoped they might make a profit and also enjoy a way of life that was not as physically exhausting as farming.

Restoring Wetlands

Restoring and creating wetlands reflected important changes in attitudes toward the environment. Up to the late 1980s, wetlands were regarded by most observers as natural features. They could be destroyed or damaged by human activities such as draining, dredging, filling, or water pollution, but prevailing opinion held that they could not be artificially reproduced or reconstructed. Waterlogged patches might be formed inadvertently as a result of mining subsidence, stream diversion, or highway construction across intermittent watercourses but such mires were unwanted accidents. It was a novel idea to design and construct wetlands that performed desired functions in ways that imitated their natural counterparts. The concept of restoration was implicit in the Swampbuster provisions of the 1985 Food Security Act. Lands set aside from crop production might attract grants from the USDA Soil Conservation Service for conversion to wetlands. A wider policy objective, announced in 1988, "to achieve no overall net loss of the nation's remaining wetlands base and to create and restore wetlands, where feasible, to increase the quantity and quality of the nation's wetland resource base" clearly intended that people should be encouraged to make new wetlands.[114]

Over much of the Midwest thousands of new wetlands were created around the edges of small farm ponds. It has been estimated that between 1955 and 1975 nearly two million acres were added to the area of wetlands in the United States as a result of digging ponds at a rate of about 50,000 per year.[115] From 1975 to 1985, an

113. Jon Margolis, Small farmers lose illusions: A bitter harvest lies ahead, *St. Louis Post-Dispatch* (4 September 1994) 21.

114. National Wetlands Policy Forum, *Protecting America's wetlands: An action agenda* (Conservation Foundation, Washington, D.C. 1988).

115. U.S. Office of Technology Assessment, *Wetlands: Their use and regulation* (OTA-0-206, Washington, D.C. 1984) 92, 94.

estimated further 800,000 acres were added.[116] Abandoned gravel pits, quarries, and irrigation ponds were also colonized by a wide variety of wetland plant communities. In 1993, a second edition of Mitsch and Gosselink's *Wetlands* contained a new chapter on wetland creation and restoration describing "the latest principles, approaches, techniques and design aspects for constructing or restoring wetlands."[117]

A small amount of new wetland was gained from restoration and creation financed through voluntary agreements under the Conservation Reserve Program instituted by the 1985 Food Security Act. Of a total of 90,000 acres gained nationwide up to 1992, about 43,000 acres were located in the upper Midwest.[118] Profiles of some projects were recorded by Jon Kusler and Mary Kentula in 1990. Forming nesting and rearing places for duck breeding, propagating aquatic plants for game fish, conducting research into wildlife habitats on slurry ponds, establishing marsh vegetation on shores of reservoirs, providing feeding grounds for waterfowl, creating new cranberry bogs, constructing artificial marshes for sewage treatment were cited as purposes for restoring wetlands.[119] Willard and others acknowledged that "the best sites have historically supported wetlands" but that many new wetlands were developed on upland sites.[120]

Mitigation and Mitigation Banks

Wetland restoration, creation, or enhancement planned expressly to compensate for permitted losses of wetlands was termed "mitigation" in considering applications for dredging and filling under Section 404 of the 1972 Water Pollution Control Act and later legis-

116. T. E. Dahl and C. E. Johnson, *Wetlands status and trends in the coterminous United States mid 1970s to mid 1980s* (U.S. Fish and Wildlife Service, Washington, D.C. 1991).

117. William J. Mitsch and James G. Gosselink, *Wetlands,* 2d ed. (Van Nostrand Reinhold, New York 1993), x, 577–615.

118. J. Josephson, Status of wetlands, *Environmental Science* 26 (1992) 422.

119. Garrett G. Hollands, Regional analysis of the creation and restoration of kettle and pothole wetlands; Daniel A. Levine and Daniel E. Willard, Regional analysis of fringe wetlands in the Midwest: Creation and restoration; Daniel E. Willard, Vicki M. Finn, Daniel A. Levine, and John E. Klarquist, Creation and restoration of riparian wetlands in the agricultural Midwest, all in Kusler and Kentula, *Wetland creation and restoration,* 1990, 281–98, 299–325, 327–50.

120. Willard et al., Riparian wetlands, in Kusler and Kentula, *Wetland creation and restoration,* 1990, 329.

lation. Guidelines agreed among the U.S. Army Corps of Engineers, Fish and Wildlife Service, Soil Conservation Service, Environmental Protection Agency, and state agencies laid down rules for the conduct of the Corps' public interest reviews. In 1980, the U.S. Environmental Protection Agency insisted that the rules be tightened and applications to destroy special aquatic sites be rejected. Because of increased workloads and interagency conflicts, district engineers resorted to compensatory replacement mitigation as an expedient for reconciling differences of opinion. Mitigation promised no net loss of wetlands and appeared to offer some environmental benefits. It facilitated the taking of small tracts of wetlands for highway construction, urban development, and harbor works providing that alternative sites were made to replace their functions.[121] Success or failure of the procedure depended on effective monitoring and skillful management of schemes.

Many doubts were expressed about the success of mitigation projects. Motives of applicants were open to suspicion. It was likely that some developers would attempt to obtain permits to dredge and fill by offering token gestures of restitution and would fail to keep their promises. There were many reasons why newly created wetlands might not be properly managed over a sufficiently long period to establish sustained habitat development. In the early 1990s, practical experience and the science base on restoration and creation had not advanced far, especially in managing the complex hydrology of prairie potholes.[122] Few projects had clearly defined specific goals, designs and construction were inadequately monitored, and long-term assessment and manipulation of hydrological regimes were neglected. It was particularly difficult to create new wetlands that functioned in the same way as wetlands they replaced.

The greatest concentration of wetland restoration in the United States between 1987 and 1991 took place in the prairie pothole region in southern Minnesota and northern Iowa (fig. 8.5). These projects were carried out by interrupting tile lines, plugging drainage ditches, or blocking natural drainage channels with earthen dams. Most basins were planted with a cover crop such as brome (*Bromus*

121. William L. Kruczynski, Mitigation and the Section 404 program: A perspective; Kruczynski, Options to be considered in preparation and evaluation of mitigation plans, both in Kusler and Kentula, *Wetland creation and restoration,* 1990, 549–54, 555–70.
122. Kusler and Kentula, *Wetland creation and restoration,* 1990, xvii, xx.

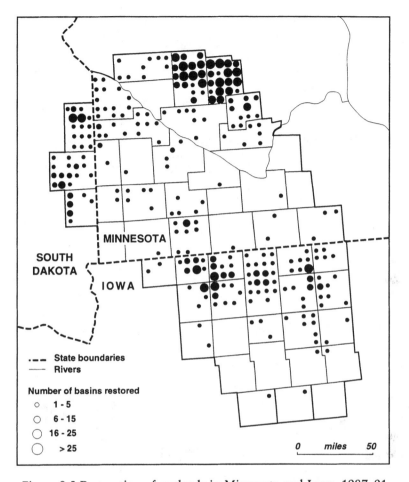

Figure 8.5 Restoration of wetlands in Minnesota and Iowa, 1987–91

Symbols represent the number of wetland restorations by township completed between 1987 and 1991 in the southern prairie pothole region. *Source:* S. M. Galatowitsch and A. G. van der Valk, Natural revegetation during restoration of wetlands in the southern prairie pothole region of North America, in Bryan D. Wheeler et al. (eds.), *Restoration of temperate wetlands* (John Wiley, Chichester 1995) 130.

inermis) or switchgrass (*Panicum virgatum*), and few attempts were made to introduce native wetland plants. S. M. Galatowitsch and A. G. van der Valk analyzed vegetation composition at 62 semipermanent wetlands whose natural drainage had been restored.[123] The

123. S. M. Galatowitsch and A. G. van der Valk, Natural revegetation during restoration

composition of vegetation in the first year after reflooding was determined primarily by how and when a wetland was drained.
Wetlands drained by open ditches contained more emergent species
than tile-drained lands, because ditches often had "relict or refugial
populations of emergent species."[124] Three years after reflooding,
restored wetlands had acquired approximately half the number of
wetland plant species present in nearby natural wetlands and only a
quarter of the species listed in published floras of high-quality pristine wetlands. All reflooded prairie wetlands lacked perimeter zones
of wet prairie and sedge meadow vegetation.[125] Colonization by
plants took place slowly and wetland hydrological systems took still
longer to establish. Duplication of all characteristics of natural wetlands was unattainable owing to the complexity and variation in
natural as well as in artificial systems.

Closer monitoring and correction of later developments were
more easily achieved in large than in small tracts of newly created
wetlands. In urban settings where many dredge-and-fill operations
took place on short lengths of waterfront it was difficult to find replacement sites in the immediate vicinity. Demands for alternative
solutions were met by setting up different types of "mitigation
banks." Some developers were taxed for damaging or destroying
wetlands and the proceeds were deposited in a habitat restoration
fund that paid for flooding an out-of-town site. Another type of mitigation bank put together a number of different mitigation contracts
to create a large, multipurpose tract of wetland. In a third type of
mitigation bank, a large area was first purchased by a group of subscribers and converted to wetland in anticipation of mitigation
agreements. Credits were withdrawn by individuals to compensate
for wetland losses when permits were granted.[126] An ingenious
method for creating large areas of wetlands was pioneered at St.
Charles and Des Plaines in northeast Illinois by John Ryan, president of Land and Water Resources, Inc.[127] Real-estate developers,

---

of wetlands in the southern prairie pothole region of North America, in Bryan D. Wheeler,
Susan C. Shaw, Wanda J. Fojt, and R. Allan Robertson (eds.), *Restoration of temperate
wetlands* (John Wiley, Chichester 1995) 129–42.

124. Ibid., 134, 138.

125. Ibid., 138–41.

126. Kusler and Kentula, *Wetland creation and restoration,* 1990, 418, 508.

127. William K. Stevens, Restored wetlands could ease threat of Mississippi floods, *New
York Times* (8 August 1995) C1, C4.

who were required to create an acre or more of wetland for every acre they drained, deposited small mitigation warrants in Ryan's mitigation bank. The bank applied bundles of these warrants to purchase and restore large, consolidated tracts of wetland. To maintain eligibility of projects under Section 404 of the Clean Water Act, the company had to submit evidence every three months to federal inspectors that the mitigation bank was actually functioning as a wetland, especially in its water flow. Both the St. Charles and Des Plaines projects demonstrated how rapidly wetlands could be restored, through largely spontaneous processes of regeneration. As soon as drainage tiles were broken, water came back and was covered by a big bloom of algae. When the algae sank, cattails grew so thickly it seemed they would push out all other plants, but muskrats soon trimmed back the cattails, and seeds carried by water and birds produced a diverse wetland habitat for a great variety of wildlife. The process required a small amount of supervision and funds were placed in trusts to pay for future maintenance. Besides contributing toward reduction of flood risks, restored wetlands prevented much sediment and chemical pollution from reaching streams. Above all, restored wetlands attracted visitors for nature study and relaxation.

On 6 March 1995, the U.S. Army Corps of Engineers, Environmental Protection Agency, Department of Agriculture, Fish and Wildlife Service, and Department of Commerce issued a joint guidance for the establishment, use, and operation of mitigation banks. The overall goal of a bank was to be the establishment of a "self-sustaining, functioning aquatic system." Mitigation credits created by banks were to be based on wetland values existing in a bank at the time credits were granted. Credit might be given for "preservation" when accompanied by creation, restoration, or enhancement of a wetland tract or, exceptionally, in stand-alone contexts. Banks were to be planned and developed to address resource needs within particular watersheds and had to have a "banking instrument." Collectively, the signatory federal agencies to a banking instrument would comprise a Mitigation Bank Review Team.[128] In the aftermath of the 1993 floods, many federal agencies, taking a lead from the president, were deeply concerned with wetland

128. Association of State Wetland Managers, Washington update (Washington, May 1995).

restoration while, at the same time, Congress was striving to cut public expenditure and curb government intervention.

Congress attempted to restrict powers exercised by federal and state agencies in regulating uses of wetlands. Wetland regulations were subject to three principal constitutional limitations: the due process clause, governing proper procedures and valid public purposes for imposing regulations; the equal protection clause, securing equal treatment for individuals similarly situated; and the taking clause, ensuring that no private property be taken for public use without just compensation. Upholders of private property rights sought to extend entitlement to "taking" compensation for losses and reductions in property values resulting from government actions. Authorities that caused lands to be flooded by damming streams or pulling down levees had to pay full compensation, but an owner might also be justified in claiming compensation for loss of development value on a wetland site where government decreed that no building was to be built. If compensation was allowed, it had to be determined whether it was for a portion of land directly affected by the regulation or for a larger holding that suffered regulatory blight. Other attempts to thwart the aims of the Environmental Protection Agency and Fish and Wildlife Service were proposals requiring officials to assess liabilities for takings that might be incurred in carrying out agencies' policies.[129] In 1995, a comprehensive wetlands conservation and management bill introduced a package of highly controversial provisions, including a much narrower delineation for the Section 404 regulatory system, a broader range of exemptions, and compensation payments for landowners who suffered 20% reductions in property values as a result of impositions of Section 404 restrictions. The overall effect of this legislation was to ease controls over wetland development. While Congress favored allowing developers more freedom, the administration remained committed to wetland preservation and restoration. Following reelection, President Clinton announced tougher measures to prevent further losses of wetlands.

In wetland districts, owners of lands subject to occasional flooding, drained lands, or "nonwetlands" received widely differing opinions on how their resources should be used or conserved.

129. Barbara Moulton, Takings legislation: Protection of property rights or threat to the public interest? *Environment* 37 (1995) 44–45.

Many local residents wanted farming to prosper, fields to be neatly cultivated, soil erosion and water pollution prevented. Many visitors wanted lakes and wetlands to be maintained for outdoor recreation and aquatic plants, fish, and wildlife to be protected. Country landowners viewed as producers of food and creators of wealth were expected to behave differently from those regarded as caterers for tourists and custodians of natural environments. Few wetland residents felt satisfied playing stereotyped roles assigned to them by others. Farmers no longer responded directly to market forces. Prices of farm products were fixed by large corporations, supported by government subsidies. Land use, water use, chemical use, animal welfare, and operators' health and safety were regulated. Financial arrangements were regulated. Conservationists, for their part, no longer acted as pure biological scientists and protectors of endangered species and natural habitats. They had to meet demands for access by hunters, fishermen, hikers, and canoeists. Wetlands were managed to increase production of game and timber, reduce incidence of floods and storms, prevent fires and water pollution, eliminate harmful insects and weeds, and improve landscape beauty. Roads, parking lots, picnic sites, and tourist accommodations were built.

Both on farms and in nature reserves, outside interests exercised extensive powers and were growing stronger. Some wetland owners felt they were no longer free to make their own private decisions about the use of land, its management, and disposal of their property. Some thought that government agencies and corporate organizations had taken away their independence and were regulating wetlands in furtherance of alien public objectives or for the benefit of distant company stockholders. A few believed that individual self-help and local self-government were being usurped by increasingly remote authorities and that power ought to be reasserted by grassroots representative bodies. Many others followed the lead taken by the administration in seeking stricter environmental protection.

9

Changing Wetland Images
and Values

What wetlands are now or were in the past reflects the way that people perceive or perceived them. A passage in Henry David Thoreau's *Journal* analyses the relation of object to percept: "It is in vain to dream of a wildness distant from ourselves. There is none such. It is the bog in our brain and bowels, the primitive vigor of Nature in us, that inspires that dream."[1] How people dreamed of, described, and depicted wetlands, how they transformed, protected, or attempted to restore them, reflected images in their minds. Mental images varied in content, intensity, and association and changed over time. Recollections of the past and evidence from historical records were reread and reshaped in the light of images in the minds of historians. Phenomena and their attributes were repeatedly redefined, reclassified, and reevaluated. In making decisions about draining, farming, protecting, or restoring wetlands, changing value systems were crucially important. Conflicts between utilitarian, productivist, socially just, ecological, aesthetic, and intrinsic values led to changes in policy, terminology, and management practices. Images, values, and action formed continually changing patterns.

Changing Images of Wetlands

The original federal land survey in the early nineteenth century plotted over 10% of land in the Midwest as swamps and marshes; in the 1980s, the federal Fish and Wildlife Service estimated that wetlands occupied less than 5% of the surface. In the course of little more than a century wet prairies were written off as "nonwetlands" and northern peatlands were reclassified as "palustrine emergent wet-

---

1. Henry David Thoreau, *Journal,* 30 August 1856, cited in Simon Schama, *Landscape and memory* (Harper Collins, London 1995) 578; also Robert R. Rothwell (ed.), *Henry David Thoreau: An American landscape* (New York 1991) 126–27.

lands." Tile drainage converted stiff, poorly drained clays and silts into dry cultivated soils, whereas most peat soils remained water-logged. In the late nineteenth century, wetlands were described and mapped by soil scientists in the service of agriculture; in the late twentieth century, wetland scientists in the service of conservation identified wetlands by their hydrological and ecological characteristics. Methods of identifying, ordering, and interpreting wetland features changed; scientific research on this topic switched from terrestrial to aquatic associations. In earlier studies, soils were identified by their derivations from glacial deposits or accumulations of organic detritus, and drainage conditions were associated with textures and permeability of parent materials; in later studies, waterlogging was associated with inflows and outflows of surface water and groundwater and growth of hydric plants. A revolution in the direction of scientific inquiry followed a shift in value judgments, and wetland values changed more radically than day-to-day management practices on waterlogged sites.

Some ecologists and wetland scientists deliberately avoided places disturbed by human activities. Frederic Clements remarked that in settled regions, "it is exceedingly difficult or entirely impossible to strike a balance between stability and change, and it becomes imperative to turn to regions much less disturbed by man, where climatic control is still paramount."[2] Edward Maltby described areas he valued as "the last truly wild and untouched places on earth."[3] Some ecologists thought tallgrass prairies were maintained by summer droughts and periodic burning and that fires were ignited by lightning rather than firebrands. Some northern swamps were observed to occupy basins dammed by tree trunks and branches. Obstructions were presumed to be caused by windthrown trees or built by beavers rather than humans. In the late twentieth century, the part played by human activity in modifying wetlands was examined closely and found to be important and, in places, to be a dominant process. New awareness and new attitudes were concerned with human exploitation of animals, plants, water, and soil and attendant problems of depletion, pollution, and erosion.

Without written records it is difficult to guess what native

2. F. E. Clements, Nature and structure of climax, *Journal of Ecology* 24 (1936) 252–84.

3. Edward Maltby, *Waterlogged wealth: Why waste the world's wet places?* (Earthscan, London 1986) 9.

Americans thought about wetlands in pre-Columbian times. Because villages and fields were situated close to lakes and marshes it may be inferred that resources of an aquatic environment were highly valued. Midwest tribes hunted and gathered fish, waterfowl, large mammals, wild rice, and a multitude of other plants. They also raised mounds or ridges on which they grew corn, beans, and squash. After the arrival of Europeans, native populations at first declined sharply, then recovered. Contacts were established in the seventeenth century through expansion of the fur trade. European traders and missionaries began to describe and interpret native American behavior and attitudes. Natives were regarded as being passively dependent on the seasonal fruitfulness of wetlands, taking no heed for scarcity in times ahead. On the other hand, many tribes were reported to be skilled in curing and storing fish and meat as well as cultivating gardens and fields and storing seeds and grain. Some observers surmised that native Americans took from the environment no more than they needed for their own subsistence and did not deplete natural resources. On the other hand, many willingly or unwillingly participated in trapping beaver almost to the point of extinction and regularly started fires that burned down tree seedlings. All questions concerning native occupance of wetlands were erased from memory when white American pioneers entered the Midwest.

During the latter half of the twentieth century, renewed interest in the history of native Americans revived old debates about passivity, frugality, and respect for nature and raised fresh questions about harvesting many different products from the same areas of wetland. Such questions prompted research into systems of intercropping, fallowing, and selective felling of mixed-age, mixed-species forests. Attempts were made to learn about people's ideas through examining their practices. At the same time, myths abounded. Some writers venerated native Americans as "ecological saviors, endowed with tribal wisdom undimmed over millennia."[4] In the Midwest, after a century of removals, too few saviors survived to exert an appreciable influence on ecological heritage.

From 1787 onward, a newly independent American nation formally claimed possession of territory northwest of the Ohio River

4. David Lowenthal, *Possessed by the past: The heritage crusade and the spoils of history* (Free Press, New York 1996) 81.

and in 1803 purchased the Louisiana territory west of the Mississippi River, thereby doubling the surface extent of the United States. Many newcomers were repelled by their first encounter with swamps and marshes which were difficult to cross on foot and almost impossible to travel through on horseback or by wagon. Visitors were attacked by swarms of bloodsucking insects and stricken with dreaded swamp fevers. At the beginning of the nineteenth century, cattlemen cut wild hay and used the land for seasonal grazing, but permanent settlement by farmers took place very slowly. Corn-growing began on small wet prairies in Ohio during the first quarter of the nineteenth century and spread fitfully northward and westward. Most purchasers of land were more intent on making profits from quick sales than settling down and investing additional funds in building, sodbusting, fencing, and, most expensively, in draining. Many seeking family farms at minimum cost shunned wet prairies in favor of drier lands further west.

First entries on large prairies in northern Indiana, Illinois, Iowa, and Minnesota were made by large landowners. Late arrivals might obtain tenant holdings or, as a last resort, work as hired hands. Opportunities to climb the agricultural ladder were few, arduous, and dearly won. Expansion of agriculture was led by capital investment in technological innovation. In the mid-nineteenth century, new methods of building farmhouses and barns were developed, and steel plows for breaking prairie sod and barbed wire for fencing stock were invented. The coming of railroads and introduction of drainage tiles were keys to opening wet prairies. Railroads crossed large, open grasslands in Illinois, Iowa, and Minnesota between 1850 and 1870, enticing prospective settlers to break new ground. From 1870 to 1920, ditching and tile draining transformed these lands into the most productive farms in the corn belt. Wet prairies rapidly lost their persistent, strongly negative image. In a remarkably short time they appeared to rise from rags to riches. Rising crop yields and rapidly increasing land values expressed a mood of soaring confidence. Few investors stopped to calculate exact returns on money expended on land and drainage works. Had they done so, few would have been able to justify the high expectations held around 1910.

Enthusiasm for draining induced some to hope that cutover northern peatlands might be reclaimed for agriculture. Railroads and sawmills were built in the lumbering era. Peat soils were easy to cultivate and rich in nitrogen, but in low-lying basins the growing season was too short for most crops apart from hay and oats. Adding to physical difficulties caused by peat shrinkage, deficiencies in phosphates and potash, and early and late frosts, drainage in northern peatlands was badly timed and poorly organized. When tractors and other motor vehicles replaced horses, demand for hay and oats fell. Many drainage enterprises were launched on far too large a scale. Promoters bought too much land, borrowed too much money, and failed to attract settlers with necessary skills and stamina. In the financial crisis in the late 1920s, farmers went bankrupt and failures of drainage enterprises to redeem their bonds cast shadows over the dependability of public authority loan stocks. Lands in arrears on capital repayments, interest charges, and unpaid taxes reverted to public ownership. Cultivation ceased, settlers abandoned their homes, and land fell derelict. In the 1930s, agriculture was replaced by conservation of wetland habitats for wildlife, reforestation, and outdoor recreation. Conservation measures were social and political responses to fears about environmental degradation and resource depletion. Government projects attempted to remedy injuries caused by land use practices that had failed.

Changing Wetland Values

In 1949, a new land ethic was enunciated by Aldo Leopold.[5] Ownership of wetlands or other wild lands was conceived as more like joining a community than buying or selling a commodity for personal consumption. Northern peatland communities could not afford to support many human members. People equipped with ditch-diggers, plows, automobiles, and power boats were able to damage and destroy growth of aquatic vegetation and disrupt wildlife habitats; their positive contributions were less predictable. Leopold's ethic appealed to the consciences of many for whom "active

5. Aldo Leopold, *A Sand County almanac* [Oxford University Press, New York 1949] (Ballantine Books, New York 1990) xviii, 237–64, esp. 240: "A land ethic changes the role of *Homo sapiens* from conqueror of the land community to plain member and citizen of it. It implies respect for his fellow-members, and also respect for the community as such." Leopold's land ethic rejected exclusively economic solutions to land-use problems. On 262, he wrote: "Examine each question in terms of what is ethically and esthetically right, as well as what is economically expedient. A thing is right when it tends to preserve the integrity, stability and beauty of the biotic community. It is wrong when it tends otherwise."

cherishing of nature is accounted good and its wanton destruction
evil," but the concept of belonging to an ecological community that
included nonhumans appeared to be flawed.[6] Leopold's critics
picked on the assumption that wetland owners might share their
property with ducks, reeds, peat, and water. A notion that nonhuman members of wetland communities, including both predators
and prey, enjoyed rights or owed each other duties seemed absurd.
John Passmore exposed the fallacy with an apt illustration: "Bacteria
and men do not recognise mutual obligations nor do they have
common interests."[7] Humans were unable to communicate directly
with ecological systems or their components although some reso-
nances might be gained from empathetic contacts or by inventing
imaginative fables in which animals performed and spoke like humans.[8] Leopold himself did not claim natural rights for particular
organisms. He was concerned about the survival of associations of
species and ecosystemic processes. As a professional game manager,
he knew that healthy functioning of biotic communities was maintained by intelligent culling of herds.[9]

Leopold's land ethic challenged Midwesterners and others to
consider seriously the principle that humans were a part of nature
and did not stand apart from it. Later thinkers, jurists, theologians,
Gaians, deep ecologists, and animal rights activists proposed alternative models for normative behavior in relation to environment.[10]
New values were proclaimed. The limits of humanity were extended. Poor people in rural areas thought they were more likely
than wealthy residents in suburban neighborhoods to have toxic
wastes dumped in their lakes and wetlands. Environmental justice
demanded that poor as well as rich should be accorded equal protection against such threats. Increasing numbers of people protested

6. John Passmore, *Man's responsibility for nature: Ecological problems and Western traditions* (Duckworth, London 1974) 186.

7. Ibid., 116.

8. Niklas Luhmann, translated by John Bednarz, *Ecological communication* (Polity Press,
Cambridge 1989) 15–35; problems of relating ecological communication to economics, law,
and science are discussed 51–83.

9. Roderick Frazier Nash, *The rights of nature: A history of environmental ethics*
(University of Wisconsin Press, Madison 1989) 71; Susan L. Flader, *Thinking like a moun-
tain: Aldo Leopold and the evolution of an ecological attitude toward deer, wolves, and forests*
(University of Missouri Press, Columbia 1974).

10. I. G. Simmons, *Interpreting nature: Cultural construction of the environment* (Routledge, London 1993) 117–42, critically reviews several of these philosophies.

against pain and suffering inflicted on sentient fellow creatures. Such injuries were described in human terms as "cruelty" to animals. Sympathy extended to animals once classed as vermin. Humane methods had to be found for trapping and killing wolves, coyotes, and other predators. Civilized people were expected to show care and respect for nature.

A fundamental departure from anthropocentric, instrumental, utilitarian valuations of environment, wherein actions were judged right or wrong by reference to whether they contributed to human welfare, occurred when nature was recognized as having intrinsic value for its own sake. Flowers in bloom or birds in song were no less beautiful when not witnessed by human observers; rocks and organisms were no less important to the functioning of ecosystems for not being useful to human economies. Many people reached a conclusion that living things were their own justification for being; their value was independent of any worth they might have for humans. The spread of environmental consciousness was reflected in a transformation of the Sierra Club. In 1960, it was a largely regional organization of 15,000 mountaineers. By 1967, membership had grown to nearly 60,000 and the club had become a national voice for an emerging environmental movement. In 1971, membership passed 100,000; in 1981, it rose to 250,000; and in 1991, it approached 500,000.[11] Environmental concerns focused on perceived duties to prevent extinction of rare species and wilderness areas. A new biocentric ethic fired a campaign to establish wildlife refuges. The Wilderness Act of 1964 provided the first legislative guarantees of freedom from human interference for wildlife and ecosystems. A few people regarded this as a first step toward granting to nonhuman members of biotic communities rights to life, liberty, and the pursuit of happiness.[12] In 1966, Congress passed a tentative Endangered Species Protection Act and in 1973, this was replaced by a much broader and stronger act.[13] The objectives and scope of these measures were tested in lawsuits but deep ecologists thought that people ought not to waste their energies quarreling about rights

11. Roderick Nash, *Wilderness and the American mind,* 3d ed. ([1967]; Yale University Press, New Haven 1982) x. The preface to the third edition discusses the fortunate timing of the appearance of this book in a period of growing scholarly, political, and popular interest in ecology and ideas of wilderness.
12. Nash, *Rights of nature,* 1989, 85.
13. Ibid., 172–79.

but should try to learn how to live in harmony with one another in their natural environment. The environmental movement also had activists taking direct action, breaking laws, liberating animals from laboratories, obstructing the clearance of ancient forests, resisting the damming of natural streams. Few incidents took place in the Midwest. In 1978, on the Minnesota prairies, a group of farmers known as the "Bolt Weevils" dismantled electrical power lines and blocked survey and construction work.[14] Awareness of environmental issues was raised by newspaper articles and television broadcasts. Documentary films on wilderness, wildlife, and natural history gained popularity. At Midwestern universities, enrollments in courses on ecology, earth sciences, environmental studies, and conservation showed steady increases. Over a period of more than twenty years, from about 1965 to the end of the 1980s, the extension of rights of nature gradually changed the way people thought about wetlands in the Midwest.

While possession of land offered prospects for making profits, most private owners were unwilling to surrender their proprietary rights. When buying and selling land and extracting resources became unprofitable, opportunities arose to protect soils, water, and wildlife. Conservation measures were imposed directly on lands that fell into the hands of public bodies, and legislation conferred powers on government agencies to regulate uses of land by private individuals to safeguard threatened habitats and prevent destructive exploitation. In practice, efforts to reestablish wetlands were not invariably well managed. Leopold complained that in central Wisconsin, ditches were plugged wholesale and "a counter-epidemic of reflooding set in." This was good for ducks, "but not so the thickets of scrub popple, and still less the maze of new roads that inevitably follow governmental conservation."[15] During the 1950s and 1960s, agriculture enjoyed a long spell of prosperity, and resistance to restrictions on draining stiffened. At the same time, outcries against water pollution and soil erosion and calls to halt wetland losses rose to crisis pitch.

In 1969, the media urged radical reform, public demonstrations were held, and Congress passed a National Environmental Policy

14. Ibid., 191, cites the *New York Times* (10 January 1978) 25, 30.
15. Leopold, *Sand County almanac,* 1990, 107.

Act.[16] In effect, government assumed a leading responsibility for land management, taking a special role in protecting nonhuman entities in biotic communities. Karl Butzer commented: "The United States farmer expects outside assistance to tackle any environmental problems, and environmental hazards have become equated with governmental responsibility."[17] Government agencies charged with planning, regulating, and subsidizing land use multiplied and pursued divergent goals. In the 1960s, farmers on wet prairies applied to the USDA for grants to drain wet patches, received subsidies for increased outputs of corn and other farm produce, obtained loans for machinery and improvements, and became ever more dependent on money from government sources. By a variety of payments and allowances, producers were encouraged to add to growing surpluses of food that world markets could no longer absorb. Owners of undrained wetlands, on the other hand, were restrained by increasingly stringent regulations and ultimately by penalties from draining sloughs. Through the 1980s, under stress from financial crisis, family upheaval, drought, and flood, clashes between farmers and government agents occurred more frequently. Family farmers and other private landowners were alarmed by their individual weakness, isolation, and declining powers. Government executives attempted to reconcile their differences and coordinate their actions as much to secure their own survival as bureaucrats as to protect wetlands. For many people outside as well as in government employment, saving remaining wetlands appeared to be top policy priority.

Important questions were raised by attempts to restore wetlands. Were restorations authentic and natural or were they cultivated water gardens? In creating them, plants, fish, and animals might be imported from distant sites. Unwanted species might be excluded or exterminated. Water levels and plant communities might be altered to prevent wetlands evolving into upland forests. Other questions concerned the purposes of restoration and conservation. Were wetlands intended to serve as resorts for outdoor recreation and, if so,

16. Marvin W. Mikesell, Geography as the study of environment: An assessment of some old and new commitments, in Ian R. Manners and Marvin W. Mikesell (eds.), *Perspectives on environment*, Commission on College Geography Publ. 13 (Assoc. Amer. Geog., Washington, D.C. 1974) 1–23.

17. Karl W. Butzer, Accelerated soil erosion: A problem of man-land relationships, in Manners and Mikesell, *Perspectives on environment*, 1974, 74.

what facilities were required by visitors? Were wetlands to be restored as romantic protests against excesses of high-tech agriculture,
to recall spirits of vanished wilderness and, if so, how might sensa-
tions of primeval chaos and solitude be recaptured? Conservation of
wildlife habitats also raised questions of access and restrictions upon
access. Were there still progressive thinkers who were repelled by
dangers lurking in insalubrious, watery wastes and scientitists who
opposed efforts to preserve or reinstate wetlands? How might divergent objectives be pursued in what were designated as scarce and
diminishing environmental resources?

An obstacle to reaching a consensus about appropriate uses for
wetlands was deep uncertainty about whether wetland ecosystems
were tough and resilient systems that would regenerate sponta-
neously or whether they were fragile and sensitive to slight human
interference. Until late in the nineteenth century, it was widely assumed that forces of nature were immensely stronger than those of
humans. It seemed unlikely that environments could be damaged
gravely or permanently by human activity. In the 1920s, Frederic
Clements's view that the delicate balance of nature was upset by the
most transient human presence gained wide acceptance. Only in
their pristine state did ecosystems attain stability and maximum diversity.[18] Since that time, Midwesterners have been warned repeatedly about environmental crises, in the 1930s, in the 1960s, and
again in the 1980s. They experienced fears that wetlands, profoundly altered by past activities, might neither recover from the
most recent assaults nor be saved from further injury as long as humans were present. Yet clearly there was little chance that humans
would go away or commit suicide in order that whooping cranes
might return to their feeding grounds in northern bogs. Giovanna Di
Chiro remarked: "Trying to solve the historical problems of particular capitalist and patriarchal dominations of nature by removing
'humans as contaminants' does not move us closer to a sustainable
future."[19] People had to take responsibility for reinventing mixed-
species communities in which they could continue to have a place
to live.

18. Frederic Clements cited in William Cronon, The trouble with wilderness; or getting
back to the wrong nature, *Environmental History* 1 (1996) 17–18.

19. Toward a conclusion, in William Cronon (ed.), *Uncommon ground: Toward reinventing nature* (W. W. Norton, New York 1995) 453.

Beyond many uncertainties surrounding the future course of their development, it was clear that without human interference and adjustment, natural processes of growth and decay would continue to transform wetland communities; some would change into forests and some would disappear altogether. There was no way that living ecosystems could be preserved like museum exhibits; their changing character had to be managed and continually adjusted. It was, first of all, important to acknowledge that wetlands were culturally constructed; they were representations of reality. It was also necessary to understand that they contained, often unnoticed, an extraordinary amount of history; they were profoundly modified by past intentions and activities. For better or worse, survival of wetlands in the Midwest depended, and still depends, on changing human attitudes, on new ideas, on informed decisions, and on constructive actions.

BIBLIOGRAPHY

Ackerknecht, Erwin H. *Malaria in the upper Mississippi valley, 1760–1900* (Baltimore 1945).

Adamus, P. R., and L. T. Stockwell. *A method for wetland functional assessment.* 2 vols. U.S. Department of Transport, Federal Highway Administration Report FHWA-IP-82-23 (Washington, D.C. 1983).

Ade, George. Prairie kings of yesterday. *Saturday Evening Post* 204 (4 July 1931) 14, 76.

Ahrens, Marjorie H. A contribution to the history of land administration in Minnesota: The origins of the Red Lake Game Preserve. M.A. thesis, Geography, University of Minnesota 1987.

Albert, A. R. Status of organic soil use in Wisconsin. *Proc. Soil Sci. Soc. of Amer.* 10 (1945) 275–78.

Albert, A. R., and O. R. Zeasman. *Farming muck and peat in Wisconsin.* Univ. Wisc. Coll. Agric. Circular 456 (Madison 1953).

Alway, F. J. Agricultural value and reclamation of Minnesota peat soils. *Minn. Univ. Bull.* 188 (1920).

Anderson, James B. White River: Historical influences observed in Morgan County. *Indiana Mag. Hist.* 43 (1947) 249.

Anderson, Sherwood. *Winesburg, Ohio* (B. W. Huebsch, New York 1919).

Andrews, Richard N. L. (ed.). *Land in America: Commodity and natural resource* (D. C. Heath, Lexington, Mass. 1979).

Argersinger, Peter H., and Jo Ann Argersinger. The machine breakers: Farmworkers and social change in the rural Midwest of the 1870s. *Agric. Hist.* 58 (1984) 393–410.

Ashby, Wallace. Problems of the new settler on reclaimed cutover land. *Agric. Engineering* 5 (February 1924) 27–29.

Atack, Jeremy. The agricultural ladder revisited: A new look at an old question with some data for 1860. *Agric. Hist.* 63 (1989) 1–25.

Axelrod, D. I. Rise of the grassland biome, central North America. *Bot. Rev.* 51 (1985) 163–202.

Baerwald, Thomas J. Forces at work on the landscape. In Clark, *Minnesota,* 1989, 30–48.

Bailey, Warren. *The one-man farm* (USDA Economic Research Service, Washington, D.C. 1973).

Baird, Robert. *View of the valley of the Mississippi* (N. S. Tanner, Philadelphia 1832).

Baker, Gladys L., Wayne D. Rasmussen, Vivian Wiser, and Jane M. Porter. *Century of service: The first 100 years of the United States Department of Agriculture* (USDA, Washington, D.C. 1963).

Baker, O. E. Agricultural regions of North America, part 4: The corn belt. *Econ. Geog.* 3 (1927) 447–65.

Barlowe, Raleigh. *Public landownership in the lake states. In Mich. Agric. Expt. Sta. Spec. Bull.* 351 (1948).

____. Changing land use and policies: The lake states. In Flader (ed.), *Great Lakes forest,* 1983, 156–76.

Barnes, C. P. Land resource inventory in Michigan. *Econ. Geog.* 5 (1929) 22–35.

Barrows, Harlan H. *Geography of the middle Illinois valley.* In *Ill. State Geol. Surv. Bull.* 15 (1910).

Beardsley, Levi. *Reminiscences: Personal and other incidents* (New York 1852).

Beck, Lewis C. *A gazetteer of the states of Illinois and Missouri* (Charles and George Webster, Albany 1823).

Bellrose, F. C., and N. M. Trudeau. Wetlands and their relationship to migrating and winter populations of waterfowl. In D. D. Hook et al. (eds.), *The ecology and management of wetlands,* vol. 1 (Croom Helm, London 1988) 183–94.

Benson, Lee. The historical background of Turner's Frontier essay. *Agric. Hist.* 25 (1951) 66–77.

Benton, Thomas Hart. *Thirty years' view,* vol. 1 (New York 1858).

Bernabo, J. C., and T. Webb III. Changing patterns in the Holocene pollen record of north-eastern North America: A mapped summary. *Quaternary Research* 8 (1977) 64–96.

Berry, W. J. The influence of natural environment in north central Iowa. *Iowa Journ. Hist. and Politics* 25 (1927) 280–94.

Billington, Ray Allen. *Westward expansion: A history of the American frontier* (Macmillan, New York 1949).

Black, J. D., and L. C. Gray, *Land settlement and colonization in the Great Lakes states.* USDA Bulletin 1295 (GPO, Washington, D.C. 1925).

Blegen, Theodore C. *Minnesota: A history of the state* (University of Minnesota Press, Minneapolis 1963).

Bliss, Eugene F. (trans. and ed.). *Diary of David Zeisberger, a Moravian missionary among the Indians of Ohio* (R. Clarke, Cincinnati 1885).

Blodget, Lorin. *Climatology of the United States* (J. B. Lippincott, Philadelphia 1857).

Bly, Carol. *Letters from the country* (Harper and Row, New York 1981).

Bogart, Ernest L., and Charles M. Thompson. *The industrial state, 1870–*

1893. Vol. 4 of C. W. Alvord (ed.), *The centennial history of Illinois* (Chicago 1922).

Boggess, Arthur Clinton. *The settlement of Illinois, 1778–1830* (Chicago Historical Society, Chicago 1908).

Bogue, Allan G. *From prairie to corn belt: Farming on the Illinois and Iowa prairies in the nineteenth century* (University of Chicago Press, Chicago 1963).

Bogue, Allan G., and Margaret B. Bogue. "Profits" and the frontier land speculator. *Journ. Econ. Hist.* 17 (1957) 1–24.

Bogue, Margaret Beattie. The Swamp Land Act and wet land utilization in Illinois, 1850–1890. *Agric. Hist.* 25 (1951) 169–80.

____. *Patterns from the sod: Land use and tenure in the Grand Prairie, 1850– 1900* (Illinois State Hist. Soc., Springfield 1959).

Bond, J. W. *Minnesota and its resources* (Redfield, New York 1854).

Borchert, John R. The climate of the central North American grassland. *Annals Assoc. Amer. Geog.* 40 (1950) 1–39.

____. *Minnesota's changing geography* (University of Minnesota Press, Minneapolis 1959).

Bourdo, E. A. A review of the General Land Office survey and of its use in quantitative studies of former forests. *Ecology* 37 (1956) 754–68.

Bourdo, Eric A., Jr. The forest the settlers saw. In Flader (ed.), *Great Lakes forest*, 1983, 3–16.

Bradbury, I. K., and J. Grace. Primary production in wetlands. In A. J. P. Gore (ed.), *Ecosystems of the world, 4A; Mires: Swamp, bog, fen, and moor, general studies* (Elsevier, Amsterdam 1983) 285–310.

Bradbury, John. *Travels in the interior of America in the years 1809, 1810, and 1811* (Liverpool 1817).

Brady, Nyle C. *The nature and properties of soils.* 8th ed. (Macmillan, New York 1974).

Brigham, A. P. The development of wheat culture in North America. *Geog. Journ.* 35 (1910) 42–56.

Brigham, C. H. The lumber region of Michigan. *North Amer. Rev.* 107 (July 1868) 97–98.

Brown, Dwight A. Early nineteenth-century grasslands of the midcontinent plains. *Annals Assoc. Amer. Geog.* 83 (1993) 589–612.

Brown, Ralph H. *Historical geography of the United States* (Harcourt, Brace, New York 1948).

Brown, Samuel R. *The western gazetteer; or emigrant's directory* (Auburn, N.Y. 1817).

Burns, Bert Earl. Artificial drainage in Blue Earth County, Minnesota. Ph.D. diss., Geography, University of Nebraska 1954.

Bush, Clive. 'Gilded backgrounds': Reflections on the perception of space and landscape in America. In Mick Gidley and Robert Lawson-Peebles (eds.),

Views of American landscapes (Cambridge University Press, Cambridge 1989) 13–30.

Buttel F. H., and L. Busch. The public agricultural research system at the crossroads. *Agric. Hist.* 62 (1988) 292–312.

Butzer, Karl W. Accelerated soil erosion: A problem of man-land relationships. In Manners and Mikesell, *Perspectives on environment*, 1974, 57–78.

————. The Indian legacy in the American landscape. In Conzen (ed.), *The making of the American landscape*, 1990, 27–50.

————. The Americas before and after 1492: An introduction to current geographical research. *Annals Assoc. Amer. Geog.* 82 (1992) 345–68.

Buzzard, Robert G. Redtop production of south-east Illinois. *Annals Assoc. Amer. Geog.* 20 (1930) 24.

Caird, James. *Prairie farming in America* (New York 1859).

Calef, Wesley, and Robert Newcomb. An average slope map of Illinois. *Annals Assoc. Amer. Geog.* 43 (1953) 305–16.

Cannon, Brian Q. Immigrants in American agriculture. *Agric. Hist.* 65 (1991) 17–35.

Carman, Harry J. English views of Middle Western agriculture, 1850–1870. *Agric. Hist.* 8 (1934) 3–19.

Carman, J. Ernest. *The Mississippi valley between Savanna and Devonport, Illinois. Ill. State Geol. Surv. Bull.* 13 (1909).

Carrier, Lyman, and Katherine S. Bort. The history of Kentucky bluegrass and white clover in the United States. *Journ. Amer. Soc. Agronomy* 8 (1916) 256–66.

Carter, Harvey L. Rural Indiana in transition, 1850–1860. *Agric. Hist.* 20 (1946) 107–21.

Cass, Lewis B. Considerations on the present state of the Indians and their removal to the west of the Mississippi. *North American Review* 66 (1830) 5.

Cassell, Mark S. *A pale blank area: Archaeology and context of agrarian development in Wisconsin's central sands, 1860–1940* (Mississippi Valley Archaeology Center, University of Wisconsin–La Crosse 1995).

Chamberlin, T. C. *Geology of Wisconsin, Survey of 1873–1879*, vol. 1 (Madison 1883); vol. 2 (Madison 1877).

Changnon, Stanley A. (ed.). *The great flood of 1993: Causes, impacts, and responses* (Westview Press, Boulder 1996).

Chase, Stuart, *Rich land, poor land: A study of waste in the natural resources of America* (AMS Press, New York 1969).

Childs, Ebenezer. Recollections of Wisconsin since 1820. *Wisc. Hist. Coll.* 4 (1859) 175–77.

Church, G. J. The worst is supposed to be over this weekend. *Time* (26 July 1993) 27–33.

Clark, Clifford (ed.). *Minnesota in a century of change: The state and its people since 1900* (Minn. Hist. Soc. Press, St. Paul 1989).

Clark, Edwin II. The United States. In Kerry Turner and Tom Jones (eds.), *Wetlands: Market and intervention failures* (Earthscan, London 1991) 39–72.

Clark, J. I. *Farming the cutover: The settlement of northern Wisconsin* (Wisconsin State Hist. Soc., Madison 1956).

Clawson, Marion, R. Burnell Held, and Charles Stoddard. *Land for the future* (Resources for the Future, Baltimore 1960).

Clements, F. E. *Plant succession: An analysis of the development of vegetation.* Carnegie Institution Publ. 242 (Washington, D.C. 1916).

____. Nature and structure of climax. *Journal of Ecology* 24 (1936) 252–84.

Cochrane, Willard W. *Farm prices* (University of Minnesota Press, Minneapolis 1958).

Collins, Scott L. Introduction: Fire as a natural disturbance in tallgrass prairie ecosystems. In Collins and Wallace (eds.), *Fire in North American tallgrass prairies,* 1990, 3–7.

Collins, Scott L., and Linda L. Wallace (eds.). *Fire in North American tallgrass prairies* (University of Oklahoma Press, Norman 1990).

Committee on Characterization of Wetlands, William M. Lewis (chair). *Wetlands: Characteristics and boundaries* (National Research Council, Washington, D.C. 1995).

Committee on Land Use and Forestry. *Forest land use in Wisconsin* (Madison 1932).

Comstock, Gary (ed.). *Is there a moral obligation to save the family farm?* (Iowa State University Press, Ames 1987).

Conzen, Michael P. The spatial effects of antecedent conditions upon urban growth. M.A. thesis, Geography, University of Wisconsin, Madison 1968.

____. The maturing urban system in the United States, 1840–1900. *Annals Assoc. Amer. Geog.* 67 (1977) 88–108.

____ (ed.). *The making of the American landscape* (Unwin Hyman, Boston 1990).

Conzen, Michael P., Thomas A. Rumney, and Graeme Wynn (eds.). *A scholar's guide to geographical writing on the American and Canadian past,* University of Chicago Geography Research Paper 235 (University of Chicago Press, Chicago 1993).

Cooper, Clare C. The role of railroads in the settlement of Iowa: A study in historical geography. M.A. thesis, Geography, University of Nebraska 1958.

Corliss, Carlton J. *Main line of mid-America: The story of the Illinois Central* (Creative Age Press, New York 1950).

Cottam, Grant. The phytosociology of an oak wood in southern Wisconsin. *Ecology* 30 (1949) 171–287.

Cowardin, Lewis M., Virginia Carter, Francis C. Golet, and Edward T. LaRoe, *Classification of wetlands and deepwater habitats of the United States* (U.S. Fish and Wildlife Service, Washington, D.C. 1979) reprinted 1992.

Cronon, William. *Changes in the land: Indians, colonists, and the ecology of New England* (Hill and Wang, New York 1983).

____. *Nature's metropolis: Chicago and the great west* (W. W. Norton, New York 1991).

____. A place for stories: Nature, history, and narrative. *Journ. Amer. Hist.* 78 (1992) 1347–76.

____. The trouble with wilderness; or getting back to the wrong nature. *Environmental History* 1 (1996) 7–28.

____ (ed.). *Uncommon ground: Toward reinventing nature* (W. W. Norton, New York 1995).

Crosby, Alfred W. *The Columbian exchange: Biological and cultural consequences of 1492* (Greenwood, Westport 1972).

____. Ecological imperialism: The overseas migration of western Europeans as a biological phenomenon. In Donald Worster (ed.), *The ends of the earth: Perspectives on modern environmental history* (Cambridge University Press, Cambridge 1988) 103–17.

Curry-Roper, Janel M., and John Bowles. Local factors in changing land-tenure patterns. *Geog. Rev.* 81 (1991) 443–56.

Curry-Roper, Janel M., and Carol Veldman Rudie. Hollandale: The evolution of a Dutch farming community. *Focus* 40 (1990) 13–18.

Curtis, J. T. *The vegetation of Wisconsin* (University of Wisconsin Press, Madison 1959).

____. The modification of mid-latitude grasslands and forests by man. In Thomas, *Man's role,* 1956, 721–36.

Cushing, Edward J. Problems in the Quaternary phytogeography of the Great Lakes region. In H. E. Wright and David G. Frey (eds.), *The Quaternary of the United States* (Princeton University Press, Princeton 1965) 403–16.

Dahl, T. E. *Wetlands losses in the United States, 1780s to 1980s* (U.S. Fish and Wildlife Service, Washington, D.C. 1990).

Dahl, T. E., and C. E. Johnson. *Wetlands status and trends in the coterminous United States mid 1970s to mid 1980s* (U.S. Fish and Wildlife Service, Washington, D.C. 1991).

Damman, A. W. H., and T. W. French. *The ecology of peat bogs of the glaciated northeastern United States: A community profile.* U.S. Fish and Wildlife Service, Biol. Report 85 (7.16) (Washington, D.C. 1987).

Dana, Edmund. *A description of the Bounty Lands in the state of Illinois* (Looker, Reynolds, Cincinnati 1819).

Dana, Samuel Trask, John H. Allison, and Russell N. Cunningham. *Minnesota lands: Ownership, use, and management of forest and related lands* (American Forestry Association, Washington, D.C. 1960).

Danhof, Clarence H. The fencing problem in the eighteen-fifties. *Agric. Hist.* 18 (1944) 168–86.

Davidson, Osha Gray. *Broken heartland: The rise of America's rural ghetto* (Doubleday, New York 1990).

Davis, Charles A. *Peat: Essays on its origin, uses, and distribution in Michigan.* Report State Board of Geol. Survey Michigan for 1906 (East Lansing 1907) 97–395.

Davis, Charles M. The High Plains of Michigan. *Papers Mich. Acad. Sci. Arts and Letters* 21 (1936) 339.

Deasy, George F. The tourist industry in a "north woods" county. *Econ. Geog.* 25 (1949) 240–59.

____. Agriculture in Luce County, Michigan, 1880 to 1930. *Agric. Hist.* 24 (1950) 29–42.

DeBoer, Stanley G. *The deer damage to forest reproduction survey.* Wisc. Cons. Dept. Publ. 340 (Madison 1947).

Deer, Robert E. A Menominee perspective. In Flader, *Great Lakes forest,* 1983, 113–18.

Delwiche, E. S. *Opportunities for profitable farming in northern Wisconsin.* In *Univ. Wisc. Agric. Expt. Sta. Bull.* 196 (Madison 1910).

Denevan, William M. The pristine myth: The landscape of the Americas in 1492. *Annals Assoc. Amer. Geog.* 82 (1992) 369–85.

____ (ed.). *The native population of the Americas in 1492* (University of Wisconsin Press, Madison 1992).

Denys, Nicolas. *The description and natural history of the coasts of North America (Acadia).* Edited and translated by W. F. Ganong. Publication of the Champlain Society 2 (Toronto 1908).

Destler, Chester McArthur. Agricultural readjustment and agrarian unrest in Illinois, 1880–1896. *Agric. Hist.* 21 (1947) 104–16.

Dick, W. B. A study of the original vegetation of Wayne County, Michigan. *Papers Mich. Acad. Sci., Arts and Letters* 22 (1936) 329–34.

Dickens, Charles. *American notes for general circulation.* Edited by John S. Whitley and Arnold Goldman ([1842] Penguin Classics, Harmondsworth 1985).

____. *Martin Chuzzlewit.* Edited by P. N. Furbank ([1843–44] Penguin Classics, Harmondsworth 1986).

Diedrick, R. T. The agricultural value of wet soils in the upper Midwest. In B. Richardson (ed.), *Selected proceedings of the Midwest conference on wetland values and management* (Freshwater Society, St. Paul, Minn. 1981) 97–106.

Dodge, Stanley D. Sequent occupance on an Illinois prairie. *Bull. Geog. Soc. Philadelphia* 29 (1931) 205–9.

Doolittle, William E. Agriculture in North America on the eve of contact: A reassessment. *Annals Assoc. Amer. Geog.* 82 (1992) 386–401.

Drache, Hiram M. Midwest agriculture: Changing with technology. *Agric. Hist.* 50 (1976) 290–302.

Drake, Daniel. *A systematic treatise, historical, etiological, and practical on the principal diseases of the interior valley of North America, as they appear in Caucasian, African, Indian, and Esquimaux varieties of its population* (W. B. Smith, Cincinnati 1850).

du Creux, François, *The history of Canada or New France, 1625–1658,* vol. 1. Edited by James B. Conacher. Publication of the Champlain Society 30 (Toronto 1951).

Duffy, John, *The healers: A history of American medicine* (University of Illinois Press, Urbana 1979).

Dunbar, G. S. Henry Clay on Kentucky bluegrass, 1838. *Agric. Hist.* 51 (1977) 520–23.

Durand, Loyal. Wisconsin cranberry industry. *Econ. Geog.* 18 (1942) 159–82.

Easterbrook, Gregg. Making sense of agriculture: A revisionist look at farm policy. In Comstock (ed.), *Obligation to save the family farm,* 1987, 3–30.

Ellis, Albert G. Upper Wisconsin country. *Wisc. Hist. Coll.* 3 (1857) 435–52.

Ellis, W. S. The Mississippi: River under siege. *Water: National Geographic Special Edition* (November 1993) 90–105.

Ellsworth, Henry W. *Valley of the upper Wabash, Indiana, with hints on its agricultural advantages: Plan of a dwelling, estimates of cultivation, and notices of labor saving machines* (1838).

Ervin, D. E. Some lessons about the political-economic effects of set-aside: The United States' experience. In British Crop Protection Council, *Set aside: Proceedings of a symposium held at Cambridge University, 15–18 September 1992,* BCPC Monograph 50 (Farnham, Surrey 1992) 3–12.

Esarey, Logan (ed.). *Messages and letters of William Henry Harrison.* Indiana Historical Collections 9 (Indianapolis 1922).

Federal Emergency Management Agency. Federal programs for property acquisition, relocation, and elevation (FEMA leaflet, Chicago, 20 December 1993).

Fenneman, Nevin M. *The lakes of southeastern Wisconsin.* In *Wisc. Geol. Nat. Hist. Surv. Bull.* 8 (Madison 1902).

Fentem, Arlin D. Cash feed grain in the corn belt. Ph.D. diss., Geography, University of Wisconsin 1974.

Finley, Dean, and J. E. Potzger. Characteristics of the original vegetation in some prairie counties of Indiana. *Butler Univ. Bot. Studies* 10 (1952) 114–18.

Finley, Robert W. *Original vegetation cover of Wisconsin, compiled from United States General Land Office notes* (U.S. Department of Agriculture, Forest Service, St. Paul, Minn. 1976).

Fishlow, Albert. *American railroads and the transformation of the ante-bellum economy* (Harvard University Press, Cambridge 1965).

Flader, Susan L. *Thinking like a mountain: Aldo Leopold and the evolution of an ecological attitude toward deer, wolves, and forests* (University of Missouri Press, Columbia 1974).

—— (ed.). *The Great Lakes forest: An environmental and social history* (University of Minnesota Press, Minneapolis 1983).

Flake, L. D. Wetland diversity and waterfowl. In Greeson et al., *Wetland values and functions*, 1979, 312–19.

Flower, George. *History of the English settlement in Edwards County.* Collections of the Chicago Historical Society, vol. 1 (Fergus, Chicago 1882).

Fogel, Robert William. *Railroads and American economic growth: Essays in econometric history* (Johns Hopkins Press, Baltimore 1964).

Fornari, Harry D. U.S. grain exports: A bicentennial overview. In Vivian Wiser (ed.), *Two centuries of American agriculture* (Agric. Hist. Soc., Washington, D.C. 1976) 137–50.

Fowler, Melvin. Middle Mississippian agricultural fields. *American Antiquity* 34 (1969) 365–75.

——. *The Cahokia atlas: A historical atlas of Cahokia archaeology.* Studies in Illinois Archaeology 6 (Ill. Hist. Preservation Agency, Springfield 1989).

Frayer, W. E. *Status and trends of wetlands and deepwater habitats in the coterminous United States, 1970s to 1980s* (Michigan Technological University, Houghton 1991).

Frayer, W. E., T. J. Monahan, D. C. Bowden, and F. A. Graybill. *Status and trends of wetlands and deepwater habitats in the coterminous United States, 1950s to 1970s* (Department of Forest and Wood Science, Colorado State University, Fort Collins 1983).

French, Henry F. *Farm drainage* (A. O. Moore, New York 1859).

Friedberger, Mark. The farm family and the inheritance process: Evidence from the corn belt, 1870–1950. *Agric. Hist.* 57 (1983) 1–13.

Fries, Robert F. *Empire in pine: The story of lumbering in Wisconsin, 1830–1900* (State Hist. Soc. Wisconsin, Madison 1951).

Fritz, Ronald, Roger Suffling, and Thomas Ajit Younger. The influence of the fur trade, famine, and forest fires on moose and woodland caribou populations in northwestern Ontario from 1786 to 1911. *Environmental Management* 17 (1993) 477–89.

Fuller, George N. *Economic and social beginnings of Michigan* (Michigan Historical Publications, Lansing 1916).

Galatowitsch, S. M., and A. G. van der Valk. Natural revegetation during restoration of wetlands in the southern prairie pothole region of North America. In Wheeler et al. (eds.), *Restoration of temperate wetlands,* 1995, 129–42.

Gallagher, James P. Prehistoric field systems in the upper Midwest. In William I. Woods (ed.), *Late prehistoric agriculture: Observations from the Midwest,* Studies in Illinois Archaeology 7 (Illinois Historic Preservation Agency, Springfield 1993).

Gallagher, James P., and Robert F. Sasso. Investigations into Oneota ridged field agriculture on the northern margin of the prairie peninsula. *Plains Anthropologist* 32 (1987) 141–51.

Galloway, Gerald E. The Mississippi basin flood of 1993. Paper prepared for Workshop on reducing the vulnerability of river basin energy, agriculture, and transportation systems to floods. Foz do Iguaçu, Brazil (29 November 1995) 14.

Garland, Hamlin. *A son of the middle border* (Macmillan, New York 1917).

Gartner, W. G. The Hulbert Creek ridged fields: Pre-Columbian agriculture near the Dells, Wisconsin. M.A. thesis, Geography, University of Wisconsin, Madison 1992.

Gates, David M., C. H. D. Clark, and James T. Harris. Wildlife in a changing environment. In Flader, *Great Lakes forest,* 1983, 52–80.

Gates, F. C. A bog in central Illinois. *Torreya* 11 (1911) 205–11.

Gates, Paul Wallace. The promotion of agriculture by the Illinois Central Railroad, 1855–1870. *Agric. Hist.* 5 (1931) 64.

——. *The Illinois Central Railroad and its colonization work* (Harvard University Press, Cambridge 1934).

——. *The Wisconsin pine lands of Cornell University* (Cornell University Press, Ithaca 1944).

——. Hoosier cattle kings. *Indiana Magazine of History* 44 (March 1948) 1–24.

——. Cattle kings in the prairies. *Miss. Valley Hist. Rev.* 35 (December 1948) 379–412.

——. *The farmer's age: Agriculture 1815–1860.* Econ. Hist. of the United States III (Harper and Row, New York 1960).

——. *Landlords and tenants on the prairie frontier: Studies in American land policy* (Cornell University Press, Ithaca 1973).

Geisler, Charles. Ownership: An overview. *Rural Sociology* 58 (1993) 532–46.

Gerhard, Frederick. *Illinois as it is* (Chicago 1857).

Gillespie, G. L. Examination of a route for a canal from Lake Michigan to the Wabash River, Indiana. In U.S. Congress, House of Representatives, *Report of the Chief Engineers,* 44th Cong., 2d Sess., Ex. Doc. 1, pt. 2, 1744 (Washington, D.C. 1876) 454–63.

Gjerde, Jon. The effect of community on migration: Three Minnesota townships, 1885–1905. *Journ. Hist. Geog.* 5 (1979) 403–22.

Glaser, P. H. *The ecology of patterned boreal peatlands of northern Minnesota: A community profile.* U.S. Fish and Wildlife Service, Report 85 (7.14) (Washington, D.C. 1987).

Gleason, H. A. The vegetational history of the Middle West. *Annals Assoc. Amer. Geog.* 12 (1932) 39–85.

Goc, Michael J. The Wisconsin dust bowl. *Wisconsin Magazine of History* 73 (1990) 163–201.

Goldstein, Jon H. *Competition for wetlands in the Midwest: An economic analysis* (Resources for the Future, Washington, D.C. 1971).

Golet, F. C., and J. S. Larson. *Classification of freshwater wetlands in the glaciated Northeast.* U.S. Fish and Wildlife Service, Resource Publ. 116 (Washington, D.C. 1974).

Good, R. E., D. F. Whigham, and R. L. Simpson (eds.). *Freshwater wetlands: Ecological processes and management potential* (Academic Press, New York 1978).

Goolsby, Donald A., William A. Battaglin, and E. Michael Thurman. *Occurrence and transport of agricultural chemicals in the Mississippi River basin, July through August 1993.* Circular 1120-C (U.S. Geological Survey, Washington, D.C. 1993).

Gosselink, J. G., and R. E. Turner, The role of hydrology in freshwater wetland ecosystems. In Good et al., *Freshwater wetlands,* 1978, 63–78.

Greeson, P. E., J. R. Clark, and J. E. Clark (eds.). *Wetland functions and values: The state of our understanding* (Amer. Water Resources Assoc., Minneapolis 1979).

Gregory, John G. *West central Wisconsin: A history,* vol. 1 (Indianapolis 1933).

Grodinsky, Julius. *Transcontinental railway strategy, 1869–1888: A study of businessmen* (University of Pennsylvania Press, Philadelphia 1962).

Grotewold, Andreas. *Regional changes in corn production in the United States from 1909 to 1949.* Department of Geography Research Paper 40 (University of Chicago, Chicago 1955).

Grue, C. E., M. W. Time, G. A. Swanson, S. M. Borthwick, and L. R. Deweese. *Agricultural chemicals and the quality of the prairie pothole wetlands for adult and juvenile waterfowl: What are the concerns?* National Symposium on Protection of Wetlands from Agricultural Impacts (Colorado State University, Fort Collins 1988).

Gustafson, A. F., C. H. Guise, W. J. Hamilton, and H. Ries. *Conservation in the United States* (Comstock Publishing, Ithaca 1939).

Hadwiger, Don F. *The politics of agricultural research* (University of Nebraska Press, Lincoln 1982).

Hamilton, Stanislaus M. (ed.). *The writings of James Monroe,* vol. 1 (G. P. Putnam, New York 1898).

Harding, Benjamin. *A tour through the western country, AD 1818 & 1819* (New London 1819).

Harkin, Duncan A. The significance of the Menominee experience in the forest history of the Great Lakes region. In Flader (ed.), *Great Lakes forest,* 1983, 96–112.

Harl, Neil E. The financial crisis in the United States. In Comstock (ed.), *Obligation to save the family farm,* 1987, 112–29.

Harris, Thaddeus M. *The journal of a tour into the territory northwest of the Alleghany Mountains made in the spring of 1803* (Boston 1805).

Harrison, Gordon. *Mosquitoes, malaria, and man: A history of hostilities since 1880* (John Murray, London 1978).

Hart, John Fraser. The Middle West. In Hart (ed.), *Regions of the United States* (Harper and Row, New York 1972) 258–82.

____. *The look of the land* (Prentice-Hall, Englewood Cliffs, N.J. 1975).

____. Resort areas in Wisconsin. *Geog. Rev.* 74 (1984) 193.

____. Change in the Corn Belt. *Geog. Rev.* 76 (1986) 51–72.

____. *The land that feeds us* (W. W. Norton, New York 1991).

____. Part-ownership and farm enlargement in the Midwest. *Annals Assoc. Amer. Geog.* 81 (1991) 66–79.

Hartman, W. A., and J. D. Black. *Economic aspects of land settlement in the cutover region of the Great Lakes states*. USDA Circular 160 (GPO, Washington, D.C. 1931).

Hartnett, Sean. The land market on the Wisconsin frontier: An examination of landownership processes in Turtle and La Prairie townships, 1839–1890. *Agric. Hist.* 65 (1991) 38–77.

Hartshorne, Richard. A classification of the agricultural regions of Europe and North America on a uniform statistical basis. *Annals Assoc. Amer. Geog.* 25 (1935) 99–120.

Haswell, John R. Drainage in the humid regions. In U.S. Department of Agriculture, *Soils and men*, 1938, 723–36.

Hayden, A. A botanical survey in the Iowa lake region of Clay and Palo Alto Counties. *Iowa State Coll. Journ. Sci.* 17 (1943) 277–416.

Hays, Samuel P. *Conservation and the gospel of efficiency: The Progressive conservation movement, 1890–1920* (Harvard University Press, Cambridge 1959).

Hayter, Earl W. *The troubled farmer, 1850–1900: Rural adjustment to industrialism* (Northern Illinois University Press, De Kalb 1968).

Hedges, James B. The colonization work of the Northern Pacific. *Mississippi Valley Hist. Rev.* 13 (1926) 311–42.

Heinselman, Miron L. Forest sites, bog processes, and peatland types in the glacial Lake Agassiz region, Minnesota. *Ecol. Monographs* 33 (1963) 327–74.

____. Landscape evolution and peatland types and the Lake Agassiz Peatlands Natural Area, Minnesota. *Ecol. Monograph* 40 (1970) 235–61.

Helgeson, Arlan C. The promotion of agricultural settlement in northern Wisconsin, 1880–1925. Ph.D. diss., History, University of Wisconsin 1951.

____. *Farms in the cutover: Agricultural settlement in northern Wisconsin* (Wisconsin State Hist. Soc., Madison 1962).

Henderson, Caspar. Chainsaw massacre? Not in Wisconsin. *Independent* (4 March 1996) 18.

Hendrickson, Paul. Those who are no longer with us. In Comstock (ed.), *Obligation to save the family farm*, 1987, 47–53.

Henlein, Paul C. *Cattle kingdom in the Ohio valley, 1783–1860* (University of Kentucky Press, Lexington 1959).

Henry, William A. *Northern Wisconsin: A handbook for the homeseeker* (Democrat Printing Co., Madison 1896).

Hewes, Leslie. Some features of early woodland and prairie settlement in a central Iowa county. *Annals Assoc. Amer. Geog.* 40 (1950) 40–57.

____. The northern wet prairie of the United States: Nature, sources of information, and extent. *Annals Assoc. Amer. Geog.* 41 (1951) 307–23.

____. Drained land in the United States in the light of the Drainage Census. *Professional Geographer* 5 (1953) 6–12.

Hewes, Leslie, and P. E. Frandson. Occupying the wet prairie: The role of artificial drainage in Story County, Iowa. *Annals Assoc. Amer. Geog.* 42 (1952) 24–50.

Hibbard, Benjamin Horace. *A history of the public land policies* (Peter Smith, New York 1939).

Hicks, John D. The western Middle West, 1900–1914. *Agric. Hist.* 20 (1946) 65–77.

Higbee, Edward C. *American agriculture: Geography, resources, conservation* (John Wiley, New York 1958).

Hilliard, Sam B. A robust new nation, 1783–1820. In Mitchell and Groves (eds.), *North America*, 1987, 149–71.

Hollands, Garrett G. Regional analysis of the creation and restoration of kettle and pothole wetlands. In Kusler and Kentula, *Wetland creation and restoration*, 1990, 281–98.

Hotchkiss, George W. *History of the lumber and forest industry of the Northwest* (G. W. Hotchkiss, Chicago 1898).

Hough, B., and A. Bourne. *Map of the state of Ohio from actual survey* (Philadelphia 1815).

Hough, Franklin. *Report on Forestry,* vol. 3 (GPO, Washington, D.C. 1882).

Hough, Walter. *Fire as an agent in human culture.* U.S. National Museum Bulletin 139 (Washington, D.C. 1926).

Howe, H. *Historical collections of Ohio* (Bradley and Anthony, Cincinnati 1848).

Hudson, John C. *Making the corn belt: A geographical history of Middlewestern agriculture* (Indiana University Press, Bloomington 1994).

Huels, Frederick W. *The peat resources of Wisconsin.* In *Wisc. Geol. Nat. Hist. Surv. Bull.* 45 (Madison 1915).

Hunter, Dianna. *Breaking hard ground: Stories of the Minnesota Farm Advocates* (Holy Cow! Press, Duluth 1990).

Hurst, James Willard. *Law and economic growth: The legal history of the lumber industry in Wisconsin, 1836–1915* (Belknap Press, Harvard University, Cambridge 1964).

____. The institutional environment of the logging era in Wisconsin. In Flader (ed.), *Great Lakes forest,* 1983, 137–55.

Interagency Floodplain Management Review Committee, Scientific Assessment and Strategy Team. *A blueprint for change: Science for floodplain management in the 21st century* (SAST, Washington, D.C. 1994).

Ives, L. J., Jr. The natural vegetation of Lorain County, Ohio. M.A. thesis, Oberlin College 1947.

Jacobs, Wilbur R. *Dispossessing the American Indian: Indians and whites on the colonial frontier* (Scribner, New York 1972).

_____. The Indian and the frontier in American history: A need for revision. *Western Historical Quarterly* 4 (1973) 43–56.

Jahoda, Gloria. *The trail of tears: The American Indian removals, 1813–1855* (Allen and Unwin, London 1976).

Jakle, John A. *Images of the Ohio valley: A historical geography of travel, 1740 to 1860* (Oxford University Press, New York 1977).

Jefferson, Thomas. *Writings.* 20 vols. (Thomas Jefferson Memorial Association, Washington, D.C. 1903–4).

Jennings, Jesse D. *Prehistory of North America* (McGraw Hill, New York 1968).

Jesness, Oscar B., Reynolds I. Nowell, and associates. *A program for land use in northern Minnesota: A type study in land utilization* (University of Minnesota Press, Minneapolis 1935).

Johnson, Arthur M., and Barry E. Supple. *Boston capitalists and western railroads: A study in nineteenth century railroad investment process* (Harvard University Press, Cambridge 1967).

Johnson, Hildegard Binder. The location of German immigrants in the Middle West. *Annals Assoc. Amer. Geog.* 41 (1951) 1–41.

_____. *Order upon the land: The U.S. rectangular land survey and the upper Mississippi country* (Oxford University Press, New York 1976).

Johnston, John. Draining. *Trans. New York State Agric. Soc.* 15 (Albany 1855) 257–59.

Jolly, Robert W. *1993 Iowa Farm Finance Survey* (Iowa State University Extension Service, Ames 1993).

Jones, E. R. A sane plan for marsh development. *Fourth Report, Wisconsin State Drainage Association Proceedings* (January 1919–December 1920) 48–54.

_____. Land drainage in Wisconsin. Press statement for State Department of Engineering and College of Agriculture, 1923. [Records of State Drainage Engineer, Madison.]

_____. Central Wisconsin Drainage Districts 1924, Reports and Memoranda. [Records of State Drainage Engineer, Madison.]

_____. Preliminary plan for putting permanent prosperity into the drainage districts and counties of central Wisconsin, submitted to the Attorney General, 1924. [Records of State Drainage Engineer, Madison.]

_____. Keeping faith with the Swamp Land Fund. *Seventh Report, Wisconsin State Drainage Association* (Madison, March 1927).

____. *Marsh problem old*. Report released 13 March 1928. [Records of State Drainage Engineer, Madison.]

Jones, E. R., and B. G. Packer. Drainage district farms in central Wisconsin. *Univ. Wisc. Agric. Expt. Sta. Bull.* 358 (Madison, October 1923).

Jones, E. R., and O. R. Zeasman. An outlet drain for every farm. *Univ. Wisc. Agric. Expt. Sta. Bull.* 351 (Madison, December 1922).

____. Drain wet fields. *Univ. Wisc. Agric. Expt. Sta. Bull.* 365 (Madison, June 1924).

Jordan, Terry G. Between the forest and the prairie. *Agric. Hist.* 38 (1964) 205–16.

____. *North American cattle-ranching frontiers: Origins, diffusion, and differentiation* (University of New Mexico Press, Albuquerque 1993).

Josephson, J. Status of wetlands. *Environmental Science* 26 (1992) 422.

Kaatz, Martin R. The Black Swamp: A study in historical geography. *Annals Assoc. Amer. Geog.* 45 (1955) 1–35.

Kane, L. Selling the cutover lands in Wisconsin. *Business History Review* 28 (1954) 236–47.

Kantrud, H. A., G. L. Krapu, and G. A. Swanson. *Prairie basin wetlands of the Dakotas: A community profile*. U.S. Fish and Wildlife Service, Biol. Report 85 (Washington, D.C. 1989).

Kay, Jeanne. Wisconsin Indian hunting patterns, 1634–1836. *Annals Assoc. Amer. Geog.* 69 (1979) 402–18.

____. Preconditions of natural resource conservation. *Agric. Hist.* 59 (1985) 124–35.

Keating, William H. *Narrative of an expedition to the source of St. Peter's River, Lake Winnepeek, Lake of the Woods, & c. performed in the year 1823 under the command of Stephen H. Long* [1824], vol. 1 (Ross and Haines, Minneapolis 1959).

Keillor, Garrison. *Lake Wobegon days* (Penguin, Harmondsworth 1986).

Kennedy, Joseph C. G. *Preliminary report on the eighth census, 1860*. House of Representatives, 37th Cong., 2d Sess. Ex. Doc. 116 (GPO, Washington, D.C. 1862).

Kenney, F. R., and W. L. McAtee, The problem: Drained areas and wildlife habitats. In U.S. Department of Agriculture, *Soils and men*, 1938, 77–83.

Kenoyer, Leslie A. Ecological notes on Kalamazoo County, Michigan, based on the original land survey. *Papers Mich. Acad. Sci., Arts and Letters* 11 (1929) 215.

____. Forest distribution in southwest Michigan as interpreted from the original land survey, 1826–1832. *Papers Mich. Acad. Sci., Arts and Letters* 19 (1933) 107–111.

King, J. E. Late quaternary vegetational history of Illinois. *Ecol. Monogr.* 51 (1981) 43–62.

Kinkel, Kenneth E. A hydroclimatological assessment of the rainfall. In Changnon, *The great flood of 1993*, 1996, 52–67.

Klippart, John. *The principles and practice of land drainage.* 3d ed. (R. Clarke, Cincinnati 1888).

Knudsen, G. J. *History of beaver in Wisconsin* (Wisconsin Conservation Department, Madison 1963).

Kohlmeyer, Fred W. *Timber roots: The Laird, Norton story, 1855–1905* (Winona Hist. Soc., Winona 1972).

Komarek, E. V. Fire ecology: Grasslands and man. *Tall Timbers Fire Ecology Conference Proceedings* 4 (1965) 169–220.

———. Fire: And the ecology of man. *Tall Timbers Fire Ecology Conference Proceedings* 6 (1967) 143–70.

Krech, Shepard III (ed.). *Indians, animals, and the fur trade: A critique of "Keepers of the game"* (University of Georgia Press, Athens 1981).

Kroodsma, D. E. Habitat values for nongame wetland birds. In Greeson et al., *Wetland values and functions,* 1979, 320–29.

Kruczynski, William L. Mitigation and the Section 404 program: A perspective. In Kusler and Kentula, *Wetland creation and restoration,* 1990, 549–54.

———. Options to be considered in preparation and evaluation of mitigation plans. In Kusler and Kentula, *Wetland creation and restoration,* 1990, 555–70.

Kusler, Jon A., and Mary E. Kentula. *Wetland creation and restoration: The status of the science* (Island Press, Washington, D.C. 1990).

Kutzbach, J. E., and H. E. Wright. Simulation of the climate of 18,000 yr BP: Results for North America/North Atlantic/European sector. *Quaternary Sci. Rev.* 4 (1986) 147–87.

Ladin, Jay. Mortgage credit in Tippecanoe County, Indiana, 1865–1880. *Agric. Hist.* 41 (1967) 37–43.

Lapham, Increase A. *Wisconsin: Its geography and topography* (Milwaukee 1846).

Larsen, J. A. *Ecology of the northern lowland bogs and conifer forests* (Academic Press, New York 1982).

Larson, Agnes M. *History of the white pine industry in Minnesota* (University of Minnesota Press, Minneapolis 1949).

Larson, G. A., G. Roloff, and W. E. Larson. A new approach to marginal agricultural land classification. *Journ. Soil and Water Conservation* 43 (1987) 103–5.

Lasley, Paul. The crisis in Iowa. In Comstock (ed.), *Obligation to save the family farm,* 1987, 99–101.

———. *Iowa farm and rural life poll: 1993 summary report* (Iowa State University Extension Service, Ames 1993).

Lee, Gerhard B. Soil Region J: Soils of stream bottoms and major wetlands. In Francis D. Hole (ed.), *Soils of Wisconsin,* in *Wisc. Geol. Nat. Hist. Surv. Bull.* (Madison 1976), 115–22.

Leitch, J. A., and L. E. Danielson. *Social, economic, and institutional incen-*

tives to drain or preserve prairie wetlands (Department of Agricultural and Applied Economics, University of Minnesota, St. Paul 1979).

Lemon, James T. *The best poor man's country: A geographical study of early southeastern Pennsylvania* (Johns Hopkins Press, Baltimore 1972).

Leopold, Aldo. *A Sand County almanac* [Oxford University Press, New York 1949] (Ballantine Books, New York 1990).

Leverett, Frank. *Surface geology and agricultural conditions of the southern peninsula of Michigan.* In *Mich. Geol. and Biol. Surv. Bull.* 9 (Lansing 1912).

Levine, Daniel A., and Daniel E. Willard. Regional analysis of fringe wetlands in the Midwest: Creation and restoration. In Kusler and Kentula, *Wetland creation and restoration,* 1990, 299–325.

Lewis, Sinclair. *Main Street* (Harcourt Brace, New York 1920).

Lieth, H. Primary production of the major vegetation units of the world. In H. Lieth and R. H. Whitaker (eds.), *Primary productivity of the biosphere* (Springer-Verlag, New York 1975) 203–15.

Lindeman, R. L. The developmental history of Cedar Creek Lake, Minnesota. *Amer. Midl. Naturalist* 25 (1941) 101–12.

Linderman, Frank, *Plenty-coups: Chief of the Crows* ([1930] reprint Lincoln 1962).

Lindsey, A. A. The Indiana of 1816. In A. A. Lindsey (ed.), *Natural features of Indiana* (Indiana Acad. Sci., Indianapolis 1966) x–xxix.

Lockwood, James H. Early times and events in Wisconsin. *Wisc. Hist. Coll.* 2 (1856) 130–41.

Lokken, Roscoe L. *Iowa: Public land disposal* (State Hist. Soc. of Iowa, Iowa City 1942).

Lothrop, J. S. *Champaign County directory, 1870–1871* (Chicago 1871).

Lovell, W. George. Heavy shadows and black night: Disease and depopulation in colonial Spanish America. *Annals Assoc. Amer. Geog.* 82 (1992) 426–63.

Lovins, Amory, L. Hunter, and Marty Bender, Energy in agriculture. In W. Jackson, W. Berry, and B. Colman (eds.), *Meeting the expectations of the land* (North Point Press, San Francisco 1984).

Lowenthal, David. *Possessed by the past: The heritage crusade and the spoils of history* (Free Press, New York 1996).

Luhmann, Niklas. *Ecological communication.* Translated by John Bednarz (Polity Press, Cambridge 1989).

Mackintosh, Jette. Ethnic patterns in Danish immigrant agriculture: A study of Audubon and Shelby counties, Iowa. *Agric. Hist.* 64 (1990) 59–77.

____. Migration and mobility among Danish settlers in southwest Iowa. *Journ. Hist. Geog.* 17 (1991) 165–89.

Mairson, Alan. The great flood of '93. *National Geographic* 185 (January 1994) 42–81.

Malcolm, Andrew. *Final harvest: An American tragedy* (Signet Books, New York 1987).

Malin, James C. The grassland of North America: Its occupance and the challenge of continuous reappraisals. In Thomas, *Man's role,* 1956, 350–66.

Maltby, Edward. *Waterlogged wealth: Why waste the world's wet places?* (Earthscan, London 1986).

Manners, Ian R., and Marvin W. Mikesell (eds.). *Perspectives on environment.* Commission on College Geography Publ. 13 (Assoc. Amer. Geog., Washington, D.C. 1974).

Marbut, C. F. Soils of the United States. Pt. 3 of O. E. Baker (ed.), *Atlas of American Agriculture* (Washington, D.C. 1935).

Margolis, Jon. Small farmers lose illusions: A bitter harvest lies ahead. *St. Louis Post-Dispatch* (4 September 1994) 21.

Marschner, Francis J. *The original vegetation of Minnesota: Compiled from U.S. General Land Office survey notes* [1930] (North Central Forest Experiment Station, St. Paul, Minn. 1974).

Marsden, Terry, Richard Munton, Sarah Whatmore, and Jo Little. Towards a political economy of capitalist agriculture: A British perspective. *International Journ. Urban and Regional Research* 4 (1986) 513–15.

Marsh, George Perkins. *Man and nature,* ed. David Lowenthal ([1864]; Belknap, Harvard University Press, Cambridge 1965).

Martin, A. C., N. Hotchkiss, F. M. Uhler, and W. S. Bourn. *Classification of wetlands of the United States.* U.S. Fish and Wildlife Service, *Spec. Sci. Rep. Wildlife* 20 (Washington, D.C. 1953).

Martin, Calom, *Keepers of the game: Indian-animal relationships and the fur trade* (University of California Press, Berkeley 1978).

Martin, Lawrence. *The physical geography of Wisconsin.* In *Wisc. Geol. Nat. Hist. Surv. Bull.* 36 (Madison 1932).

Mattingly, Rosanna L., Edwin E. Herricks, and Douglas M. Johnston. Channelization and levee construction in Illinois: Review and implications for management. *Environmental Management* 17 (1993) 781–95.

McCorvie, Mary R., and Christopher L. Lant. Drainage district formation and the loss of Midwestern wetlands, 1850–1930. *Agric. Hist.* 67 (1993) 13–39.

McCrory, S. H. Historical notes of land drainage in the United States. *Proc. Amer. Soc. Civil Engineers* 53 (1927) 1628–36.

McIntyre, Wallace E. Land utilization of three typical upland prairie townships. *Econ. Geog.* 25 (1949) 260–74.

McManis, Douglas R. *The initial evaluation and utilization of the Illinois prairies, 1815–1840.* Department of Geography Research Paper 94 (University of Chicago, Chicago 1964).

McMurry, K. C. The use of land for recreation. *Annals Assoc. Amer. Geog.* 20 (1930) 7–20.

McNall, P. E., H. O. Henderson, A. R. Albert, and W. R. Abbott. Farming in the central sandy area of Wisconsin. *Univ. Wisc. Agric. Expt. Sta. Bull.* 497 (1952).

Mead, W. R. A Finnish settlement in central Minnesota. *Acta Geographica* 13 (Helsinki 1954) 3–16.

Meinig, D. W. *The shaping of America: A geographical perspective on 500 years of history.* Vol. 1, *Atlantic America, 1492–1800* (Yale University Press, New Haven 1986).

————. *The shaping of America: A geographical perspective on 500 years of history.* Vol. 2, *Continental America, 1800–1867* (Yale University Press, New Haven 1993).

Menzel, B. W. Agricultural management practices and the integrity of inter-stream biological habitat. In F. W. Schaller and G. W. Bailey (eds.), *Agricultural management and water quality* (American Water Resources Association, Minneapolis 1978).

Meyer, Alfred H. The Kankakee "marsh" of northern Indiana and Illinois. *Papers Mich. Acad. Sci., Arts and Letters* 21 (1935) 359–96.

————. Circulation and settlement patterns of the Calumet region of northwest Indiana and northeast Illinois: The first stage of occupance; the Pottawatomie and the fur trader, 1830. *Annals Assoc. Amer. Geog.* 44 (1954) 245–74.

Michigan Department of Natural Resources, Recreation Division. *Michigan's 1987–88 Recreation Action Program* (East Lansing 1988).

Mikesell, Marvin W. Geography as the study of environment: An assessment of some old and new commitments. In Manners and Mikesell, *Perspectives on environment,* 1974, 1–23.

Miller, George J. Some geographic influences in the settlement of Michigan and in the distribution of its population. *Bull. Amer. Geog. Soc.* 45 (1913) 321–48.

Minnesota Department of Natural Resources, *Minnesota peatlands* (St. Paul 1978).

Minnesota Outdoor Recreation Resources Commission. *Acquisition of wild-life land in Minnesota (Wetland Program)* (State Capitol, St. Paul 1965).

Mitchell, D. W. *Ten years in the United States: Being an Englishman's views of men and things in the North and South* (London 1862).

Mitchell, Robert D., and Paul A. Groves (eds.). *North America: A historical geography of a changing continent* (Hutchinson, London 1987).

Mitchell, Robert D., and Milton B. Newton. *The Appalachian frontier: Views from the East and the Southwest.* Historical Geography Research Series 21 (Cheltenham 1988).

Mitchell, S. Augustus. *Illinois in 1837* (Mitchell, Philadelphia 1837).

Mitsch, W. J. Interactions between a riparian swamp and a river in southern Illinois. In R. R. Johnson and J. F. McCormick (tech. coords.), *Strategies for the protection and management of floodplain wetlands and other ri-*

parian ecosystems, U.S. Forest Service General Technical Report WO-12 (Washington, D.C. 1979).

Mitsch, William J., and James G. Gosselink. *Wetlands,* 2d ed. (Van Nostrand Reinhold, New York 1993).

Mitsch, W. J., and B. C. Reeder. Nutrient and hydrologic budgets of a Great Lakes coastal freshwater wetland during a drought year. *Wetlands Ecology and Management* 1 (1992) 211–23.

Mitsch, W. J., W. Rust, A. Behnke, and L. Lai. *Environmental observations of a riparian ecosystem during flood season.* University of Illinois Water Resources Center Research Report 142 (Urbana 1979).

Moehlman, Arthur Henry. The Red River of the north. *Geog. Rev.* 25 (1935) 79–91.

Moline, Robert T. The modification of the wet prairie in southern Minnesota. Ph.D. diss., Geography, University of Minnesota 1969.

____. *The citizen and water attitudes in southern Minnesota.* Office of Water Research and Technology Project B-042-Minnesota (Minneapolis 1974).

____. Cultural modification of wet prairie landscapes. In Thomas J. Baerwald and Karen L. Harrington (eds.), *A.A.G. '86 Twin Cities field trip guide* (Assoc. Amer. Geog., Minneapolis 1986) 194–203.

Mooney, James. *The aboriginal population of America north of Mexico.* Edited by J. R. Swanton. Smithsonian Institution, Miscellaneous Collection 80 (GPO, Washington, D.C. 1928).

Moore, Peter D. Soils and ecology: Temperate wetlands. In Williams, *Wetlands,* 1990, 95–114.

Moulton, Barbara. Takings legislation: Protection of property rights or threat to the public interest? *Environment* 37 (1995) 44–45.

Munton, R. J. C., T. Marsden, and S. Whatmore. Technological change in a period of agricultural adjustment. In P. Lowe, T. K. Marsden, and S. J. Whatmore (eds.), *Technological change and the rural environment* (Fulton, London 1990) 104–26.

Murray, Stanley N. Railroads and the agricultural development of the Red River valley of the north, 1870–1890. *Agric. Hist.* 31 (1957) 64–66.

Naiman, R. J., T. Manning, and C. A. Johnston. Beaver population fluctuations and tropospheric methane emissions in boreal wetlands. *Biogeochemistry* 12 (1991) 1–15.

Nash, Gary B. *Red, white, and black: The peoples of early America* (Prentice-Hall, Englewood Cliffs, N.J. 1982).

Nash, Roderick Frazier. *Wilderness and the American mind.* 3d ed. ([1967]; Yale University Press, New Haven 1982).

____. *The rights of nature: A history of environmental ethics* (University of Wisconsin Press, Madison 1989).

____ (ed.). *The American environment: Readings in the history of conservation* (Addison-Wesley, Reading, Mass. 1976).

Nass, David L. The rural experience. In Clark, *Minnesota,* 1989, 129–54.

Nassauer, Joan Iverson. The aesthetics of horticulture: Neatness as a form of care. *HortScience* 23 (1988) 973–77.

____. Agricultural policy and aesthetic objectives. *Journ. Soil and Water Conservation* 44 (1989) 384–87.

National Wetlands Policy Forum. *Protecting America's wetlands: An action agenda* (Conservation Foundation, Washington, D.C. 1988).

Natural Resources Defense Council. *Land use controls in the United States: A handbook on the legal rights of citizens* (Dial Press, New York 1977).

Nellis, Duane. Agricultural externalities and the environment in the United States. In I. R. Bowler, C. R. Bryant, and M. D. Nellis (eds.), *Contemporary rural systems in transition: Agriculture and environment,* vol. 1 (CAB International, Wallingford 1992).

Newell, F. H. What may be accomplished by reclamation. *Annals Amer. Acad. Pol. and Soc. Sci.* 33 (1909) 658–63.

Nichols, David A. *Lincoln and the Indians: Civil War policy and politics* (University of Missouri Press, Columbia 1978).

Novitzki, R. P. Hydrologic characteristics of Wisconsin's wetlands and their influence on floods, stream flow, and sediment. In Greeson et al., *Wetland function and values,* 1979, 377–88.

Ogawa, H., and J. W. Male. Simulating the flood mitigation role of wetlands. *Journ. Water Resources Planning and Management* 12 (1986) 114–28.

Oliver, W. *Eight months in Illinois: With information to emigrants.* March of America facsimile series 81 ([1843]; University Microfilms, Ann Arbor, Mich. 1966).

Ostergren, Robert. A community transplanted: The formative experience of a Swedish immigrant community in the upper Middle West. *Journ. Hist. Geog.* 5 (1979) 189–212.

Overton, Richard C. *Burlington west: A colonization history of the Burlington Railroad* (Harvard University Press, Cambridge 1941).

Owen, C. R., and H. M. Jacobs. Wetland protection as land-use planning: The impact of Section 404 in Wisconsin, U.S.A. *Environmental Management* 16 (1992) 345–53.

Owen, David Dale. *Report of a geological survey of Wisconsin, Iowa, and Minnesota: And incidentally a portion of Nebraska territory* (Lippincott, Grambo, Philadelphia 1852).

Paddock, Joe, Nancy Paddock, and Carol Bly. *Soil and survival* (Sierra Club Books, San Francisco 1986).

Page, Brian, and Richard Walker. From settlement to Fordism: The agro-industrial revolution in the American Midwest. *Econ. Geog.* 67 (1991) 281–315.

Page, John Lorence. *Climate of Illinois* (University of Illinois Press, Urbana 1949).

Palmer, Ben. *Swamp land drainage with special reference to Minnesota.* University of Minnesota Studies in Social Sciences 5 (Minneapolis 1915).

Parker, G. R., and G. Schneider. Biomass and productivity of an alder swamp in northern Michigan. *Canadian Journ. Forest Research* 5 (1975) 403–9.

Parrett, C., N. B. Melcher, and R. W. James. *Flood discharges in the upper Mississippi River basin 1993.* U.S. Geological Survey Circular 1120-A (Washington, D.C. 1993).

Passmore, John. *Man's responsibility for nature: Ecological problems and Western traditions* (Duckworth, London 1974).

Pattison, William D. *Beginnings of the American rectangular land survey system, 1784–1800.* Department of Geography Research Paper 50 (University of Chicago, Chicago 1957).

Paullin, Charles O., and John K. Wright. *Atlas of the historical geography of the United States* (Carnegie Institute and American Geographical Society, Washington, D.C. 1932).

Pavelis, G. A. (ed.). *Farm drainage in the United States: History, status, and prospects.* U.S. Department of Agriculture, Economic Research Service, Misc. Publ. 1455 (Washington, D.C. 1987).

Peck, John M. *A gazetteer of Illinois in three parts* (R. Goudy, Jacksonville 1834).

Perkins, W. LeRoy, The significance of drain tile in Indiana. *Econ. Geog.* 7 (1931) 380–89.

Peterson, Harold F. Some colonization projects of the Northern Pacific Railroad. *Minnesota History* 10 (1929) 127–44.

Philippi, Nancy S. *Revisiting flood control: An examination of federal control policy in light of the 1993 flood event on the upper Mississippi River* (Wetlands Research, Chicago 1994).

____. *Spending federal flood control dollars: Three case studies of the 1993 Mississippi River floods* (Wetlands Research, Chicago 1995).

Pickels, G. W., and F. B. Leonard. *Engineering and legal aspects of land drainage in Illinois.* In *Ill. State Geol. Surv. Bull.* 42 (1928).

Poggi, Edith Muriel, *The prairie province of Illinois.* Ill. Studies in Soc. Sci. 19 (University of Illinois Press, Urbana 1934).

Potter, Clive, Paul Burnham, Angela Edwards, Ruth Gasson, and Bryn Green. *The diversion of land: Conservation in a period of farming contraction* (Routledge, London 1991).

Power, Richard Lyle. Wet lands and the Hoosier stereotype. *Miss. Valley Hist. Rev.* 22 (1935) 33–48.

____. *Planting corn belt culture: The impress of the upland Southerner and Yankee in the old Northwest* (Indiana State Hist. Soc., Indianapolis 1953).

Prince, Hugh. Floods in the upper Mississippi River basin, 1993: Newspapers, official views, and forgotten farmlands. *Area* 27 (1995) 118–26.

____. A marshland chronicle, 1830–1960: From artificial drainage to outdoor recreation in central Wisconsin. *Journ. Hist. Geog.* 21 (1995) 3–22.

Pyne, Stephen. *Fire in America: A cultural history of wildland and rural fire* (Princeton University Press, Princeton 1982).

Quade, Henry W., et al. *The nature and effects of county drainage ditches in south central Minnesota.* In *Water Resources Research Center Bulletin* 105 (University of Minnesota, Minneapolis 1980).

Ralph, George A. *Report on topographical and drainage survey of swamp and marshy lands owned by the state of Minnesota* (State Drainage Commission of Minnesota, Crookston 1907).

Rasmussen, Wayne D. (ed.). *Readings in the history of American agriculture* (University of Illinois Press, Urbana 1960).

Reader, R. J. Primary production in northern bog marshes. In Good et al., *Freshwater wetlands,* 1978, 53–62.

Reiners, W. A. Structure and energetics of three Minnesota forests. *Ecological Monographs* 42 (1972) 71–94.

Reinhardt, Hazel H. Social adjustments to a changing environment. In Flader (ed.), *Great Lakes forest,* 1983, 205–19.

Report of the Forestry Commission of the State of Wisconsin (Democrat Printing Co., Madison 1898).

Report of the state forester of Wisconsin, 1909–10 (Madison 1910).

Rice, John G. The role of culture and community in frontier prairie farming. *Journ. Hist. Geog.* 3 (1977) 155–75.

Richardson, C. J. Primary productivity values in freshwater wetlands. In Greeson et al., *Wetland function and values,* 1979, 131–45.

Richardson, J. L., and J. L. Arndt, What use prairie potholes? *Journ. Soil and Water Conservation* 44 (1989) 196–98.

Richter, Conrad, *The trees* (A. A. Knopf, New York 1940).

Riley, Thomas J., and Glen Freimuth. Fields systems and frost drainage in prehistoric agriculture of the upper Great Lakes. *American Antiquity* 44 (1979) 27–85.

Risser, Paul G. Landscape processes and the vegetation of the North American grassland. In Collins and Wallace, *Fire in tallgrass prairies,* 1990, 133–46.

Robbins, Roy M. *Our landed heritage: The public domain, 1776–1936* (Princeton University Press, Princeton 1942).

Rogin, Leo. *The introduction of farm machinery in its relation to the productivity of labor in the agriculture of the United States during the nineteenth century.* University of California Publications in Economics 9 (University of California Press, Berkeley 1931).

Rolvaag, Ole. *Giants in the earth* (Harper, New York 1927).

Rose, John K. Delavan Prairie: An Illinois corn belt community. *Journ. Geog.* 32 (1933) 1–13.

Rosenblatt, Paul C. *Farming is in our blood: Family farms in economic crisis* (Iowa State University Press, Ames 1990).

Rostlund, Erhard. *Freshwater fish and fishing in native North America.* University of California Publ. Geography 9 (Berkeley 1952).

――――. The evidence for the use of fish as fertilizer in aboriginal North America. *Journ. Geog.* 56 (1957) 222–28.

Roth, Filibert. *On the forestry conditions of northern Wisconsin.* In *Wisc. Geol. Nat. Hist. Surv. Bull.* 1 (Madison 1898).

Rothwell, Robert R. (ed.). *Henry David Thoreau: An American landscape* (New York 1991).

Royce, C. C. Indian land cessions in the United States. *Eighteenth Annual Report of the Bureau of American Ethology,* pt. 2, *1896–7* (Washington, D.C. 1899).

Salamon, Sonya. *Prairie patrimony: Family, farming, and community in the Midwest* (University of North Carolina Press, Chapel Hill 1992).

――――. Culture and agricultural land tenure. *Rural Sociology* 58 (1993) 580–98.

Sampson, Homer C. An ecological survey of the prairie vegetation of Illinois. *Ill. Nat. Hist. Surv. Bull.* 13, art. 16 (1921) 523–77.

Satz, R. N. *American Indian policy in the Jacksonian era* (University of Nebraska Press, Lincoln 1975).

Sauer, Carl O. *Geography of the upper Illinois valley and history of its development.* In *Ill. State Geol. Surv. Bull.* 27 (1916).

――――. Grassland climax, fire, and man. *Journ. Range Management* 3 (1950) 16–21.

――――. *Agricultural origins and dispersals* (American Geog. Soc., New York 1952).

――――. The agency of man on earth. In Thomas, *Man's role,* 1956, 49–69.

――――. *Seventeenth century North America* (Turtle Island, Berkeley 1980).

Schafer, Joseph. *A history of agriculture in Wisconsin.* Wisconsin Domesday Book 1 (State Hist. Soc. Wisconsin, Madison 1922).

――――. *Wisconsin Domesday Book: Town studies,* vol. 1 (State Hist. Soc., Madison 1924).

――――. *Four Wisconsin counties: Prairie and forest.* Wisconsin Domesday Book 2 (State Hist. Soc., Madison 1927).

Schama, Simon. *Landscape and memory* (Harper Collins, London 1995).

Schmaltz, Norman J. Michigan's land economic survey. *Agric. Hist.* 52 (1978) 229–46.

――――. The land nobody wanted: The dilemma of Michigan's cutover lands. *Michigan History* 67 (1983) 32–40.

Schob, David E. Sodbusting on the upper Midwestern frontier, 1820–1860. *Agric. Hist.* 47 (1973) 47–56.

――――. *Hired hands and plowboys: Farm labor in the Midwest, 1815–60* (University of Illinois Press, Urbana 1975).

Schorger, A. W. The beaver in early Wisconsin. *Trans. Wisconsin Acad.* 54 (1965) 147–79.

Sclick, W. J. *The theory of underdrainage*. In *Iowa Engin. Expt. Sta. Bull.* 50 (1918).

Scott, Roy V. *The agrarian movement in Illinois, 1880–1896* (University of Illinois Press, Urbana 1962).

——. Land use and American railroads in the twentieth century. *Agric. Hist.* 53 (1979) 683–703.

Sears, P. B. The natural vegetation of Ohio: I. A map of the virgin forest. *Ohio Journ. Sci.* 25 (1925) 139–49.

——. The natural vegetation of Ohio: II. The prairies. *Ohio Journ. Sci.* 26 (1926) 128–46.

Severson, Robert F. The source of mortgage credit for Champaign County, 1865–1880. *Agric. Hist.* 36 (1962) 150–55.

Shaler, N. S. Fresh water morasses of the United States. *United States Geological Survey Tenth Annual Report,* pt. 2 (Washington, D.C. 1890).

Shames, Deborah (ed.). *Freedom with reservation: The Menominee struggle to save their land and people* (National Committee to Save the Menominee People and Forests, Washington, D.C. 1972).

Shaw, Earl. Swine production in the corn belt of the United States. *Econ. Geog.* 12 (1936) 359–72.

Shaw, S. P., and C. G. Fredine. *Wetlands of the United States: Their extent and their value to waterfowl and other wildlife.* U.S. Fish and Wildlife Service Circular 39 (Washington, D.C. 1956).

Shery, Robert W. The migration of a plant: Kentucky bluegrass followed settlers to the New World. *Natural History* 74 (1965) 43–44.

Shetler, S. Three faces of Eden. In H. J. Viola and C. Margolis (eds.), *Seeds of change: A quincentennial commemoration* (Smithsonian Institution, Washington, D.C. 1991) 225–47.

Shimek, B. The prairies. *Lab. Nat. Hist. Univ. Iowa Bull.* 6 (1911) 169–240.

Shjeflo, J. B. Evapotranspiration and the water budget of prairie potholes in North Dakota. *U.S. Geol. Surv. Prof. Paper* 585-B (1968).

Short, C. W. Observations on the botany of Illinois, more especially in reference to the autumnal flora of the prairies. *Western Journ. of Medicine and Surgery* n.s. 3 (1845) 185–98.

Shortridge, James R. *The Middle West: Its meaning in American culture* (University of Kansas Press, Lawrence 1989).

Simmons, Ian G. *Interpreting nature: Cultural construction of the environment* (Routledge, London 1993).

Sitterley, J., and J. Falconer. Better land utilization for Ohio. *Ohio State Univ. Agric. Expt. Sta. Bull.* 108, Dept. of Rural Economy, mimeo (Columbus 1938).

Sluyter, Andrew. Intensive wetland agriculture in Mesoamerica: Space, time, and form. *Annals Assoc. Amer. Geog.* 84 (1994) 557–84.

Smiley, Jane. *A thousand acres* (Flamingo, London 1992).

Smith, Guy Harold. The relative relief of Ohio. *Geog. Rev.* 25 (1935) 272–84.

Smith, Huron H. *Ethnobotany of the Menomini Indians*. Bulletin of the Public Museum of the City of Milwaukee 1923 (reprinted by Greenwood Press, Westport 1970).

Society of Wetland Scientists. *1995 membership directory and handbook*. Supplement to *Wetlands* 15 (1995).

Soper, E. K. *The peat deposits of Minnesota*. In *Minn. Geol. Surv. Bull.* 16 (1919).

Spencer, J. E., and Ronald J. Horvath. How does an agricultural region originate? *Annals Assoc. Amer. Geog.* 53 (1963) 74–92.

Spindler, George, and Louise Spindler. *Dreamers without power: The Menomini Indians* (Holt, Rinehart and Winston, New York 1971).

Staats, J. Riley. The geography of the central sand plain of Wisconsin. Ph.D. diss., Geography, University of Wisconsin 1933.

Stallings, J. H. *Soil conservation* (Prentice-Hall, Englewood Cliffs 1957).

State Conservation Commission of Wisconsin. *Twenty-first biennial report* (Madison 1949).

State of Minnesota. *Laws*. (State of Minnesota, St. Paul) 1881–1933.

Stevens, William K. Restored wetlands could ease threat of Mississippi floods. *New York Times* (8 August 1995) C1, C4.

Stewart, Lowell O. *Public land surveys: History, instructions, methods* (Collegiate Press, Ames, Iowa 1935).

Stewart, Omer C. Burning and natural vegetation in the United States. *Geog. Rev.* 41 (1951) 317–20.

____. Fire as the first great force employed by man. In Thomas, *Man's role*, 1956, 115–33.

Stewart, R. E., and H. A. Kantrud, *Classification of natural ponds and lakes in the glaciated prairie region*. U.S. Fish and Wildlife Service Resource Publ. 92 (Washington, D.C. 1971).

Stout, A. B., A biological and statistical analysis of the vegetation of a typical wild hay meadow. *Wisc. Acad. Sci., Arts and Letters* 17 (1914) 438.

Straus, Murray A. Societal needs and personal characteristics in the choice of farm, blue collar, and white collar occupations by farmers' sons. *Rural Sociology* 29 (1964) 408–25.

Suffling, Roger. Catastrophic disturbance and landscape diversity: The implications of fire control and climate change in subarctic forests. In Michael R. Moss (ed.), *Landscape ecology and management*, Proceedings of the First Symposium of the Canadian Society for Landscape Ecology and Management (Polyscience Publications, Montreal 1988) 111–20.

Suffling, Roger, and Ron Fritz. The ecology of a famine: Northwestern Ontario in 1815–17. In C. R. Harington (ed.), *The year without a summer? World climate in 1816* (Canadian Museum of Nature, Ottawa 1992) 203–17.

Swander, Mary. Iowa, colored blue. *New York Times Magazine* (19 September 1993) 36–94.

Swierenga, Robert P. *Pioneers and profits: Land speculation on the Iowa frontier* (Iowa State University Press, Ames 1968).

Tansley, A. G. *The British Islands and their vegetation* (Cambridge University Press, Cambridge 1939).

Tauxe, Caroline. The myth of the family farm: An essay on hegemonic process in American society. *Dialectical Anthropology* 17 (1992) 291–318.

Terasmae, J. Postglacial history of Canadian muskeg. In N. W. Radforth and C. O. Brawner (eds.), *Muskeg and the northern environment in Canada* (University of Toronto Press, Toronto 1977) 9–30.

Thomas, Christine L. One hundred and twenty years of citizen involvement with the Wisconsin Natural Resources Board. *Environmental History Review* 15 (1991) 61–81.

Thomas, David. *Travels through the western country in the summer of 1816* (Auburn, N.Y. 1819).

Thomas, W. L. (ed.). *Man's role in changing the face of the earth* (University of Chicago Press, Chicago 1956)

Thompson, John G. The rise and decline of the wheat growing industry in Wisconsin. *Univ. Wisc. Bull.* 292 (1909).

Thomson, Gladys Scott. *A pioneer family: The Birkbecks in Illinois, 1818–1827* (London 1953).

Thornthwaite, C. W., John R. Mather, and Douglas B. Carter. *Three water balance maps of eastern North America* (Resources for the Future, Washington, D.C. 1958).

Thrower, Norman J. W. *Original survey and land subdivision: A comparative study of the form and effect of contrasting cadastral surveys.* AAG Monograph (Rand McNally, Chicago 1966).

Thwaites, F. T. *Outline of glacial geology* (Madison 1953).

Tiner, Ralph W. *Wetlands of the United States: Current status and recent trends.* U.S. Fish and Wildlife Service, National Wetlands Inventory (Washington, D.C. 1984).

____. How wet is a wetland? *Great Lakes Wetlands* 2 (1991) 1–7.

Tiner R. W., and B. O. Wilen, *The U.S. Fish and Wildlife Service National Wetlands Inventory project* (U.S. Fish and Wildlife Service, Washington, D.C. 1983).

Titus, H. *The land nobody wanted.* In *Mich. Agric. Expt. Sta. Bull.* 332 (1945).

Tobin, Graham, and Burrell E. Montz. *The great Midwestern floods of 1993* (Harcourt Brace, Orlando 1994).

Towar, J. D. *Peat deposits of Michigan.* In *Mich. State Agric. Coll. Expt. Sta. Bull.* 181 (East Lansing 1900).

Transeau, E. N. Precipitation types of the prairie and forested regions of the central states. *Annals Assoc. Amer. Geog.* 20 (1930) 44–45.

____. The prairie peninsula. *Ecology* 16 (1935) 423–37.

Turner, Frederick Jackson. The character and influence of the Indian trade in

Wisconsin: A study of the trading post as an institution (Johns Hopkins University Studies in Historical and Political Science, Baltimore 1891); reprinted in *The early writings of Frederick Jackson Turner,* introduction by Fulmer Mood (University of Wisconsin Press, Madison 1938) 87–181.

____. The significance of the frontier in American history. *American Historical Association Annual Report* (1893) 199–227; reprinted in *The early writings of Frederick Jackson Turner,* with introduction by Fulmer Mood (University of Wisconsin Press, Madison 1938) 183–229.

Turner, L. M. Grassland in the flood plains of Illinois rivers. *Amer. Midland Nat.* 15 (1934) 770–80.

Tweeten, Luther. Has the family farm been treated unjustly? In Comstock (ed.), *Obligation to save the family farm,* 1987, 222–25.

Twining, Charles E. The lumbering frontier. In Flader (ed.), *Great Lakes forest,* 1983, 121–36.

Ubelaker, Douglas H. The sources and methodology for Mooney's estimates of North American Indian populations. In Denevan, *Native population,* 1992, 243–88.

Ullsperger, H. W. Report on Portage County Drainage District, 1916. [Records of State Drainage Engineer, Madison.]

United States. *Statutes at large* (Act of 4 September 1841) 457 [Preemption Act].

____. *Statutes at large* 9 (Act of 28 September 1850) 519 [Swamp Land Act].

U.S. Army Corps of Engineers. *Federal manual for identifying and delineating jurisdictional wetlands* (Washington, D.C. 1989).

____. *Floodplain management assessment* (Washington, D.C. 1995).

U.S. Bureau of Census. *Fourteenth census of the United States, 1920: Agriculture,* vol. 7, *Irrigation and drainage* (Washington, D.C. 1922).

____. *Fifteenth census of the United States, 1930,* vol. 4, *Drainage of agricultural lands* (Washington, D.C. 1932).

____. *Sixteenth census of the United States, 1940: Drainage of agricultural lands* (Washington, D.C. 1942).

____. *Seventeenth census of population, 1950* (Washington, D.C. 1952).

____. *Census of agriculture, 1950,* vol. 4, *Drainage of agricultural lands* (Washington, D.C. 1952).

____. *Census of agriculture, 1959,* vol. 4, *Drainage of agricultural lands* (Washington, D.C. 1962).

____. *Census of agriculture, 1969* (Washington, D.C. 1969).

____. *Census of agriculture, 1978: Drainage* (Washington, D.C. 1978).

U.S. Census Office. *Eighth census,* vol. 1, *Population of the United States, 1860* (Washington, D.C. 1864).

____. *Ninth census: Productions of agriculture, 1870* (Washington, D.C. 1874).

____. *Tenth census: Productions of agriculture, 1880* (Washington, D.C. 1883).

____. *Eleventh census of population, 1890: Extra Census Bulletin*, vol. 2 (Washington 20 April 1891).

____. *Twelfth census, 1900*, vol. 5, *Agriculture* (Washington, D.C. 1902).

U.S. Congress. *General acts of Congress respecting the sale and disposition of the public lands with instructions issued from time to time by the Secretary of the Treasury and Commissioner of the General Land Office and official opinions of the Attorney General on questions arising under the land laws* (Gales and Seaton, Washington, D.C. 1838).

U.S. Congress. *House Reports of Committees*. 40th Cong., 2d Sess. 25 (1868).

U.S. Congress, House of Representatives. 22d Cong., 2d Sess., Ex. Doc. 2 (1832) 1, 10–11.

____. 26th Cong., 2d Sess., *Congressional Globe* (4 January 1841) appendix, 19.

U.S. Department of Agriculture. *Soils and men: Yearbook of agriculture, 1938* (Washington, D.C. 1938).

____. *What the conservation provisions of the Food, Agriculture, Conservation, and Trade Act mean to you* (Washington, November 1991).

U.S. Department of Agriculture and Soil Conservation Service. *Drainage of agricultural land 1967* (Washington, D.C. 1968).

U.S. Department of Interior, Fish and Wildlife Service. National wetlands inventory. Unpublished data, St. Petersburg, Fla. 1983.

____. *Wetlands: Meeting the president's challenge* (Washington, D.C. 1990).

U.S. Executive Office of the President. *Sharing the challenge: Floodplain management into the 21st century* (Washington, D.C. 1994).

U.S. Office of Technology Assessment. *Wetlands: Their use and regulation* (OTA-0-206, Washington, D.C. 1984).

U.S. Secretary of the Interior. *The impact of federal programs on wetlands*, vol. 1 (Washington, October 1988).

U.S. Soil Conservation Service. *Soil taxonomy: A basic system of soil classification for making and interpreting soil surveys*. USSCS Agricultural Handbook 436 (Washington, D.C. 1975).

U.S. Soil Conservation Service in cooperation with National Technical Committee for Hydric Soils. *Hydric soils of the United States* (Washington, D.C. 1987).

University of Minnesota Center for Urban and Regional Affairs. *Thematic map: Available wetlands for bioenergy purposes* (University of Minnesota, St. Paul 1981).

van der Valk, A. G., and C. B. Davis. Primary production of prairie glacial marshes. In R. E. Good, D. F. Whigham, and R. L. Simpson (eds.), *Freshwater wetlands: Ecological processes and management potential* (Academic Press, New York 1978) 21–37.

Vestal, Arthur G. Preliminary account of the forests in Cumberland County, Illinois. *Trans. Ill. Acad. Sci.* 12 (1919) 240.

____. Why the Illinois settlers chose forest lands. *Trans. Illinois State Acad. Sci.* 32 (1939) 85–87.

Visher, S. S. *The climate of Indiana* (Bloomington 1944).

Vogel, John N. *Great Lakes lumber on the Great Plains: The Laird, Norton Company in South Dakota* (University of Iowa Press, Iowa City 1992).

von Tungeln, George H. *A rural social survey of Orange Township, Black-hawk County, Iowa.* In *Iowa Agric. Expt. Sta. Bull.* 184 (Ames 1918).

____. *A rural social survey of Lone Tree Township, Clay County, Iowa.* In *Iowa Agric. Expt. Sta. Bull.* 193 (Ames 1920).

von Tungeln, George H., and Harry L. Eells. *Rural social survey of Hudson, Orange, and Jesup consolidated school districts, Blackhawk and Bucha-nan counties, Iowa.* In *Iowa Agric. Expt. Sta. Bull.* 224 (Ames 1924).

von Tungeln, George H., E. L. Kirkpatrick, C. R. Hoffer, and J. F. Thaden. *The social aspects of rural life and farm tenantry, Cedar County, Iowa.* In *Iowa Agric. Expt. Sta. Bull.* 217 (Ames 1923).

Waller, Robert. *The bridges of Madison County* (Warner, New York 1992).

Walsh, Margaret. From pork merchant to meat packer: The Midwestern meat industry in the mid-nineteenth century. *Agric. Hist.* 56 (1982) 127–37.

Walters, William D., Jr., and Floyd Mansberger. Initial field location in Illinois. *Agric. Hist.* 57 (1983) 289–96.

Wang, Jen Yu. The phytoclimate of Wisconsin. M.S. thesis, Meteorology, University of Wisconsin 1955.

Ward, David (ed.). *Geographic perspectives on America's past* (Oxford University Press, New York 1979).

Warntz, William. An historical consideration of the terms "corn" and "corn belt" in the United States. *Agric. Hist.* 31 (1957) 40–45.

Waselkov, Gregory. Prehistoric agriculture in the central Mississippi valley. *Agric. Hist.* 51 (1977) 513–19.

Washington, George. *The writings of George Washington,* vol. 27. Edited by John C. Fitzpatrick (GPO, Washington, D.C. 1938).

Watterson, Arthur Weldon. *Economy and land use patterns in McLean County, Illinois.* Department of Geography Research Paper 17 (University of Chicago, Chicago 1950).

Weaver, John C. Changing patterns of cropland use in the Middle West. *Econ. Geog.* 30 (1954) 1–47.

____. Crop-combination regions in the Middle West. *Geog. Rev.* 44 (1954) 175–200.

____. Crop-combination regions for 1919 and 1929 in the Middle West. *Geog. Rev.* 44 (1954) 560–72.

Weaver, John E. *North American prairie* (Johnsen, Lincoln, Neb. 1954).

____. *Prairie plants and their environment* (University of Nebraska Press, Lincoln 1968).

Weaver, Marion. *History of tile drainage (in America prior to 1900)* (M. M. Weaver, Waterloo, N.Y. 1964).

Webb, Walter Prescott. *The Great Plains* (Ginn and Co., Boston 1931).

Webb, Thompson III, Patrick J. Bartlein, and John E. Kutzbach. Climate change in eastern North America during the past 18,000 years: Comparisons of pollen data with model results. In W. F. Ruddiman and H. E. Wright (eds.), *The geology of North America: North America and adjacent oceans during the last deglaciation* (Geol. Soc. of America, Boulder 1987) 447–62.

Weidman, Samuel. *Preliminary report on the soils and agricultural conditions of north central Wisconsin.* In *Wisc. Geol. Nat. Hist. Surv. Bull.* 11 (Madison 1903).

Weidman, Samuel, and Alfred R. Schultz. *The underground and surface water supplies of Wisconsin.* In *Wisc. Geol. Nat. Hist. Surv. Bull.* 35 (Madison 1915).

Weller, M. W. *Freshwater marshes: Ecology and wildlife management,* 2d ed. (University of Minnesota Press, Minneapolis 1987).

Weller, M. W., and C. S. Spatcher, *Role of habitat in the distribution and abundance of marsh birds.* In *Iowa State Univ. Agric. and Home Econ. Expt. Sta. Spec. Report* 43 (Ames, Iowa 1965).

Wessel, Thomas R. Agriculture, Indians, and American history. *Agric. Hist.* 50 (1976) 9–20.

Wheeler, Bryan D., Susan C. Shaw, Wanda J. Fojt, and R. Allan Robertson (eds.). *Restoration of temperate wetlands* (John Wiley, Chichester 1995).

Whitbeck, Ray Hughes. Economic aspects of the glaciation of Wisconsin. *Annals Assoc. Amer. Geog.* 3 (1913) 71.

____. *The geography and industries of Wisconsin.* In *Wisc. Biol. Nat. Hist. Surv. Bull.* 26 (Madison 1913).

White, Richard. Native Americans and the environment. In W. R. Swagerty (ed.), *Scholars and the Indian experience: Critical reviews of recent writing in the social sciences* (Indiana University Press, Bloomington 1984).

____. "Are you an environmentalist or do you work for a living?": Work and nature. In Cronon, *Uncommon ground,* 1995, 171–85.

White House, Office on Environmental Policy. *Protecting America's wetlands: A fair, flexible, and effective approach* (GPO, Washington, D.C. 1993).

Whitney, Gordon G. An ecological history of the Great Lakes forest of Michigan. *Journ. of Ecology* 75 (1987) 667–84.

____. *From coastal wilderness to fruited plain: A history of environmental change in temperate North America, from 1500 to the present* (Cambridge University Press, Cambridge 1994).

Whitson, A. R. *Soil Survey of Adams County.* In *Wisc. Geol. Nat. Hist. Surv. Bull.* 61D (Madison 1924).

____. *Soils of Wisconsin.* In *Wisc. Geol. Nat. Hist. Surv. Bull.* 68 (Madison 1927).

Whitson, A. R., W. J. Geib, Guy Conrey, W. C. Boardman, and Clinton B.

Post. *Soil survey of Wood County.* In *Wisc. Geol. Nat. Hist. Surv. Bull.* 52B (Madison 1918).

Whitson, A. R., W. J. Geib, T. J. Dunnewald, and Lewis P. Hanson. *Soil survey of Portage County.* In *Wisc. Geol. Nat. Hist. Surv. Bull.* 52C (Madison 1918).

Whitson, A. R., W. J. Geib, L. R. Schoenmann, C. A. Leclair, O. E. Baker, and E. B. Watson. *Soil survey of Juneau County.* In *Wisc. Geol. Nat. Hist. Surv. Bull.* 38 (Madison 1914).

Whitson, A. R., and E. R. Jones. *Drainage conditions of Wisconsin.* In *Univ. Wisc. Agric. Expt. Sta. Bull.* 146 (Madison 1907).

Whitson, A. R., and F. J. Sievers. *The development of marsh soils.* In *Univ. Wisc. Agric. Expt. Sta. Bull.* 205 (Madison 1911).

Wilen, B. O. Fact sheets and information (National Wetlands Inventory, St. Petersburg, Fla. 1991).

Wilken, D. F., et al. *Fifty-eighth annual summary of Illinois farm business records* (University of Illinois, College of Agriculture, Urbana 1983).

Willard, Daniel E., Vicki M. Finn, Daniel A. Levine, and John E. Klarquist. Creation and restoration of riparian wetlands in the agricultural Midwest. In Kusler and Kentula, *Wetland creation and restoration,* 1990, 327–50.

Williams, Michael. *Americans and their forests: A historical geography* (Cambridge University Press, Cambridge 1989).

_____. Agricultural impacts in temperate wetlands. In Williams, *Wetlands,* 1990, 181–216.

_____. The human use of wetlands. *Progress in Human Geography* 15 (1991) 10.

_____. The relations of environmental history and historical geography. *Journ. Hist. Geog.* 20 (1994) 3–21.

_____ (ed.). *Wetlands: A threatened landscape* (Blackwell, Oxford 1990).

Winsor, Roger A. Environmental imagery of the wet prairie of east central Illinois, 1820–1920. *Journ. Hist. Geog.* 13 (1987) 375–97.

Winter, T. C. Hydrologic studies of wetlands in the northern prairie. In A. G. van der Valk (ed.), *Northern prairie wetlands* (Iowa State University Press, Ames 1989) 16–54.

Wisconsin Conservation Department. Progress report on drainage. Press release 15 March 1950. [Records of State Drainage Engineer, Madison.]

_____. *Wisconsin lakes.* Publ. 218–51 (Madison 1951).

_____. *Wisconsin trout streams.* Publ. 213–51 (Madison 1951).

Wisconsin Department of Agricultural Administration. Inquiry into indebtedness of drainage districts, 1923. [State Historical Society Archives, Madison.]

Wisconsin Department of Agricultural Administration, Immigration Division Records, 1920–30, Proceedings of drainage district conferences 1923–25. [State Historical Society Archives, Madison, Series 9/1/1, 3.]

Wooten, H. H., and L. A. Jones. The history of our drainage enterprises. In

USDA, *Water: Yearbook of Agriculture, 1955* (Washington, D.C. 1955) 478–91.

Worster, Donald (ed.). *Nature's economy: A history of ecological ideas* (Cambridge University Press, Cambridge 1977).

Wunderlich, Gene. The land question: Are there answers? *Rural Sociology 58* (1993) 547–59.

Wyman, Walker, and Lee Prentice. *The lumberjack frontier: The life of a logger in the early days on Chippeway* (University of Nebraska Press, Lincoln 1969).

INDEX

The University of Chicago
GEOGRAPHY RESEARCH PAPERS

Titles in Print

127. GOHEEN, PETER G. *Victorian Toronto, 1850 to 1900: Pattern and Process of Growth.* 1970. xiii + 278 pp.

132. MOLINE, NORMAN T. *Mobility and the Small Town, 1900–1930.* 1971. ix + 169 pp.

136. BUTZER, KARL W. *Recent History of an Ethiopian Delta: The Omo River and the Level of Lake Rudolf.* 1971. xvi + 184 pp.

152. MIKESELL, MARVIN W., ed. *Geographers Abroad: Essays on the Problems and Prospects of Research in Foreign Areas.* 1973. ix + 296 pp.

181. GOODWIN, GARY C. *Cherokees in Transition: A Study of Changing Culture and Environment Prior to 1775.* 1977. ix + 207 pp.

186. BUTZER, KARL W., ed. *Dimensions of Human Geography: Essays on Some Familiar and Neglected Themes.* 1978. vii + 190 pp.

194. HARRIS, CHAUNCY D. *Annotated World List of Selected Current Geographical Serials, Fourth Edition. 1980.* 1980. iv + 165 pp.

206. HARRIS, CHAUNCY D. *Bibliography of Geography. Part II: Regional. Volume 1. The United States of America.* 1984. viii + 178 pp.

207–208. WHEATLEY, PAUL. *Nagara and Commandery: Origins of the Southeast Asian Urban Traditions.* 1983. xv + 472 pp.

209. SAARINEN, THOMAS F.; DAVID SEAMON; and JAMES L. SELL, eds. *Environmental Perception and Behavior: An Inventory and Prospect.* 1984. x + 263 pp.

210. WESCOAT, JAMES L., JR. *Integrated Water Development: Water Use and Conservation Practice in Western Colorado.* 1984. xi + 239 pp.

213. EDMONDS, RICHARD LOUIS. *Northern Frontiers of Qing China and Tokugawa Japan: A Comparative Study of Frontier Policy.* 1985. xi + 209 pp.

216. OBERMEYER, NANCY J. *Bureaucrats, Clients, and Geography: The Bailly Nuclear Power Plant Battle in Northern Indiana.* 1989. x + 135 pp.

217–218. CONZEN, MICHAEL P., ed. *World Patterns of Modern Urban Change: Essays in Honor of Chauncy D. Harris.* 1986. x + 479 pp.

222. DORN, MARILYN APRIL. *The Administrative Partitioning of Costa Rica: Politics and Planners in the 1970s.* 1989. xi + 126 pp.

223. ASTROTH, JOSEPH H., JR. *Understanding Peasant Agriculture: An Integrated Land-Use Model for the Punjab.* 1990. xiii + 173 pp.

224. PLATT, RUTHERFORD H.; SHEILA G. PELCZARSKI; and BARBARA K. BURBANK, eds. *Cities on the Beach: Management Issues of Developed Coastal Barriers.* 1987. vii + 324 pp.

225. LATZ, GIL. *Agricultural Development in Japan: The Land Improvement District in Concept and Practice.* 1989. viii + 135 pp.

227. MURPHY, ALEXANDER B. *The Regional Dynamics of Language Differentiation in Belgium: A Study in Cultural-Political Geography.* 1988. xiii + 249 pp.

228–229. BISHOP, BARRY C. *Karnali under Stress: Livelihood Strategies and Seasonal Rhythms in a Changing Nepal Himalaya.* 1990. xviii + 460 pp.

230. MUELLER-WILLE, CHRISTOPHER. *Natural Landscape Amenities and Suburban Growth: Metropolitan Chicago, 1970–1980.* 1990. xi + 153 pp.

231. WILKINSON, M. JUSTIN. *Paleoenvironments in the Namib Desert: The Lower Tumas Basin in the Late Cenozoic.* 1990. xv + 196 pp.

232. DUBOIS, RANDOM. *Soil Erosion in a Coastal River Basin: A Case Study from the Philippines.* 1990. xii + 138 pp.

233. PALM, RISA, AND MICHAEL E. HODGSON. *After a California Earthquake: Attitude and Behavior Change.* 1992. xii + 130 pp.

234. KUMMER, DAVID M. *Deforestation in the Postwar Philippines.* 1992. xviii + 179 pp.

235. CONZEN, MICHAEL P., THOMAS A. RUMNEY, AND GRAEME WYNN. *A Scholar's Guide to Geographical Writing on the American and Canadian Past.* 1993. xiii + 751 pp.

236. COHEN, SHAUL EPHRAIM. *The Politics of Planting: Israeli-Palestinian Competition for Control of Land in the Jerusalem Periphery.* 1993. xiv + 203 pp.

237. EMMETT, CHAD F. *Beyond the Basilica: Christians and Muslims in Nazareth.* 1994. xix + 303 pp.

238. PRICE, EDWARD T. *Dividing the Land: Early American Beginnings of Our Private Property Mosaic.* 1995. xviii + 410 pp.

239. PAPADOPOULOS, ALEX G. *Urban Regimes and Strategies: Building Europe's Central Executive District in Brussels.* 1996. xviii + 290 pp.

240. KNOWLES, ANNE KELLY. *Calvinists Incorporated: Welsh Immigrants on Ohio's Industrial Frontier.* 1997. xxiv + 330 pp.

241. PRINCE, HUGH. *Wetlands of the American Midwest: A Historical Geography of Changing Attitudes.* 1997. xiv + 395 pp.